King of Ragtime

KING OF RAGTIME

Scott Joplin and His Era

Edward A. Berlin

New York   Oxford
OXFORD UNIVERSITY PRESS
1994

Oxford University Press

Oxford   New York   Toronto
Delhi   Bombay   Calcutta   Madras   Karachi
Kuala Lumpur   Singapore   Hong Kong   Tokyo
Nairobi   Dar es Salaam   Cape Town
Melbourne   Auckland   Madrid

and associated companies in
Berlin   Ibadan

Library of Congress Cataloging-in-Publication Data
Berlin, Edward A.
King of Ragtime : Scott Joplin and His Era / Edward A. Berlin.
p.   cm.   Includes bibliographical references and index.
ISBN 0-19-508739-9
1. Joplin, Scott, 1868–1917.
2. Composers—United States—Biography.
I. Title.   ML410.J75B5   1994
780'.92—dc20   93-28318

ILLUSTRATION CREDITS
Fisk University Library's Special Collections: Queen City Cornet Band (p. 21), The Maple Leaf Club (p. 39),
and Scott Joplin (p. 190).
State Fair Community College Ragtime Archives, Sedalia, Missouri: Maple Leaf Club business card (p. 39).
Pierpont Morgan Library: *Maple Leaf Rag* contract (p. 55). Owned by James J. Fuld.
Trebor Jay Tichenor: Williams and Walker print (p. 61).
State Historical Society of Missouri: Eleanor Stark (p. 73), Tom Turpin (p. 91), and Scott Joplin photograph
from the *St. Louis Globe-Democrat* (p. 95).
Patricia Lamb Conn: Joseph F. Lamb (p. 174).
Maple Leaf at beginning of sidebars, by Paul Pascal.
Author's photograph by the Department of Art and Photography, Queensborough Community College.

9 8 7 6 5 4 3 2

Printed in the United States of America
on acid free paper

Dedicated to the pioneers of the ragtime revival,
those who rekindled interest in
a great American music:
Wally Rose, Rudi Blesh,
Harriet Janis, and Max Morath

Preface

Music publishing turned a corner in the 1890s. It became more daring, more colorful, and far more profitable. Marketing of popular music reached unprecedented heights and Tin Pan Alley was born.

Much popular music had a maudlin cast. The immense success of Charles K. Harris's *After the Ball* in 1892 demonstrated the public's love of pathos. Songwriters responded to the demand with such ballads as *She May Have Seen Better Days* (1894), *Those Wedding Bells Shall Not Ring Out* (1896), *Take Back Your Gold* (1897), and *She Is More To Be Pitied Than Censured* (1898).

But while the public gloried in this self-indulgent and extreme sentimentality, a brash, abrasive, and decidedly zippier style was moving to center stage. This was ragtime, both an irrepressible, swinging dance music and a colloquial, cynical song form. In its vocal guise, it was the antithesis of the precious, saccharine parlor song, telling not of excessive sensibilities, self-sacrifice, and honor, but of black men and women in ludicrous, and frequently demeaning, situations.

Ragtime was at first dismissed as a curious expression of black Americans and stage minstrels, a vulgar fad that would quickly disappear. But as it evolved and adapted itself to the white majority, it took root and became sufficiently broad to embrace both Scott Joplin's *Maple Leaf Rag* and Irving Berlin's *Everybody's Doing It*. Ragtime dominated American popular music for the two decades from the late 1890s until the late 1910s and was this country's first uniquely national style of music. Out of it came succeeding generations of popular song and jazz.

Ragtime receded from public consciousness during the 1920s and 1930s. By the year 1941, it had been dormant for two decades and was mostly forgotten. Occasionally, people of older generations used the term as a synonym for swing or jazz. A few record collectors recognized ragtime as a distinct style, but relegated it to a minor role in the evolution of American music, that of a precursor, a stage-setter for jazz.

In the 1940s, jazz was on the move. A new generation of jazz musicians was looking for its voice, seeking an alternative to big-band swing. In New York, inventive young musicians developed bebop, something entirely new. In California, musicians oriented toward the old tried to refresh jazz by returning to its roots. In the process, they discovered ragtime. These were the first stirrings, the awakening of a music that had faded in 1917.

By 1949, ragtime had become a subset of "traditional jazz," also referred to as "Dixieland." Rudi Blesh, an outspoken advocate for the earlier styles of jazz, reissued this music on his small Circle Records label. In his book *Shining Trumpets* (1946),[1] he proclaimed the virtues of older, purer, and blacker jazz and decried the commercial sounds of white-oriented swing. He also attacked black musicians who, by his standards, had betrayed their African-American roots.

Finding that music historians had overlooked ragtime, Blesh collaborated with Harriet Janis, his partner at Circle Records, to fill the gap. In less than a year-and-a-half, they turned out a masterpiece: *They All Played Ragtime*. The book was both the first social history of the ragtime world and the first biography of Scott Joplin.

Blesh and Janis wrote a book that overflows with enthusiasm, one that has thrilled and inspired legions of fans in the decades following its publication. They had the foresight and good sense to go beyond library research, to interview more than one hundred surviving participants of the ragtime scene. They told a tale that had never before been recounted, told it passionately and with authority. The reader, in finishing the volume, knew the story of ragtime and the story of Scott Joplin.

For all its deserved praise, the book has serious flaws. Some shortcomings are obvious. Since it was written quickly, its organization is maddeningly haphazard. The authors, lacking formal historical training or a sense for scholarly niceties, had little concern for presenting evidence. Where they offered documentation, they frequently confused their sources. When they lacked information, they had no qualms about fictionalizing.

Less obvious is a problem concerning the book's most distinctive feature: the testimony of those who were still able to bear witness. This testimony is invaluable, a source that cannot be duplicated today. For seeking out the original ragtimers and preserving their words, a task that occurred to no one else, Blesh and Janis are forever honored.

But testimony is not an unmixed blessing. Skilled investigators quickly learn that those being interviewed, in a misdirected effort to help, may slant their report to what they believe the interviewer wants to hear. Or in probing their memories for events of some forty or more years earlier, they may make mistakes, perhaps confuse people and events, compress years.

Take the example of Eubie Blake (1883–1983). Blake was a major player and witness of the ragtime years. In the last twenty years of his life, as interest in ragtime increased, he was called upon innumerable times to recall stories. To the end of his long life his mental faculties remained remarkably acute, and no one would accuse him of dishonesty. But he was also a show-man, interested in a good response. As he repeated stories, the accounts frequently changed somewhat, becoming more elaborate and entertaining.

Even when a story remained essentially constant, a seemingly insignifi-cant detail might vary, revealing the frailty of memory. In recounting his meeting with Scott Joplin, Blake's story of what occurred always remained the same. But the year of the meeting frequently changed, varying between 1907 and 1915. From his vantage point, the events of the meeting were of consequence, not the years. It was all so long ago. What difference do a few years make, one way or the other?

To the historical researcher trying to establish a chronology, the differ-ence can be critical. The lesson is that we should learn what we can from testimony but retain a healthy skepticism and continue to search for more evidence. Do not assume that having been on the scene enables a witness to relate events with absolute fidelity.

To the average, less critical ragtime fan, the Blesh-Janis book satisfied a need for information and context; for some, it intensified a love for ragtime. As Chapter Thirteen relates, the following for ragtime grew in the 1950s, 1960s, and most dramatically in the 1970s; *They All Played Ragtime* was a significant factor in that growth, especially during the first two of those decades. Fan clubs and newsletters that emerged during that period added a few details and discoveries to the Joplin biography, but the tale Blesh and Janis told, and the interpretations and points of view they offered, remained essentially unaltered. Numerous articles, biographies, and histories of rag-time repeated the Joplin story as Blesh and Janis had laid it out.

My own interest in ragtime began with Scott Joplin. I wrote *Ragtime. A Musical and Cultural History* (1980) in part to provided a context with which to understand Joplin's music. After I finished the book my appreciation for Joplin deepened, but I had no plans to write a book about him.

The present book developed out of a discussion with the late Martin Williams, the great jazz writer and, at that time, acquistions editor for the Smithsonian Institution Press. Williams had proposed a small book to sum-marize the then-current knowledge about Joplin. I accepted the assignment provisionally, noting that I disagreed with some of the accepted conclusions about Joplin; I wanted to do a certain amount of new research to clarify ambiguities and contradictions. I felt I could fulfill the assignment rather quickly.

I should have known better. My experience with music research taught that any topic re-examimed yields new material and alternative interpreta-tions.

Still, I was unprepared for what I encountered. So much had already been written about Joplin. Yet, the most likely sources of information in public archives and newpapers had barely been examined. Dipping into a few of these newpapers proved so fruitful I had to search further. One newspaper led to another, and I ultimately read through miles of microfilm, uncovering hundreds of new details. We find Joplin in previouly unsuspected situations that give new meaning to his music and his life. In some ways he is less attractive, more blemished and troubled, making his accomplishments that much more remarkable.

The newly acquired details give us a fresh view of Joplin and a different perspective on the ragtime world he occupied. This perspective is integral to our story and our understanding, for Joplin is inseparable from the world he inhabited, from its events and its cultural and racial climate. In seeing how his colleagues struggled, how they succeeded, and how they failed, we gain a greater understanding of Joplin's efforts.

Gaps remain, for every answer that I might provide brings its own host of new questions. The paradox is that finding answers does not necessarily provide us with a greater assuredness of the truth. Increased information may undermine our biases and weaken our convictions. We come to doubt, and perhaps even discard, what we had previously accepted as true. The certainty we felt from reading Blesh and Janis slips away.

In the end, Scott Joplin, the King of Ragtime Writers, remains elusive. So many of the artifacts that might have provided answers have been destroyed or lost. But while we do not reach a goal of pristine clarity, we do come to a greater understanding and appreciation of the ragtime world and of Scott Joplin.

Malverne, N.Y.   E.A.B.
September 7, 1993

Acknowledgments

Several years ago I contributed an article to a volume published to mark the 65th birthday of musicologist H. Wiley Hitchcock. A few weeks after its publication, I received a note from Hitchcock that included the following:

> One thing that has struck me especially about your Joplin opera essay is the degree of friendly networking that goes on among the ragtimers. I've sensed it before, but your article's acknowledgments make it really explicit. How different from the atmosphere that often seems to prevail in the musicological establishment—one of fear and suspicion, territorial imperatives and jealous guarding of turf. Your corner of the scholarly world seems clear and inviting: you must enjoy being in it.

Hitchcock was right. It is a joy to be part of the ragtime community, and members of that community are remarkably forthcoming to requests for assistance. Quite early on, and throughout my study, several prominent figures in ragtime generously gave me access to materials essential to my study. Foremost among these were Max Morath, John Edward Hasse, Dick Zimmerman, and Trebor Jay Tichenor.

Others, both within and outside the ragtime world, assisted me in a variety of ways. Lynn Abbott shared information about Joplin's early years as a minstrel; Texarkana resident and jazz historian Jerry Atkins sent me years of correspondences he had had with previous Joplin investigators; Lawrence Gushee and John Graziano forwarded to me Joplin items they came across while pursuing their own research; Ian Whitcomb examined materials for me at the British Museum; pianist John Arpin told me of his conversations with Joplin student William Sullivan; Michael Montgomery, Elliott Adams, and Charles Templeton sent me copies of rare music; Robert Ault, Dennis Pash, Eli Kaufman, and Clarke Buehling demonstrated their expertise in answering questions about obscure musical instruments of the ragtime years; psychiatrist and ragtimer Terry Parrish discussed with me theories of Joplin's emotional and mental state.

I thank, also, the following for assistance, ideas, leads, information, and materials: Phyllis Anderson, Ralph Bohn, Rob DeLand, David Reffkin, Joseph Scotti, Scott DeVeaux, Naomi Edmonds, Sol Goodman, John Hancock, William Kenney, James Maher, Rose Nolen, Daniel Paget, Doug Seroff, Richard Egan, Edith Schnall, Dick Earle, Jim Wiemers, Rhonda Sizemore, Carolyn Geida, Russel Hotzler, Daniel Burke, John Vivenzio, Larry Melton, John Moore, Rev. Marvin Albright, and Julie Renberg.

Good librarians and archivists can be a major asset, and I was fortunate in coming upon many who eagerly exceeded their normal duties on my behalf. Among these were Wayne Shirley of the Music Division of the Library of Congress; Charles Roberts of the Copyright Office; Frances Barulich and John Shepherd of the Music Division of the New York Public Library; Sandra Novik (Acting Chief Librarian), Herbert Knobler, Donald Bryk, and Marcia Kovler of the Queensborough Community College Library; Donald Morton of the Sedalia Library, and Helen Darrah, who did research for me there; Roberta Hagood, who searched for Scott Joplin's presence in Hannibal, Missouri; Edward J. Russo of the Lincoln Library of Springfield, Illinois; Leta Hendricks, of the Galesburg (Illinois) Public Library; Gene DeGruson, Special Collections Librarian at the Pittsburg (Kansas) State University, and John Chambers, who read through archival newspapers for me; Mary Ann Lemon, of the Ottumwa (Iowa) Public Library; Stephanie Shulte, of the Cedar Rapids (Iowa) Public Library; Carolyn Baker, of the Beatrice (Nebraska) Public Library; Barb Bandlow, of the Fremont (Nebraska) Public Library; Beverly Elder, of the Mason City (Iowa) Public Library; Henry Rock and Norma Rock, of the New York Library of the Mormon Church; Beth Howse of the Fisk University Library, and Henry Jones who went through the Brun Campbell papers for me there; Patricia Finley of the Onondaga County Library in Syracuse, NY; Patricia Payette, of the Free Public Library of Bayonne, NJ; Libby Nemota and Kemia Hudson of the Palm Beach (Florida) County Library; Judy Neumann, Joyce Kruger, and Brian Kelley (director) of the Palm Beach Community College Library. Should I have inadvertently omitted someone from these paragraphs whom I should have recognized, and I sincerely apologize.

My main research assistant, responsible for innumerble discoveries, was Reba Berlin, my aunt. Contributing to the family effort, my father, Milton Berlin, read through years of newspapers on microfilm, and commented upon every page of my manuscript. My wife Andrée kept things in perspective, gave her usual good counsel, and assured that I remembered how to laugh at myself.

In the final stage, I had the able assistance of editors at Oxford University Press: Senior Vice President Sheldon Meyer, assistant editor Karen Wolny, and copy editor Samuel C. Plummer, a super sleuth who uncovered errors missed by everyone else, at the same time providing ingenious solutions.

This research was supported (in part) by two grants from The City University of New York PSC/CUNY Research Award Programs.

Contents

King of Ragtime

CHAPTER ONE

The Early Years, 1868–1893

SCOTT JOPLIN was the "King of ragtime writers." For fifteen years, beginning with his first ragtime publications in 1899, he composed music unlike any ever before written. The piano-playing public clamored for his music; newspapers and magazines proclaimed his genius; musicians examined his scores with open admiration.

An African American born to the first post-slavery generation, Joplin had a dream of artistic greatness. He accepted society's stated ideals and believed that with hard work, education, self-improvement, and respectability, he could overcome the obstacles of background, class, and race.

He almost succeeded. Yet despite the acclaim, he was never during his lifetime to realize the full fruits of success. Cognizant of the disparity between his achievements and his position, he prophesied that recognition would come after his death:

> [Lottie Joplin, his widow:] He would often say that he'd never be appreciated until after he was dead.

> [Edward B. Marks, a music publisher:] "Boy," he used to tell the other colored song writers, "when I'm dead twenty-five years, people are going to begin to recognize me."[1]

His quarter-century prophecy was almost exactly on the mark. In 1944, twenty-seven years after his death, the Joplin revival began with the article "Scott Joplin: Overlooked Genius."[2] Through the following quarter-century, recognition and appreciation gradually increased, culminating with unprecedented popular and critical acclaim in the 1970s. Belatedly, he achieved the pinnacle of success and respect that had eluded him during his lifetime.

The authenticated facts of Scott Joplin's life are surprisingly scant. Questionable testimony by aged individuals is one problem. Documents may also be seriously flawed. Conflicting and illogical information too often leads to uncertainty. The various problems are illustrated by the difficulty in determining Scott Joplin's date of birth.

November 24, 1868, the date long accepted as Scott Joplin's birth date, is almost certainly incorrect. The date derives from biographical information submitted in 1942 by Lottie Joplin, the composer's widow, when she joined the American Society of Composers, Authors, and Publishers (ASCAP) in his name. In the *ASCAP Biographical Dictionary* his entry begins

> Joplin, Scott, composer, pianist; b. Texarkana, Tex., Nov. 24, 1868; d. New York, April 4, 1919. ASCAP 1942. . . .[3]

Lottie demonstrates that her memory in 1942 was faulty. We can overlook her error in placing Joplin's birth in Texarkana. Perhaps Joplin, himself, did not know that there was no Texarkana when he was born, that the town where he spent most of his childhood was founded five years after his birth. But she was present when he died on April 1, 1917, yet remembered the incident as occurring more than two years later.

The information she gave of his birth is also highly questionable. For his death certificate in 1917 (see on page 239), she had given his age as 49 and his year of birth as 1868, leaving the month and day blank. If she knew his full birth date in 1942, why did she not know it in 1917? In addition, if he had died at the age of 49, his birthday would have had to be before April 2, 1868.

The U.S. Census of 1870 and of 1880 also indicate that Joplin was born before the middle of 1868. The 1870 census, taken in Davis County, Texas, on July 18, indicates that his age on his last birthday was two. Similarly the 1880 census, taken in Texarkana, Texas, on June 17 and 18, lists his age on his last birthday as twelve.

Later census listings are no help. The 1890 census was destroyed in a fire. The 1900 census (see on page 84), taken in Sedalia, Missouri, on June 6, gives his birth date as October 1872—obviously incorrect, for he had already appeared in the 1870 census. His final census entry, that of New York on April 22, 1910 (see on page 197), gives his age as 40, also in disagreement with the earlier listings.

Scott Joplin's date of birth is not the important issue. Even if he was not born on November 24, 1868, he was born within a year of that date. The larger issue is that, from the very point of his birth, the facts concerning him are elusive. In studying his life, we are continually faced with conflicting

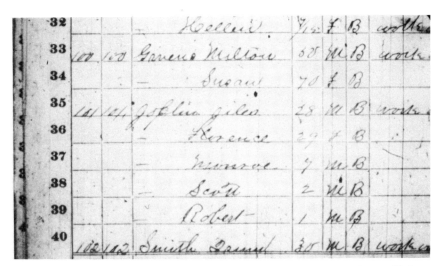

United States Census, 1870, Davis County, Post Office of Linden, Texas. Taken on July 18, 1870.

The Joplin family is on lines 35 through 39. In the columns following Scott Joplin's name, on line 38, are "2" [age on last birthday], "M" [male], "B" [black]; not shown is a column indicating he was born in Texas.

evidence. In view of this general lack of certainty, the approach in this biography is to present all the reasonable information, to discuss the options, and to suggest what seems most plausible.

A few facts are known about Scott Joplin's parents. His mother, Florence Givens, was freeborn in Kentucky around 1841.[4] His father, Giles, was born into slavery in North Carolina around 1842.[5] Scott probably was born in northeastern Texas, but just where is not known. Testimony in 1959 from Alexander Ford, a resident of Texarkana who had been a childhood friend of the Joplin children, gives us our first location for Scott Joplin: near Marshall, Texas, a town in the northeastern part of the state some 70 miles south of Texarkana.[6] In 1870, as shown in the census, the family lived about 32 miles north of Marshall on a farm owned by William Caves in Davis (now Cass) County near the current town of Linden. The family at that time was composed of Giles, a 28-year-old farm worker; his wife Florence, age 29; and the children Monroe, age 9, Scott, age 2, and Robert, age 1. Listed just before the Joplins are Milton and Susan Givens, ages 60 and 70, probably Florence's parents.

The next census, that of 1880, has the family living some 38 miles northeast of Linden in Texarkana. Straddling the border between Texas and Arkansas, Texarkana then had a population of about 3,250.[7] The Joplins lived on the 600 block of Hazel Street. The family was then comprised of Giles, age 38,

*United States Census, 1880, Texarkana, Texas, page 29, SD 2, ED 2. Taken on
June 17 and 18, 1880.*

*In the columns following Scott's name are "B" [black], "M" [male], "12"
[age on last birthday], "son," "going to school." Occupation for Giles is listed as
"Common laborer"; for Florence, "Wash & Iron."*

working as a "common laborer"; Florence, age 39, who took in laundry;
Monroe, age 19, a porter in a store; Scott, age 12; Robert, 11; Jose (correctly,
Osie), age 10; William, age 4; Johnny (correctly, Myrtle), age 3 months.

The year of the Joplins' arrival in Texarkana is not known, but is subject
to interesting speculation. According to Ford it was in 1880, right after
Myrtle was born. Blesh and Janis, however, state that Scott, at the age of
seven (which would have been in 1875), was permitted to play a piano in a
neighbor's house in Texarkana.[8] White Texarkana attorney L. Jean Cook
partially supports this assertion: Cook's mother told him of giving the young
Scott permission to play their piano while Florence cleaned the house.[9]

Scott Joplin himself left a clue suggesting the truth of the Blesh-Janis-
Cook version. His opera *Treemonisha* (1911) takes place near Texarkana and
has autobiographical hints. In the preface, Joplin states that the heroine
Treemonisha, *at the age of seven,* received her education from a nearby white
woman in exchange for her mother's housework. It would seem that Joplin
here both commemorates this significant stage in his musical education and
pays tribute to his mother for her part in it.[10]

The Joplins were a musical family: Florence played the banjo and sang,
and Giles played violin. Giles apparently taught his children the violin as

well, for Scott, Robert, and William all played that instrument. In later years, Robert and William became professional musicians, with Robert also gaining fame as a dancer and all-around vaudevillian during the first two decades of the 20th century.

Giles left the family in the early 1880s, going to live with another woman, but he maintained contact with the family. In later years, he would live with his adult children. During Scott's childhood, Florence assumed support of the family with domestic work.[11] In addition to obtaining the young Scott's use of the Cook piano, Florence managed to purchase a piano for him around 1881 or 1882.[12]

According to Zenobia Campbell, another elderly black Texarkana resident interviewed in 1959, the young Scott was serious, ambitious, and spoke of his intention to make something of himself. In addition to going to school, Scott studied with several local music teachers; among them, Mag Washington, a mulatto woman, and J. C. Johnson, who was part black and part Indian.[13] A more significant teacher was a German immigrant who has been identified as Julius Weiss.[14]

Julius Weiss was born in Germany in 1840 or 1841, graduated from a university in Saxony, and is thought to have come to the United States in the late 1860s, settling in St. Louis. In the late 1870s he was hired as private tutor to the children of Robert W. Rodgers, a founding landowner in Texarkana. Weiss taught the children German, astronomy, mathematics, and violin, and also taught music to other children in town. In later years, one of the Rodgers children wrote that Weiss provided the basis for his lifelong interest in opera.[15]

This German teacher has attained mythic proportions in the Joplin legends. Impressed with the talent of the young Scott, the teacher gave him free lessons and exposure to European art music. The evidence suggests that he had a profound influence on the young Joplin. In later life, according to Joplin's widow Lottie, Scott maintained correspondence with his former teacher, occasionally sending him gifts of money.[16] The essence of what Weiss accomplished was to impart to Scott an appreciation of music as an art as well as an entertainment. Weiss helped shape Joplin's aspirations and ambitions toward high artistic goals.

Alexander Ford told how Joplin began his musical career, at the age of 16, forming a vocal quartet with three other boys: his brother William, Wesley Kirby, and Tom Clark. However, since William would have been only eight at this time, it is doubtful that he was in the group; probably brother Robert, only a year younger than Scott, was the fourth member. The group performed in Texarkana and nearby towns. Scott also performed on the piano in dance halls and taught guitar and mandolin.[17]

Joplin left Texarkana at some point during his middle or late teens; just when is another issue on which there is conflicting evidence. Based on testimony of Zenobia Campbell, who recalled Joplin's attending Texarkana's Orr

School in 1887,[18] Haskins and Benson have him leaving home around 1888, when he was 19 or 20 years old.[19] Blesh and Janis, however, say that he left "in his early teens" and arrived in St. Louis around 1885.[20] The accepted story is that he traveled as a saloon and honky-tonk pianist through Texas, Louisiana, Missouri, Illinois, Ohio, and Kentucky before eventually settling in St. Louis.[21] Publisher John Stark's later reference to Joplin as "a homeless itinerant" would tend to support the belief that Joplin traveled extensively.[22]

Joplin's choice of St. Louis as a main base would have been significant, for that city was to become one of the earliest and most important sites of ragtime activity. John Turpin's Silver Dollar Saloon, at 425 South Twelfth Street, was reportedly the center of Joplin's early professional life.[23] Turpin's saloon first appears in the city directories in 1890, although it may have existed prior to that time. According to a newspaper report in 1912, John Turpin's son Tom, a major ragtimer and close friend of Joplin, composed the first rag, *Harlem Rag*, in 1892.[24] Joplin's presence in St. Louis in that year would put him at the scene of ragtime's emergence.

Yet no convincing evidence has ever been offered to substantiate the chronology or locales of Joplin's early travels. Lacking such evidence, we cannot credit assertions that he was in St. Louis by 1885, or by any other specific year. We must recognize that these reports are vague and can be accepted only in their generalities.

An interesting possibility is that, after leaving Texarkana, the adolescent Scott Joplin moved to Sedalia, Missouri, and went to school there. Sedalia, almost 200 miles away from St. Louis, is the town that was later to become identified with Joplin's early career as a ragtime composer. If he had spent some teenage years there, it might explain why he lived there in later life.

Regarding this story, Blesh and Janis had been told in 1949 by onetime Sedalia resident William G. Flynn that Scott Joplin had attended Sedalia's black Lincoln High School.[25] Blesh and Janis apparently felt that Flynn was mistaken, for they did not include this testimony in their book. However, Flynn's assertion should not be dismissed so casually, for there is supporting evidence.

Sedalia town directories from 1883 to 1887 list a black individual named Julius Walter Joplin, a blacksmith.[26] This suggests two possibilities: one is that Flynn remembered a black youngster in town named Joplin and later confused him with Scott; the other is that the Texarkana Joplins were related to Julius Walter Joplin (or came from the same slave-holder and assumed the same surname) and that the young Scott lived for a time with this relative. The latter hypothesis is supported by a subheading, after Joplin's rise to fame, in a 1903 isssue of the *Sedalia Times,* a black newspaper:

SCOTT JOPLIN A KING
His Latest Suit Is Catchy Rag Time Music
WAS RAISED IN SEDALIA[27]

Given the common inaccuracies of newspapers, this subheading cannot be cited as absolute proof of Joplin's early residence in Sedalia. However, the editor of the paper was W. H. Carter, who was also a musician. By 1903, Carter had been acquainted with Joplin for at least a decade and would have known if Joplin had been "raised in Sedalia."

Joplin's living in Sedalia would not invalidate the other claims of his travels and time spent in St. Louis. The travels could have occurred after, or even during, his stay in Sedalia. St. Louis, for example, was easily reached by railroad. His living in Sedalia and attending high school at this time of his life might also suggest additional dimensions of his character: even as a teenager, he was interested in continuing his formal education. The desire for higher education might have provided a reason for his moving to Sedalia in the 1890s.

The one certain glimpse we have of Joplin's early career is from the summer of 1891. It is an enlightening glimpse, but also puzzling and provocative. Scott Joplin in the summer of 1891 was a member of the Texarkana Minstrels in his hometown of Texarkana. It was because of a scandalous incident concerning the troupe that news of it has been preserved.

Since minstrelsy was the major theatrical expression for blacks of this period, it is not surprising that the young Joplin should have been involved. We do not know what part he performed in the troupe, but it is likely that he sang and played piano and other instruments, including the fiddle. He probably was not the leader, since he is not listed first in the one article that identifies the nine members.

In mid-July of 1891 (a Saturday night, either the 11th or 18th), the newly formed Texarkana Minstrels had its premier performance at Ghio's Opera House in Texarkana. A report indicates they put on a good show.[28]

They were booked for a repeat performance on July 30, a showing designed to coincide with a reunion in Texarkana of the Confederate Veterans' Association of Bowie County (Texas) and Miller County (Arkansas). They signed a contract to receive 40 percent of the gross receipts. When they saw the playbills for the show, they learned that their performance was advertised as a benefit to raise funds for a monument to Jefferson Davis, the late president of the Southern Confederacy.

Texarkana's black citizens were outraged that local black youths should dishonor their race by performing for such a cause. They tried, unsuccessfully, to dissuade the group from performing. The group was unhappy about the advertising—they claimed to have had no prior knowledge of the benefit—but nevertheless put on the show.

The incident created considerable stir and was reported in at least three newspapers, two in New Orleans and one in Indianapolis. Reading about the event in the New Orleans *Times-Democrat,* Mrs. Alice E. Albert, wife of the editor of the New Orleans-based *Southwestern Christian Advocate,* sponsored by the Methodist Freedmen's Aid Society, expressed her anger:

This is an outrageous shame! What has Jefferson Davis done for the colored people that they should want to help raise funds to help build a monument for him? The cause for which he fought and spent his life, I should say, left enough monuments on the backs of our poor fathers and mothers, to satisfy any people. . . . Why not devise plans to erect monuments to Abraham Lincoln, U.S. Grant, Wm. T. Sherman, Sumner and Sheridan instead? Such are the ones we should honor with monuments. The charitable thing for our people to do is to try to forget such men as Jeff. Davis. The white people of the South would respect us more for it. They know as well as we, that no people honestly feel like building monuments for anybody that fought to keep them in slavery, under the lash, and that sold their children, parents, and husbands and wives from each other. The thing is unnatural.[29]

Since the *Southwestern* was known throughout the South, being distributed through the network of Methodist churches and freedmen's schools, the troupers learned of Mrs. Albert's comments and responded in a letter. They were unpleasantly surprised to see Jefferson Davis's name on the playbills, but had no control over the producer's actions and did not donate any of their 40 percent of the proceeds to the fund. The newspaper editor was not satisfied:

This, to say the least, smells rather suspicious and makes it appear as if somebody was quite willing to go into partnership, into any kind of co-operation, so long as "our 40 per cent" was not at all diminished. Smooth it over as you will or may, the thing looks worse the more the young troupers seek to explain it. Under the circumstances, we are not at all surprised to learn that "the colored people of Texarkana were raging mad with them," for they profited from the use of Jeff Davis' name equally with the contractor and were accessories to the fact, in that they performed and shared the profits under that representation.

The troupe is composed of Dave Jackson, Will Dyson, Isaac Mingo, James Benson, Cary Daughtry, Scott Joplin, John Adams, Pleasant Jackson and Hugh Garner. Their action dishonors their race and curses the memories of John Brown, Abraham Lincoln, Wm. Loyd Garrison, Calvin Fairbank and the host of abolitionists that fought and bled that they might enjoy the privilege of organizing such a troupe.[30]

Discovery of this previously unknown fact about Joplin has the dual effect of providing answers and creating new questions. Since Joplin was in Texarkana in 1891, what does this say about the assertions of his being centered in St. Louis? Was he just home for a visit? Why were his brothers Will and Robert, also performers, not in the troupe? If he was still living in his hometown, was the embarrassing circumstance of the performance responsible for his leaving?

Minstrelsy

 "Black-face minstrelsy" was an invention of 19th-century America. Originated by white performers appearing with blackened faces, the genre purported to present in music, dance, and dialogue, humorous vignettes of African-American life. After the Civil War, black performers also became prominent in what was an immensely popular theatrical form. Black performers generally followed the conventions of minstrelsy: first, by further blackening their naturally dark-pigmented faces with burnt cork, and, second, by adopting the stereotypical portrayals of their race. Thus, American theatergoers received reinforcement of the perceptions of blacks as naive, slow-witted, able to speak only in a substandard dialect, and of being marvelous dancers.

While minstrelsy demanded that black performers accede to a racial self-mockery, it also provided a showcase for black performing talent. It nurtured such significant black performers as Bert Williams, George Walker, Ernest Hogan, Bob Cole, and Sissieretta Jones, and it was from these figures that black musical theater was to emerge.

Testimony places Joplin in Chicago in 1893 at the World's Columbian Exposition.[31] That Joplin should have been there is not surprising; though fair officials minimized participation by African-Americans (while, at the same time, featuring an exhibit by black Africans from Dahomey), black performers from all over the United States congregated there. The fair was the most spectacular exposition in American history up to that time, attended by some 27 million Americans, and had a profound effect on American music, art, architecture, and mores.

The fair was also a signal event in ragtime history, for, according to numerous accounts during the next two decades, it was here that ragtime surfaced from its incipient stages in black communities and became known to the wider American public. The following quotations from the ragtime years serve as typical references to this public emergence of ragtime at the fair:

> Not until the "midways" of our recent expositions stimulated general appreciation of Oriental rhythms did "rag-time" find supporters throughout the country.

> The [coon song and ragtime] fad had its origin along about 1893, the year of the Chicago World's Fair.

> It has been said that "rag-time" first appeared in our music-halls about the time of the Chicago World's Fair.

> The work [on the first published ragtime arrangements] was begun in Chicago, which still echoed with the joyousness of the World's Fair of 1893.[32]

We have been unable to find any contemporaneous record of ragtime being played at the fair itself, possibly because it had not yet found its name. But whether performed on the fairgrounds or not, it was probably heard in the nearby saloons and sporting houses that offered alternative entertainment to the public. Jesse Pickett's "Dream Rag," which in later years became known in versions by Eubie Blake, Abba Labba, Jack the Bear, Luckey Roberts, Willie "The Lion" Smith, James P. Johnson, and others, was reportedly heard first at the fair.[33]

We assume the reports are correct that Joplin led a band in Chicago, playing cornet, and that he associated with other ragtimers there, but the only specific information we have is his meeting with Otis Saunders. Saunders, a light-complexioned mulatto who could pass for white, was a pianist and singer, and he and Joplin became good friends. After the fair ended in October 1893 the two went to St. Louis and then to Sedalia, Missouri.[34]

Our recounting of Scott Joplin's early years is necessarily sparse. There is little credible information available from that period. As he attained fame newspapers began to document his career and person, bringing him gradually into focus as the story unfolds.

CHAPTER TWO

Sedalia, the Cradle
of Classic Ragtime

Sedalia has been nothing but music. . . . All my life this was called the
"musical town of the West."

Beatrice Martin (1890–1989),
lifelong black resident of Sedalia,
in a 1975 interview.

*I*T WAS WHILE LIVING IN SEDALIA, MISSOURI, that
Scott Joplin enjoyed his first real success. During his years
there, he became a published composer of ragtime, attained nationwide fame
as creator of the *Maple Leaf Rag,* established himself as "the king of ragtime
writers," and set forth on what he thought would be a fruitful career as a
composer for the lyric stage.

Known as the "Queen City of the Prairies," Sedalia is in central Missouri, 90 miles from Kansas City and 190 miles from St. Louis. It was founded
in the center of Pettis County on November 30, 1857, by General George
Radeen Smith, who, learning of plans for the Missouri Pacific Railway to go
through the area, purchased the land. He originally named the town Sedville
(for his daughter Sarah, who was nicknamed "Sed"), but changed the name
to Sedalia in October 1860. The railroad reached town on January 1, 1861,
and in 1862 Sedalia was named the county seat. Although Smith had 75
slaves when he arrived in Missouri in the 1830s, he came to oppose slavery,
and Sedalia during the Civil War was a Union military post.

In 1872, the Missouri, Kansas & Texas Railroad (known as the M. K. &
T. or "Katy") was brought to the town, and by the end of the decade the
Sedalia, Warsaw & Southern Railroad, a division of the Missouri Pacific, was

added. Extensive railroad repair yards were established to service the trains, and a railroad hospital was built for the employees.[1]

Situated in the midst of fertile farm country, the town became the commercial and transportation center of the region. In 1870, it had a population of 4,500; this increased to 14,000 by 1890.[2]

By 1883, Sedalia was a good-sized, thriving town. It had eight public schools (including the segregated black Lincoln School), five private schools, twenty Protestant and Catholic churches (including two for blacks), twenty-eight secret and benevolent societies, five paramilitary organizations, at least two musical ensembles (the eight-member Friemel's Orchestra and the fifteen-member Sedalia Silver Cornet & Reed Band), four dealers of music and musical instruments (John Stark, A. W. Perry & Son, J. W. Truxel, and Charles A. Dexter), five newspapers, nine banks and loan associations, and thirty-five saloons.[3] Wood's Opera House, at Lamine and 2nd, was built in 1884, and Smith's Opera Hall operated briefly in 1886–87 on Main near Ohio.[4] By 1888, the town had two baseball teams: the Shultz Hat Nine and the Y.M.C.A. Nine.[5] After the turn of the century there were two other organized teams: the Sedalia Blues, a white team, and the Sedalia Browns, a black team.[6]

As the region's center of commerce and transportation, Sedalia was crowded with men away from home, men with extra money in their pockets seeking diversions for their leisure time. Thus there emerged in town an ample supply of pool halls, saloons, gambling dens, dance halls, and brothels. These were mostly situated on Main Street, running east-west along the southern edge of the railroad (see map). They formed a dissolute pocket within a thriving and basically respectable community. The gambling dens preyed on the unwary, combining outright robbery with more subtle scams. The saloons were scenes of frequent drunken brawls and occasional murders. The brothels operated openly as thriving enterprises and attracted a regular supply of prostitutes who had little to fear from the law: an arrest for being "an inmate of a bawdy house" might result in a fine of five dollars or a sentence of three days in jail. A newspaper item in 1899 notes that fifteen madams went to police court to pay their monthly five-dollar fines.[7]

Citizens and property owners despaired at the conditions along Main Street, frequently referred to as "Battle Row," but periodic efforts to rid the area of vice had little long-term effect. As late as 1940, *Life* magazine referred to Sedalia as having "one of the midland's most notorious red-light districts."[8]

These establishments were viewed as a necessary evil and formed an important part of the local economy. Many also provided employment for musicians, especially the "piano thumpers." Joplin's student Arthur Marshall (1881–1968), who was working as a pianist by 1897, reported that the visiting cattlemen and wholesale grocers would make several hundred dollars on a trip and were good tippers.[9]

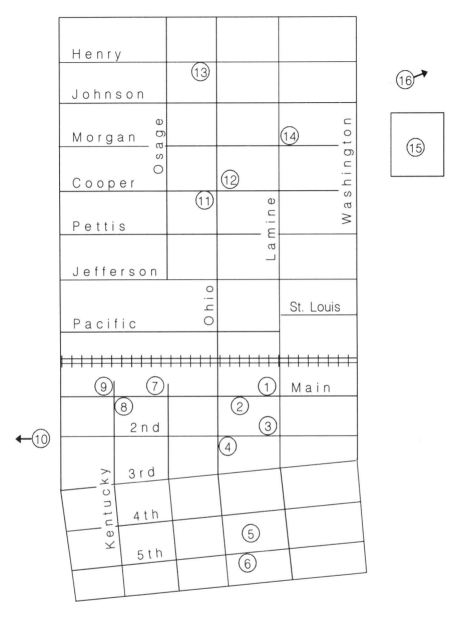

Scott Joplin's Sedalia. (1) Maple Leaf Club, 121 East Main; (2) Black 400 Club, 106 East Main; (3) Wood's Opera House, Lamine and 2nd; (4) St. Louis Clothing Store, 201–203 Ohio; (5) Courthouse; (6) John Stark & Son, 114 East 5th; (7) Ida Hathaway's brothel, 204 West Main; (8) Brothel, 219 West Main; (9) Nellie Hall's brothel, 302 West Main; (10) Liberty Park; (11) Hayden residence, 133 W. Cooper; (12) Solomon Dixon residence, 124 West Cooper; (13) Marshall residence, 135 West Henry; (14) John Wesley, Lamine and Morgan; (15) George R. Smith College; (16) Cemetery.

Prostitution in Sedalia

 Brothels and prostitution are minor themes running through the history of Sedalia. Almost daily, there were items in the newspapers about them. The houses of prostitution were known as *bawdy houses, bagnios, immoral resorts, disreputable houses,* and a number of other terms. Sometimes the brothel madams were such well-known characters that the houses were identified with their names: Trixie Lynn's place, Ida Hathaway's bagnio, and the like.

The prostitutes were referred to as *bawds, women of the town, harlots, scarlet women, girls of evil character,* and the like. They were all grouped in the same category, and few emerge from newspaper accounts with any distinctive identities. One way that a white prostitute would attract attention, although the notice was not desired, was to commit what was regarded as an offense of the most extreme depravity: consorting with a black man. Ada Carrol, "a white woman, a frail wreck," was in the company of Tom Ireland (1866–1963) and Ed Gravitt—two members of the black Queen City Cornet Band—when apprehended. The melodramatic style of the newspaper account was typical.

> Heavy steps were heard on the stairs and the woman started up in alarm. She knew to be caught in such company meant banishment from the city for her. The negroes were less concerned, . . . The woman ran to the front window, raised it and leaped out into the darkness of the night, onto the icy pavement. She fell straight down from the window ledge and her body broke on the stone steps below.

Despite the pitiful description, Carrol recovered and continued her profession. The men were at first fined thirty dollars each, but their fines were later reduced to ten dollars.[10]

Another interracial incident involved white prostitute Lottie Wright and her regular beau Emmett Cook, the Cornet Band's drummer and a singer with Joplin's Texas Medley Quartette. When Wright and Cook were caught together, they were fined a hundred dollars each, but the fines were stayed provided they leave the city. Presumably they left, but they later returned, for Cook continued working with the band. Lottie Wright, on her return, opened her own brothel and had both Arthur Marshall and Scott Joplin work for her as pianists.[11]

Anna Embree's story is one that reveals a prostitute as an individual with problems and feelings. As related in a local newspaper in 1899:

The Sad Story of a Pettis County Young Woman
A Short Life of Shame

There is a sad story in connection with this woman's life. She is but 25 years of age and was married seven years ago. . . . They had a great deal of trouble, which led to separation and divorce.

Three months ago she left her two children at the home of her parents and came to Sedalia. After her arrival here she led a life of shame and became ill and was taken to the City hospital.

While there she wrote a very touching letter to her parents, telling them of her sinful ways and begged forgiveness. She said she had been a wicked girl and thought she was going to die, and requested them to call and see her. She ended her letter by requesting that, in the event of her death, not to bury her in the potter's field.

Mr. and Mrs. Greer came to Sedalia yesterday, forgave their daughter, and took her back home, she being considerably improved.[12]

This story is unusual. More often, Sedalia's prostitutes were seen simply as anonymous merchandise, products of a system that was troublesome and evil, but nevertheless necessary.

African Americans in Sedalia were subject to many of the discriminations, restrictions, abuses, and insults common to segregated communities. On a whim, a white official could order a black baseball team to vacate the field during a scheduled game in order to accommodate white youths.[13] Some of the white newspapers referred to black Sedalians without animosity, but others seemingly missed no opportunity to disparage and mock them. Most black residences were restricted to "Lincolnville" (the part of town that was north of the railroad) and the tawdry Main Street. Efforts to move all of the town's brothels to Lincolnville were unsuccessful, but they clearly reveal community attitudes.[14] Seating in theaters was segregated, and restaurants and most saloons were restricted. Although there were black police officers and deputies, their arrest powers were restricted, and few other municipal positions went to blacks.[15]

The "color line," as segregation was frequently called, naturally rankled the African-American citizenry, but community leaders generally counseled accommodation. Typical was the advice offered by W. H. Carter, editor of the *Sedalia Times:*

If a Negro is compelled to take the farthest back seat in an opera house he should not bring further humiliation on his race by being disorderly, but should try to convince those who occupy the boxes that there is a gentleman occupying the back seat.[16]

Yet, as compared with communities of like size in Missouri, race relations in Sedalia were relatively good. Carter commented in a 1903 editorial that white Sedalians, as contrasted with white strangers in town, were generally fair in their dealings with blacks.[17] Local white citizens certainly seemed to shun discrimination's more brutal aspects. Lynchings and whippings were common elsewhere in the state, but I could find no record of such atrocities in Sedalia during the mid- to late 1890s. The *Sedalia Democrat,* a white newspaper, gave grisly accounts of lynchings in other parts of the country, an apparent effort to assure that decent Sedalians understood the barbaric nature of these incidents. In addition, African Americans had considerable freedom to direct their own lives. They were free to establish their own businesses and social institutions, including newspapers, churches, fraternal organizations, bands, baseball teams, and the like. Blacks participated in the political process and their votes were avidly sought, leading routinely to bribes at election times.[18] Blacks were not successful in their attempts to run for elected office, but several prominent black citizens, including band member Tom Ireland, physician C. S. Walden, and newspaper editor W. H. Carter, held minor positions in the local Republican party.[19]

This relatively benign attitude may have been a legacy of the town's founder. During his life, George R. Smith demonstrated considerable concern and charity for the freed slaves. After his death, his daughters Martha Smith and Sarah Cotton, in accordance with his request, donated 24 acres for the establishment of a college for blacks. On this land the Freedmen's Aid and Southern Education Society of the Methodist Church founded the George R. Smith College in 1888, dedicating it "to the moral and intellectual culture of the colored people of the west."

The college, housed in a 4-story brick and stone structure, opened in January 1894 with an enrollment of fifty-seven. It admitted both resident and nonresident students and offered Bachelor of Arts degrees in seven areas: classics, philosophy, normal (education), science, commerce, industry, and music. It also admitted nondegree students and had a college preparatory division. The *Annual Catalogue* of 1895–96 specifies no age restrictions, but recommends that students be at least 14 (16 for degree students). Students were also to be of good moral character ("Young people of bad habits are not welcome here") and had to fulfill a variety of nonsectarian requirements on chapel and church attendance and Bible study. Additional information from a 1916 college bulletin indicates prohibitions on smoking, drinking, gambling, and attendance at theaters.[20]

The academic year in 1896–97 consisted of three terms, each of ten weeks' duration. Tuition was $1.50 per month; for music students, there was an additional charge of $2.50 per month for instruction in piano, organ,

The George R. Smith College.

violin, or voice.[21] Instrumental music was taught by a blind black musician named W. G. Smith.[22] Presumably there was also instruction in music theory—there was in later years—although it is not specified in this catalogue. The college, along with its records, was destroyed by fire in 1925.

Though Smith never became an important college, it was nevertheless a beacon of hope and culture for Sedalia's black citizens. It sponsored concerts, lectures, and poetry readings, including at least one appearance by Paul Laurence Dunbar, known nationally as the "Negro Poet Laureate."[23] Education was widely believed to be the way to equality for African Americans, and Smith College was a tangible symbol of a better future.

A major leisure-time activity for Sedalians, both white and black, was music. Almost every day the newspapers noted musical events occurring in town. Bands, including military and minstrel bands, marched through the streets. Dances and balls were held several times a week, frequently at Gregg's Hall or at Liberty Park hall; sometimes the dance music was supplied by a solo piano, sometimes by an orchestra, such as Friemel's. A popular novelty dance step that is mentioned regularly was the cakewalk. It seems that everyone in Sedalia was doing the cakewalk: blacks and whites, adults and children. The acknowledged champion cakewalker of the town was Tony Williams, who also operated the Black 400 Club.

At Wood's Opera House there might be a minstrel show by a major organization such as Mahara's or Billy Kersands's, a comic opera or operetta by Victor Herbert or some other popular favorite, a musical comedy farce, a concert by march king John Philip Sousa, or a minstrel show and cakewalk put on by the local Black 400 Club. Recitals and concerts could be heard at Wood's, as well as at the town's many halls and churches. One might attend a choral concert by the Liederkranz Society, a local German-American organization; a recital by the astonishing black piano virtuoso "Blind" Boone on one of his many visits to town. Even picnics were considered incomplete without an accompanying band or vocal quartet.

One organization that played an important part in the town's musical life was the black Queen City Cornet Band, formed in December 1891. (In 1901 it changed its name to the Queen City Concert Band, but even after that date newspapers often referred to it by its original name.) The band provided music for public and private events in Sedalia and in neighboring towns. It played for celebrations, marches, Republican political rallies, concerts (many in black churches and at the George R. Smith College), dances, and train excursions. It competed with some success in musical contests against black bands throughout Missouri.[24] It occasionally traveled with the Sedalia Browns baseball team, combining musical and athletic contests.[25] Its repertory included marches, waltzes, popular songs, instrumental rags, and patriotic and sacred pieces. A dance program might include marches, selections from the opera *Carmen,* a medley of popular songs, including such works as Neil Moret's "Hiawatha" or Will Marion Cook's "On Emancipation Day," and two-steps like the Scott Joplin–Scott Hayden collaboration *Sunflower Slow Drag.* Political events naturally suggested patriotic music, while church concerts would have religious works, such as *Holy City.*[26]

A photograph from about September 1896 shows the Queen City Cornet band as a 12-piece ensemble, mostly brasses. Several of the band members were among the black community's leaders and were accomplished in other areas, as well. Clarinetist Tom Ireland was a minor politician in town. Trombonist W. H. Carter, who also played string bass in the "colored orchestra," edited the *Sedalia Times* (1893–1905). Robert O. Henderson, here shown with the baritone horn, also performed on trombone and tuba and taught the band its pieces.[27] He later attained renown in black vaudeville as a comedian. Alto hornist James Scott was also a professional organist; snare drummer Emmett Cook was a member of Joplin's Texas Medley Quartet.

Naturally, the membership of the band was not static. Henderson, Diggs, and Martin left Sedalia by 1902; Chism died in 1903. In 1903, the band announced a search to fill vacancies: "Six good Sedalia boys who want to learn band music."[28] The instrumentation is also likely to have changed since the photograph was taken.

The Colored Orchestra was a seven-member ensemble that performed mostly in churches. Its three violinists and cornetist were joined by some of the band members: Tom Ireland (clarinet), R. O. Henderson (trombone), and W. H. Carter (string bass).

There was much quartet singing in black Sedalia, and we read in the newspapers of such groups as the Lincolnville Quartette, the Cuba Quartette

*Queen City Cornet Band, September 1896. The personnel, as identified by band
member G. Tom Ireland, are: A. H. Hickman, bass horn; R. O. Henderson,
baritone horn; W. H. Carter, valve trombone; Al Wheeler, tenor horn; William
Travis, 1st cornet; James Chism, 2nd alto horn; James W. Scott, 1st alto horn;
G. Tom Ireland, solo clarinet; Nat Diggs, 1st clarinet; Ed W. Gravitt, solo
cornet; C. W. Gravitt, bass drum; Emmett Cook, snare drum; Henry Martin,
drum major; Bert Stewart, band librarian.*

(this was during the Spanish-American War), the Pork Chops Greasy Quartette (taking its title from a popular minstrel song by Irving Jones), the
Dalbyville Quartette, and the Sedalia Quartette. Smith College also had its
own male and female vocal ensembles.

As for religious music, aside from what was heard weekly in the
churches, there was the music of tent and camp meetings. One such occasion,
on the corner of Morgan and Ohio (which was in the black part of town),
featured jubilee singers, i.e., singers of spirituals, and attracted both black
and white listeners.[29]

The only type of music that seems to have drawn disapproval was that of
piano players in the Main Street honky-tonks. Even editor Carter, himself a
musician, complained of the "piano thumping" along Main Street. The dis-

reputable nature of the establishments from which this music emanated may have influenced his attitude toward the music itself. Since Carter's office was located at 120 East Main, in the midst of the honky-tonk area (and directly across the street from the Maple Leaf Club), the late-night "thumping" may have also been an unwelcome distraction from his work.[30]

The Queen City Cornet Band and its Financial Difficulties

 The Queen City Cornet Band was a popular organization in Sedalia, but its popularity did not translate into
financial well-being. Every few months, some organization put on a benefit for the band, and the diversity of the sponsors shows how wide the group's support was. To help finance the
purchase of new instruments, both the Maple Leaf Club and the
Black 400 Club held benefit dances (on December 8, 1898, and
February 27, 1899, respectively). Although black church organizations opposed virtually everything the Maple Leaf and Black 400
did, the two camps could agree on the single issue of assisting the
band. Even the most strait-laced groups in town, such as the
Ladies' Club of the Morgan Street Baptist Church, raised money for
the band.

Despite periodic assistance, the band had serious financial difficulties. The most severe occurred in October 1899, when the ac-
cumulated back rent on its rehearsal room on East Main Street
totalled $70. Seventy dollars was an enormous amount of money
and, in a town in which a five-room house might rent for $7 per
month, it is difficult to understand how the band could have accu-
mulated so much debt.

However it came about, the money was owed and the landlord
went to court, obtained a judgment against the band, and had the
constable seize its instruments and put them up for sale.

The band members were frantic. They made appeals all over
town for help. Some local officials came forward. Judge Wilkerson purchased all the instruments, thereby preventing their disper-
sion. Then the constable—the same one who had seized the
instruments—and Judge Kinsey solicited funds around town, raising $66.85. Judge Wilkerson accepted this amount for redeeming
the instruments. The band members, having their instruments returned, serenaded the judges and the constable and paraded
down Ohio Street.[31]

This story played out over a period of five weeks, with each
daily newspaper bringing another installment in the saga. It is a
revealing story, showing both the desperately poor conditions of

the black community and the relatively benevolent attitudes of Sedalia's white population.

From a present-day perspective, the story can also be a little uncomfortable with its condescension toward the black band members. The band had no recourse but to appeal to the white officials, who paternalistically demonstrated their generosity. The black musicians, in turn, showed their gratitude in a way that was expected: they responded musically and playfully, serenading their sponsors and marching through town. One might say that they confirmed a prevailing view of African Americans as being good-natured, playful, and musical, but also irresponsible and childlike, incapable of meeting financial obligations.

This understanding of the situation might well be a distortion, for it comes from reading accounts in the white newspapers. It is unfortunate there are no surviving issues of the *Sedalia Times* from that period, for editor W. H. Carter, the band's trombonist, might have presented an entirely different perspective.

As Beatrice Martin said, Sedalia was "the musical town of the West." It was a town in which Joplin could hear a vast amount of music making, a town that offered opportunities for him to work as a performer and composer, a town where he could find congenial colleagues and appreciation. It was an ideal setting for him, serving as a springboard to his career.

CHAPTER THREE

Scott Joplin in Sedalia, 1894–1898

We were playing rags before Scott Joplin came [to Sedalia], but after he came there was quite an abundant amount of playing of hot rags. . . . We played rags, but Scott had better form than any of them. After we got to playing rags and making them really go, the birth of rags started to spread through the central states.

> Sedalia resident and Scott Joplin
> student Arthur Marshall, in a
> 1949 interview.[1]

S COTT JOPLIN AND OTIS SAUNDERS, his companion from the Chicago Fair, arrived in Sedalia in 1894. Tom Ireland related in a letter to Brun Campbell in 1947 that Joplin was a visitor at the Queen City Cornet Band's rehearsals in 1894–95.[2]

Joplin first moved into the Marshall home at 135 West Henry. There, he shared a room with the family's two sons, Arthur, age 13, and his younger brother Lee.[3] Although Joplin lived with the family for only five or six months, Arthur, already seriously interested in music, became Joplin's protégé.[4] In later years, Arthur Marshall was to become a major ragtime pianist and composer.

If Joplin had a permanent address while living in Sedalia, no evidence of it has been found. After leaving the Marshalls, he lived in a house owned by John Wesley, a white farmer, on the corner of Lamine and Morgan.[5] Subsequently, he lived in at least two other residences: the 1900 census, taken on June 6, shows him at an unnumbered house on Washington Avenue belonging to Susanna Hawkins; the 1904 town directory, the only Sedalia directory in which his name appears, places him at 124 West Cooper, the home of Solomon Dixon. Joplin's transient living arrangements may be ascribed to his profession as a touring musician.

We can only speculate on why Joplin left St. Louis, a major center of the new black music, to go to Sedalia. One assertion, found in a self-published local history of Pettis County, was that Joplin first arrived in Sedalia while on a professional tour. He was then so attracted by the town's lively night life and the quality of the Queen City Cornet Band, which he heard playing at the Maple Leaf Club, that he decided to stay.[6]

If it is true that Joplin spent some adolescent years in Sedalia, moving to Sedalia would have been a return to a familiar locale. He also would have been aware of the town's employment opportunities for musicians. Since his pianistic skills were relatively modest, he may have found in Sedalia a greater acceptance of his level of performance than in the more competitive environment of St. Louis.

Another possible reason for Joplin's moving to Sedalia was the opening of George R. Smith College, a college for blacks, which he later attended. However, there is also a claim that his decision to attend came later.

Shortly after arriving in Sedalia, Joplin began attending rehearsals of the Queen City Cornet Band. He became a member for a while, playing first cornet, but left the group before September 1896, selecting a few members to form a "gig orchestra." This was a six-piece band consisting of cornet, clarinet (Tom Ireland), E-flat tuba (probably A. H. Hickman), baritone, drums (probably Emmett Cook), and Joplin at the piano. With this band, he played for local dances, both black and white.[7]

He also played solo piano for dances and other events at halls in town, in neighboring towns, and at the major black social clubs: the Maple Leaf Club and the Black 400 Social Club. Occasionally, he played piano in brothels. Arthur Marshall named two madams for whom he and Joplin worked: Lottie Wright and Nellie Hall.[8]

While using Sedalia as his main base of operations, Joplin traveled on tours and visits. It was probably late in 1894 that he left Sedalia to tour with his Texas Medley Quartette. As suggested by the quartet "We will rest awhile" in his opera *Treemonisha* (which is set near Texarkana), quartet singing must have been a part of his life since his earliest days as a musician. However, despite assertions that he had formed the Texas Medley Quartette while still an adolescent in Texarkana, the only tangible evidence of the Texas Medley name was during the 1894–95 period.

According to Marshall, the Texas Medley Quartette actually consisted of eight members, thereby being a double quartet. The members were Scott Joplin, conductor and occasionally lead voice; his brother Will Joplin, lead voice; his brother Robert Joplin, baritone; John Williams, lead or baritone; Leonard Williams, tenor; Emmett Cook, tenor; Richard Smith, bass; and Frank Bledsoe, bass.[9] No newspaper notices of the tour have been located, but the group apparently went as far east as Syracuse, New York, and must have made a favorable impression, for two businessmen from that city—neither one a music publisher—published songs by Joplin, these being his first two publications.

Joplin wrote both the words and music for the two songs. They were published for voice and piano, but Joplin probably composed the music for quartet performance, a thought that modern performers should consider. The first song was *Please Say You Will,* copyright February 20, 1895 by M. L. Mantell, a jeweler with a store at 129 North Salina Street in Syracuse. The cover of this song names Joplin as being "of the Texas Medley Quartette."[10]

The other song was *A Picture of Her Face,* copyright by Leiter Bros. July 3, 1895. Whether the separation of more than four months between copyright registrations means Joplin made a second trip to Syracuse, or resulted simply from a filing delay by the Leiter Bros., is not known.

The Leiter brothers, Louis and Herman, operated a thriving piano distributorship at 333 South Salina Street. Neither Mantell nor the Leiter Bros. had other music copyrights, so it was apparently their enthusiasm for Joplin's performances that prompted them to publish his music.[11]

Though these two songs were Joplin's first publications, they should not be considered as inconsequential first efforts by a young composer. Joplin was already 26 or 27 years old, by which time he probably had several years of music composition behind him. The quality of the songs supports this hypothesis, for they are fully mature samples of the polite, parlor-style song of the 1890s. Though not of a sufficiently distinctive character to attract special attention, they are thoroughly professional. The melodies have grace and balance, the harmonies are more imaginative and chromatic than those of most songs of the type. There are also clear efforts to have the music mirror the text. In the first song, as the courting lover unhappily confesses an indiscretion, the music goes into a minor mode. In the second song, the expressiveness of the augmented sixth chord, a relatively sophisticated construction, is used several times. These are not the fumbling attempts of a beginning composer.

It is possible that Joplin had assistance with the music. Brun Campbell reported in 1949 that Joplin was assisted by a woman in Hannibal: "There was a music store there run by a Mr. and Mrs. Morton Walker. In those years the wife, Marie Walker, was a very talented musician. She helped Joplin with some of his early compositions, in 1895–96."[12]

There was a music store in Hannibal run by S. M. Walker, so Campbell's report has some basis.[13] However, we have no information on the nature of Mrs. Walker's assistance. If it was just help in notating the music, we have to wonder why he did not get this from any of the black or white music teachers in Sedalia.

I could learn nothing else of the Walkers, and the few surviving Hannibal newspapers from this period have nothing on Joplin. Yet it would be reasonable to propose that Joplin had performed in Hannibal in 1894 or 1895 and had attracted the attention of Mrs. Walker. It would be similar to what had apparently happened in Syracuse and was a pattern that was repeated throughout his lifetime. In an era when blacks were generally dismissed with disdain by the white population, Joplin consistently made favorable impressions, attracting a degree of support.

It is likely that Joplin spent part of 1896 in Texas, since he had three pieces published in Temple, Texas. Since Temple is more than 200 miles from Texarkana, his assumed visit to Temple probably occurred during a tour rather than while visiting his family. It is also possible that he had traveled to the area to witness the Crush Collision, a planned train collision that he could have learned of from friends who worked on the M. K. & T. Railway line. Henry Jackson, for example, a local pianist who wrote the words for two later Joplin songs, was an M. K. & T. train porter.[14]

The Crush Collision was the intentional head-on collision and destruction of two locomotives on September 15, 1896. It was planned by railway official William Crush, who intended it as a public-relations device to get feuding farmers and railroaders to end their enmity. It drew a crowd of a reported 50,000 and ended in tragedy. The crews jumped clear before the

This advertisement is not for the Crush Collision, but for a similar intentional crash of two locomotives in 1900.

collision, but boilers of the two locomotives exploded, killing three spectators and injuring dozens more.[15]

A month later, Joplin's *Crush Collision March* was registered for copyright (October 15, 1896), with a dedication to the "M.K.&T. Ry." The most notable feature of this march is its program describing the crash. Foreshadowing the tragic conclusion to the incident, the music starts in the minor mode. The introduction uses a descending scalar passage that highlights the augmented second interval (C-sharp to B-flat) of the harmonic minor. The collision occurs in the interlude within the trio section,[16] with a description written into the printed score:

> The noise of the trains while running at the rate of
>   sixty miles per hour,
> Whistling for the crossing,
> Noise of the trains
> Whistle before the collision
> The collision

The roar of the trains is represented by a chromatic bass, which gets lower as the trains near. The train whistles are high-pitched, tuned a major second apart, each with accompanying grace notes, and they get higher as the trains get closer. The work demonstrates Joplin's interest in dramatic depiction, although the result is not wholly successful. Aside from difficulties of portraying on the piano the horrendous and clamorous nature of the event, Joplin subverts the drama to a formal convention of musical repeats. It is thereby antidramatic that, after the crash and the following strain of "calm" music, the crash reoccurs.

Almost obscured by the brief dramatic depiction is the piece's musical sophistication. In several sections Joplin forgoes the mechanical bass-chord alternations in favor of a melodic bass and interactions between bass and treble melodies (Ex. 3-1).

Ex. 3-1. Scott Joplin, Crush Collision March *(1896): (a) B8-16; (b) analytic reduction showing interplay between treble and bass lines.*

(b)   Reduction showing interplay between treble and bass lines.

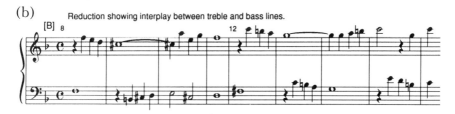

Combination March and *Harmony Club Waltz* were both registered for copyright a month later, on November 16, 1896, but they may have been published earlier, for they are listed on the cover of *Crush Collision March*. We have no clue as to the meaning of the title *Combination March;* the waltz was probably composed to honor a Harmony Club.

In each, one can note the signs of a thinking musician. For example, in *Combination March,* in measures 1–2 and 5–6 of the interlude between the A and B strains, there is an interplay between parts, similar to that in *Crush Collision.* In *Harmony Club,* strain E has slow melodic rhythms, featuring sustained and lushly harmonized third beats, accentuated so as to suggest syncopation. The following strain provides a sharp contrast with a quick, light, flowing melody (Ex. 3-2). This is as good a waltz as any composed during the period. With a significant publisher, it could have attracted the attention it deserved.

Ex. 3-2. Scott Joplin, Harmony Club Waltz *(1896): (a) D1-4; (b) E1-4.*

A constant theme in Joplin's life is a dedication to education, both as a student and as a teacher. The formative musical studies he had as a youth, under the guidance of his "German professor" Julius Weiss, apparently had a profound effect on his perceptions of music, culture, and education. For Joplin, music was not just entertainment; it was also potentially ennobling, especially the music of the "higher planes," that is, music of the European masters. In this conviction, he echoed opinions of the time. According to these opinions, it was especially important for the African American to grow

beyond "Negro music" as one manifestation of a general educational uplift-
ing, the prerequisite for acceptance into the cultural and social mainstream.

Our first glimpse of Joplin as an educator is in his relationship to Arthur
Marshall. As Marshall told it:

> He'd taken a liking to me when he roomed with me and my brother at my
> mother's home. [At] our old square piano (that was all we had to practice on)

The earliest known photograph of Scott Joplin, as it appears on the cover of
Swipesy. Cake Walk *(1900).*

Arthur Marshall, as he appears on the same cover.

> Mr. Joplin would tell me, "Go ahead and play that again—go ahead and play
> that piece again." Well, that way we began to get together and get together. . . .
> He was kind to all of us musicians that would just, as I say, 'flock' around him,
> 'cause he was an inspiration to us all. We always treated him as daddy to the
> bunch of piano players here in Sedalia.[17]

Judging by this passage, Joplin took it upon himself to guide many
young pianists in town, and they looked upon him almost as a hero, much in
the way an athlete might be considered today. Adding to the passage's credi-
bility is a parallel description of Joplin in St. Louis, given by ragtimer
Charles Thompson: "Both Joplin and [Louis] Chauvin had a string of fol-
lowers as they strutted around the district. When Joplin returned for a visit
there would be just like a parade down Market St."[18]

Another of his students was Scott Hayden, a school friend of Marshall's.
Joplin later collaborated with both Marshall and Hayden on several rags.

The one remaining student we know from this period was S. Brunson
Campbell, a 15-year-old runaway white youth who went to Sedalia to seek
out Joplin in 1899.[19] In later years, Campbell wrote much about Joplin,
recorded music supposedly in the style that Joplin had taught him, and
helped to establish a special Joplin collection at Fisk University. Campbell
could write authoritatively of Joplin during the year they were acquainted,
but his writings about Joplin during other periods are highly suspect.[20]

Scott Hayden.

Brun Campbell.

There were probably other students, as well—note the reference to "the bunch of piano players here in Sedalia"—but their names have not come down to us. In one of the *Maple Leaf Rag* stories, there is mention of a "colored boy" who was coached in playing the rag, but as he is not named this boy was apparently neither Marshall nor Hayden.

Joplin's respect for education is seen also in his enrollment, while in his late 20s, at the George R. Smith College. According to Brun Campbell, Joplin's decision to attend Smith was made in consultation with Otis Saunders and Tony Williams, the latter being Sedalia's leading black impresario, its foremost cakewalker, and head of the Black 400 Club.

> Otis Saunders, Scott Joplin and Tony Williams got into a huddle to decide Joplin's future musical career. They decided that the thing for Joplin to do was to go to a musical school there in Sedalia and learn musical arrangement and take a course in harmony. . . . He took their advice and in a brief time he left the musical school, having completed his courses. He immediately put five of his first rags down on paper. The first was his *Original Rags* in 1897 which he sold to a Kansas City, Mo. music publisher, who published it in 1898.[21]

Since the college records were destroyed in a fire, we have no accurate information as to when Joplin attended, what he studied, or whether he gained appreciably from these studies. That he attended prior to the end of 1897, as Campbell contends, is possible. Arthur Marshall is reported to have attended the college with Joplin,[22] and Marshall would have been eligible for admission in November 1895, when he turned fourteen. Their attendance could therefore have been between January 1896 and late 1897.

Presumably, as an older student, Joplin did not live on the campus and was not a degree student. Though he was an eminently respectable individual, his employment in saloons and brothels, his playing for dances, and his love for the theater could not have been looked upon kindly by the administrators of the college. It is possible, however, that even if these facts were known, the college's mission of educating willing and promising black citizens outweighed moralistic ideals.

Assertions of what Joplin learned at Smith are suspect. We are told that his lessons there enabled him to notate the complex syncopations of ragtime, and that he took "advanced courses in harmony and composition."[23]

These claims are strange, and even contradictory. How could one who had difficulty notating ragtime syncopations be prepared to take courses in advanced harmony and composition? Actually, there is nothing particularly complex about the notated rhythms of ragtime. One would think his teacher Weiss could have taught him such fundamentals, and his two song publications from 1895 suggest he was already functioning on a fairly sophisticated musical level before he entered the college. Even if he required assistance in notation from Marie Walker of Hannibal, he would not have needed to re-learn this skill at the college.

I also doubt that he could have studied advanced composition at the college. There is no notice in the newspapers of a composer teaching at Smith. And why should this small college, designed to give African Americans the most basic type of higher education, perhaps to train them as elementary school teachers, have someone qualified to teach *advanced* composition?

Joplin undoubtedly attended Smith with the anticipation that he would gain from the experience. However, with the information currently available, we have no way of knowing whether his expectations were fulfilled.

An important part of Joplin's social and musical life in the late 1890s was connected with Sedalia's two black men's clubs: the Black 400 Club, and the Maple Leaf Club. These organizations had among its members some of the town's brightest and most enterprising young black men, some of whom were among Joplin's closest associates.

In forming these clubs, the men were following a time-honored practice in town. Sedalia was a community of clubs. There were numerous fraternal and benevolent associations, including some for African Americans. There were even more specialty clubs for music (most comprised of women, but also some of men); music and dramatic performance; social service; and literary, current affairs, and philosophical discussions. There were several dance clubs, among them the Monday Night Club, the Manhattan Dancing Club, and the Autumn Leaf Club. To the annoyance (or amusement) of the white society gentlemen comprising the Autumn Leaf, their organization was sometimes confused by newspapers with the black Maple Leaf Club. These mistakes, when they occurred, were easy to detect: should one read a report of a party at the Maple Leaf Club in which ice cream and punch were served, and in which chaperones were present, one could be certain that this was actually the Autumn Leaf.[24]

Arthur Marshall stated that the Maple Leaf Club preceded the Black 400,[25] but my evidence indicates otherwise. The earliest reference is to the Black 400 in the spring of 1898.[26] At that time it was not a place, but a performing group of vaudevillians and cakewalkers led by Tony Williams. On April 19th it presented an entertainment at Wood's Opera House. That it was able to mount a production at Wood's suggests that the organization had already proved itself capable with entertainments at lesser places in town, as well as in neighboring towns. Among the competitors of the cakewalk contest on the 19th was "Doc" Brown of Kansas City, commemorated in 1899 in the music *"Doc" Brown's Cake Walk* by Charles Johnson.[27]

The event was proclaimed a brilliant success, attracting "the largest and most fashionable audience of the season." Another show was scheduled at the opera house the following month, on May 26; the advance promotion gives some idea of the show's components:

Wood's Opera House.

By special request of the society white people of Sedalia and the theater goers to repeat the cake walk of April 19, those wishing to see this grand entertainment will have another chance to see who "is the warmest baby in the bunch" . . . There will be . . . cake walk, Irish specialities, the song and dance team, and famous southern hoe-down dance, assisted by the Cuba quartette. Don't fail to see R. O. Henderson in his laughable subject, "Who Am I?" Williams and Henderson will amuse you with a character song, "Pack Your Trunk and Go."

This show, too, received a glowing review.[28]

Through the spring and summer, Tony Williams put on cakewalks in Sedalia and in neighboring towns. When members of white clubs wanted to learn this novelty dance, they would usually hire Tony Williams to instruct them.[29]

In October 1898, Tony Williams, assisted by his brother Charles, opened a club room for the Black 400 on the second floor of 106 East Main Street, above Mulford & Rawle's saloon.[30] Despite its location in "Battle Row," i.e., the section of Main Street that contained most of the town's saloons and other rough hangouts, Williams made it clear that the club would be thoroughly respectable.

It is the intention of Tony to make the club a high-toned resort, and no foolishness will be tolerated. Admission will be by card, and the manager says no toughs will be permitted to enter the premises.

The Black 400 had occupied the second story of this building (on the right).

> The colored society people will be out in numbers tonight to attend the open-
> ing ball of the season, to be given at the rooms of the "Black 400" club. Tony
> Williams, as manager of the club, will see to it that the affair is a society function
> in its strictest sense.[31]

Respectability seems to have been important to Williams, and on one occasion in the fall the club was host to Republican party politicians who were
invited to speak to black voters.

Through the fall and early winter, by which time membership in the
Black 400 had grown to 120, the club held several balls and all were reported
to have been run on a dignified level. As Tony Williams had promised,
admission was by card, such as the one issued to Miss Carrie Ireland, the
sister of clarinetist Tom Ireland, in late December 1898:

TONY WILLIAMS,
Joplin, Mo.

Tony Williams in 1900.

400 Social Club
C. E. Williams, president
106 East Main Street
Sedalia, Missouri

The management of this club would be pleased to extend their hospitality to you and anyone you may wish to bring to their Fancy Dress Calico Ball. Monday night.

Mamie Walden, pianist.
Present at door. Admission 25c.[32]

White society people were frequently invited as spectators and, from seats on the stage, would judge the contests. The music on at least two occasions in 1898, November 24 and December 26, was provided by Professor Scott Joplin, the Queen City Cornet Band also being hired on the second evening.[33]

In the meantime, the Maple Leaf Club came into being, reportedly being "composed of the best and brightest young colored people of the city." They held their first ball on Thanksgiving night, November 24, 1898. The club did not yet have its own club room and held the ball at Goodrich Hall, on East Main Street.[34]

Is it just coincidental that the date of the first Maple Leaf Club ball, November 24, is the same date that Lottie Joplin had cited as her husband's birthday? It is tempting to ascribe some significance to this convergence of dates, but we are not sure that Joplin was even present at the Maple Leaf dance. He was playing piano for the Black 400 dance that night.

The following month, on December 14, 1898, the Maple Leaf Club was incorporated. The club was run by Walker Williams and his brother Will (unrelated to the brothers Tony and Charles Williams of the Black 400), but the nominal president was H. L. Dixon. Thomas Tompkins, who was vice-president, was also one of the club's pianists. Other pianists among the club's thirty-one charter members were Scott Joplin, Joplin's pupil Arthur Marshall, Archie Chastine, John Reed, and Malford Alexander.

The purposes of the club, according to the charter papers,

> . . . is to form and maintain a club, and to maintain a club house for the purpose of advancing, by social intercourse, the bodily and mental health of such persons as might be, or thereafter become its members, and by the friendly interchange of views and discussions, advance the interests of its members; to obtain a place of common and friendly intercourse of such members with each other; to maintain a library for its members, other species of amusement and entertainments for the benefit of its members.[35]

The Maple Leaf Club seems to have followed the model of the Black 400. I have no doubt that, at its inception, the members of the Maple Leaf were sincere in setting out high goals. One of the first events the club scheduled was a masked ball for the benefit of the Queen City Cornet Band, to raise money for new instruments.[36]

The Maple Leaf soon opened its own club room, on the second floor of 121 East Main, across the street from the Black 400. A surviving business card gives us, on the reverse side, a colorful view of the club's character.[37]

The opening text on the back of the card suggests that "William's Place" may have been an alternate name, although it is never mentioned in the newspapers. Joplin played piano at the club, but his listing as "the entertainer" suggests he was more than a pianist; he probably sang and may also

The Maple Leaf Club occupied the second floor in the left side of this building.

THE MAPLE LEAF CLUB.

SEDALIA,

121 EAST MAIH ST.
W. J. WILLIAMS, PROP.

MO.

→THE GOOD TIME BOYS.←

WILLIAM'S PLACE, for Williams, E. Cook Allie Ellis, Taylor Williams. Will give a good time, for instance Master Scott Joplin, the entertainer. W. J. Williams the slow wonder said that H. L. Dixon, the cracker-jack around ladies said E, Cook, the ladies masher told Dan Smith, the clever boy, he saw L n Williams, the dude, and he said that there are others but not so good. These are the members of the "Maple Leaf Club,". Don't forget Jake Powel, the plunger and King of kitchen machanic.

Business card for the Maple Leaf Club.

have played other instruments. "E. Cook" was Emmet Cook, drummer with the Queen City Cornet Band and a singer with Joplin's Texas Medley Quartette. "L n Williams" was probably the singer Len or Lynn Williams. "Jake Powel" (correctly, "Powell") was with the Cuba Quartette, which had performed at Wood's Opera House earlier in the year. Later, with Arthur Marshall and Richard Smith, he joined Dan McCabe's Coontown 400, a minstrel company.[38]

I suspect the library mentioned in the incorporation papers was a lawyer's invention. The club definitely was not a literary society. (There was a literary society composed of black Sedalians; its meetings were held in a chapel). The club's newspaper announcements were almost all of public events such as dances, cakewalk contests, and masquerades. On one occasion, it held a special exhibition featuring the "world-wide celebrity" Jack Gentry, also known as "Black Diamond." Gentry's fame was based on his ability to hold four billiard balls "in his face." He would follow this remarkable feat by inserting into his mouth a plate of six-inch diameter.[39] The Maple Leaf Club was clearly not dedicated to high culture.

Less than a month after the Maple Leaf's incorporation, trouble began. Local black pastors complained of the Maple Leaf and the Black 400, and both clubs were ordered closed by the mayor. Tony Williams accused the mayor, a Democrat, of closing the clubs for political reasons, of trying to prevent blacks from unifying under the Republican banner in the coming election. The following day he met with the mayor and convinced him that there was nothing improper occurring at the clubs. Both clubs were then permitted to reopen.[40] (That the mayor agreed to meet with Williams and was persuaded by him demonstrates that Williams was respected in Sedalia.)

But problems did not end there. A week later, Charlie Williams announced that, due to pressure from the black churches, the Black 400 would close at 7 p.m. on Sundays to encourage its members to attend church services. That move did not satisfy the ministers, and they published an open letter accusing both clubs of gross immorality:

> . . . the Black 400 and Maple Leaf clubs are a detriment to the morals of our people— affording a loafing place for many of our girls and boys, where they drink, play cards, dance, and, we have been informed, carry on other immoral practices, too disgraceful to mention.
>
> Whereas, since we are ashamed of these iniquitous practices, and, too, having the streets filled until late at night with many of our young girls, lowering the morals of our homes, counteracting the influence of our public school, Christian college and churches; we earnestly appeal to you, the authorities of the city, to assist us in the elevation of our race by putting a stop to these nefarious club meetings.
>
> The better class of the negroes of this city are not in sympathy with such organizations and protest against their existence.

> We do not believe the statement . . . that the best white ladies and gentle-
> men of this city attend their entertainments; and we are sure that the best class
> of negroes do not attend them.
>
> By permanently putting an end to these abominable loafing places—hot
> beds of immorality—you will stop a great source of vice, create a better moral
> atmosphere for our young people, and render some of our homes happier.[41]

Charlie Williams denied the accusations, saying that his club was open
for inspection at any time, that "at two dress balls forty of the best white
people of Sedalia were in attendance as spectators," and that the club donated
part of its entertainment receipts to charities, including a charity headed by
one of the accusing pastors, a claim supported by the *Sedalia Times,* a black
newspaper.[42]

The city authorities refused to grant the pastors' request, and the clubs
continued operating. Within the week, both clubs had dancing events, and a
report noted that at the Black 400 "a large crowd of white people were given
seats of honor on the platform at the south end of the hall." In an effort to
establish legitimacy and deflect future criticisms, the Black 400 followed the
lead of the Maple Leaf and filed for incorporation. Events in following weeks,
though, gave the court reason to delay action. In the end, incorporation was
never granted.[43]

The pastors renewed their attack. Claiming their initial protests had
been too mild, they now accused the Maple Leaf and Black 400 of being
worse than any of the town's many saloons.

> We do not believe that there is a saloon in the city of Sedalia, where colored
> women and white men would be able to play cards, drink and associate togeth-
> er. We have good evidence that the Black 400 allows such.
>
> We do not believe that a saloon is allowed to sell minors intoxicating
> drinks. Minors buy all they want to drink at the Black 400 club. There is a
> saloon in the story below the hall of this club . . . and an elevator or schute for
> the saloon furnished the drink supply. . . .
>
> The better class of white citizens should not fault the negro for his crimes
> and immorality, when they allow such pest dens to run unchecked. . . . We are
> forced to acknowledge that there are too many sports, cake-walkers, loafers and
> criminals among us—all the product of such loafing dens. . . .[44]

It is difficult to appraise these conflicting claims. Respectability was an
aim of both clubs. Published reports confirm the proper deportment at the
clubs' major events. In addition, the Maple Leaf's incorporation—a rarity for
an establishment on Main Street—reflects its members' desire for recognition
and legitimacy. The clubs, as claimed, regularly invited "white society" as
spectators to their events, and while the presence of this segment of population did not guarantee good behavior, it is a condition that held weight at the

time. Both clubs included among their members some highly regarded black men of Sedalia. The Maple Leaf Club had Scott Joplin, who was praised by all the newspapers. The Black 400 Club was headed by Tony Williams, who was viewed as a highly responsible individual whose major efforts were in organizing theatrical presentations. The 400's officers included also C. S. Walden, a black physician and appointee to the Pettis County Pension Committee. It does not seem that the members of these clubs would have been a bunch of rowdies.

But, regardless of whether the ministers' protests were justified, there was some truth to their accusations. Both clubs lacked liquor licenses, but they did serve liquor. The members definitely danced, and they probably played cards and tolerated the association of black women and white men. To what extent the clubs were responsible for "immoral practices, too disgraceful to mention," is impossible to say. With at least fifteen brothels and numerous street prostitutes on Main Street, there certainly was no need for either club to sponsor illicit sex. Madams and prostitutes were regularly arrested and fined for their professions, but neither club was ever charged with similar offenses.

A clue to understanding the pastors' claims is in the final line of their second protest, in which they group cakewalkers with criminals. Some church members viewed dancing as sinful, along with drinking and card playing. While there were many saloons, gambling dens, and dance halls along Main Street, these places were clearly disreputable. The social clubs, on the other hand, had at least the veneer of respectability. This might have been the perceived danger: that the social clubs, through their basic respectability, might lure youth into what were viewed as disreputable activities.

The Ministers' Alliance and the social clubs held a meeting, at which agreements were apparently reached. Immediately afterward, several dances and concerts were scheduled at the clubs, some designed as benefits to raise money for the town's black poor.[45]

Despite these gestures of acting for the good of the community, the clubs' problems were not over. A week later, Tony Williams and another member of the Black 400 were arrested by a federal marshal for serving liquor without a license. The same evening, the club held a masquerade ball, at which the marshal was present, and the hall was packed with prominent white people who came to Tony Williams's support.

The trial was held in Clinton, some forty miles away, and the Black 400 members were defended by a former mayor of Sedalia. At the trial, Tony denied the charges, which he said were based on malice, and explained that the club's purposes were social and literary, and that it gave respectable entertainments. The club members were found innocent and Tony Williams continued running the club and organizing entertainments at Wood's and elsewhere.[46]

The Black 400 held its last dance on May 31, 1899, and then closed its doors for the summer, the hot, humid Sedalia weather making their second-floor hall too uncomfortable. (The Maple Leaf probably closed for the summer, too.) Though Tony and Charles Williams continued to present outdoor dances throughout the summer, they were disturbed at being "grossly misrepresented by the press and by the clergy" and decided to leave the club (with Dr. Walden taking over as head).[47] In August, Tony (but not Charles) settled in Joplin, Missouri, and in December he opened a 400 Club there.[48]

The hall where the Black 400 had met now reverted to its former name, D. O. H. Hall,[49] and was taken over by a new black men's organization, called simply the Social Club. This new group gave its first dance on October 25, 1899. The affair was managed by Charles Williams, had music supplied by former Joplin student Scott Hayden, and had an arrangement committee that included Howard Dixon, formerly president of the Maple Leaf Club.[50] Dixon's participation indicates that things may not have been going well across the street at the Maple Leaf Club.

The first blow-up at the Maple Leaf occurred on October 1, 1899. It was a night in which "Battle Row" lived up to its name, featuring a record number of street brawls with the use of knives and beer bottles, with women participating as enthusiastically as men. The highlight of the evening was a fight involving Arthur Marshall.

A young man named Ernst Edwards brought his favorite girl to the dance at the Maple Leaf, but Arthur took a liking to her and tried to escort her home. The two young men quarreled and went out to the street to continue their fight. There, Arthur pummeled Edwards with his cane, which was weighted. Edwards, in the words of one newspaper, "who comes of fighting stock, pulled his gun and commenced shooting, while Marshall commenced to run." Another paper reported that "from what authorities can learn, he is running yet."[51]

From here on, it was down hill for the Maple Leaf Club. On October 24, the entire night shift of the police department raided the club and arrested Walker Williams for selling liquor without a license.[52] Two months later, the club was again in the news as two women in the club room engaged in a vicious battle in which both were severely injured.[53] The end came ingloriously in January.

On January 14, 1900, members of the Maple Leaf Club invaded the Social Club and a fight ensued with knives, bottles, and one shooting, which caused a minor wound. The only reason for the fight reported by the newspapers was a rivalry that existed between the clubs. Walker Williams was cited in one paper as the leader of the invasion, but since he was not among those arrested this report might not be true. The person who used the gun was Will Hubbard, who was a charter member of the Maple Leaf but is otherwise unknown to us. Hubbard was bound over to the grand jury. The other five Maple Leafers arrested were let off with fines ranging between one and ten dollars.[54]

This last event, capping a year of protests, was more than town officials would tolerate. The black social clubs were closed on January 25, 1900.[55] The members of the Social Club protested, claiming they had done no wrong, but there are no signs that they ever again held events at 106 East Main. Charles Williams left Sedalia and joined his brother Tony in Joplin, taking Scott Hayden along to be the house pianist at their 400 Club in that town. A report has it that a teenage James Scott, later to be one of the major ragtime composers, also played at the club in 1900.[56]

In 1947 Tom Ireland, the clarinetist for the Queen City Cornet Band, commented on the Maple Leaf Club: "A young colored man Walker Williams open up a club room. . . . After Joplin arranged his 'Maple Leaf Rag' and called it the 'Maple Leaf Club.' It folded about 1900 and quit."[57]

Perhaps the brevity of Ireland's comment indicates he did not remember the club as being very important. For all its fame, the Maple Leaf Club existed for less than two years. It was not even Sedalia's dominant black men's club of the time, being secondary to the Black 400.

There is much more we would want to know about the clubs. The evidence strongly suggests that both began as reputable organizations, and that the Black 400, despite the ministers' protests and the liquor incident, remained so. The final episode, the invasion of the Social Club (or Black 400) by the Maple Leaf, is puzzling. How could the club's membership have fallen to the level of a street gang? Were these six invading members representative of the club? If so, how could Scott Joplin, an eminently respectable individual, have been a member? Or had he left the club? The Maple Leaf Club has been enshrined because of Scott Joplin's celebrated rag, but the club's fame seems otherwise unmerited.

CHAPTER FOUR

The Maple Leaf Rag, 1899–1900

But when Maple Leaf was started
　my timidity departed
I lost my trepidation
　you could taste de admiration.
Oh go 'way man
　I can hypnotize dis nation
I can shake de earth's foundation
　wid de Maple Leaf Rag.

> from *The Maple Leaf Rag. Song*
> (1904), words by Sydney Brown

*I*T WAS WITH RAGTIME that Joplin become a composer of
note. Ragtime had already started—more than a hundred rags
had been published before his earliest appeared in print—but he was on the
scene as the style was being defined. His place in ragtime history was as
composer of the idiom's most refined and sophisticated piano rags.

The time and place of ragtime's emergence cannot be pinpointed, for
incipient ragtime was known in many places throughout the nation. Its most
characteristic element is its syncopated rhythm, its ragginess, its *ragged time*.
Joplin said as much:

> "Why do you call it ragtime?" some one asked him long ago.
> "Oh!" replied Joplin, "because it has such a ragged movement. It suggests
> something like that."[1]

In their simplest, most general and stereotypical form, ragtime rhythms were syncopations in the treble—or the right hand of a piano part—set against a regular duple, march-like bass (Ex. 4-1). The syncopations were associated with the music of African Americans, and by the 1880s examples a) and b) were used on the minstrel stage to caricature black music.[2]

Ex. 4-1. Characteristic ragtime rhythms.

In the early 1890s, syncopations that were to become the stereotypical rhythms of ragtime were applied to two musical forms, the song and the march.

The ragtime song already had a generic name: "coon song." The coon song was a child of the minstrel stage, and, like much that emanated from that tradition, it caricatured and demeaned African Americans. Through the early 1890s, several coon songs appeared with syncopated rhythms. In 1986 a song appeared that explicitly connected this genre and ragtime: Ernest Hogan's song *All Coons Look Alike to Me,* which had appended to it an optional arrangement labeled "Choice Chorus, with Negro 'Rag' Accompaniment."[3]

Scott Joplin worked with both the ragtime song and the instrumental rag, but it was with the latter that he reached greatness. The instrumental rag evolved from joining the syncopated ragtime rhythms with the form and regular duple pulsations of the march. This combining of elements points to the dance origins of ragtime, for the march at this time was more than a martial form: it was also a popular dance form, known as the two-step. Even such an eminently military creation as John Philip Sousa's *The Washington Post. March* (1889) was used for dance and became known as the "Washington Post Two-Step."[4]

The form of the march, which was common also to such dances as the polka and schottische, was adapted for instrumental ragtime. This form consists of a succession of 16-measure strains, each traditionally designated in analytical diagrams with upper-case letters between double-bars. Each strain, in turn, is typically composed of four 4-measure phrases. Most often there are three strains, the first two being in the tonic key (designated with a Roman number "I"), and the last, called the "trio," being in the subdominant key (Roman numeral "IV"). The general layout, with repeats indicated with pairs of two dots (or "colons") within the double-bars, is

Trio
‖: A :‖: B :‖ A ‖: C :‖
I   IV

Added to this format was frequently a four-measure introduction, a four-measure introduction to the trio, and an interlude between the repeats of the C strain.

Trio

Intro ‖: A :‖: B :‖ A ‖ Intro2 ‖ C ‖ Interlude ‖ C ‖
     I   IV

There were many departures from these patterns. The reprise of A after B might be omitted; A might return after C, in the tonic or in the subdominant; or there may be several additional strains added and different key relationships. Rags by Joplin and the other "classic" ragtimers differed from the norm of three strains in having a fourth (D) strain.[5]

Ragtime, or some early form of it, was probably known to the pianists of the St. Louis Chestnut Valley region when Joplin is thought to have been there in the late 1880s and early 1890s. His close friend Tom Turpin is reported to have composed *Harlem Rag* in 1892. The following year, at the Chicago World's Fair, the ragtime style became known to the general public.[6]

The rags of the early part of the 1890s were not published until later years, if at all. William Krell's *Mississippi Rag* was the first instrumental rag to appear in print, in 1897, a year after the initial publication of ragtime coon songs. Later in 1897, Turpin's early *Harlem Rag* became the first instrumental ragtime publication by a black composer.[7]

By the time Joplin published his first rags, in 1899, there were more than a hundred rags in print. The vast majority used the rhythms cited in the examples cited earlier in the chapter and could be typified by the melodic types, formal designs, and rhythmic configurations found in the A strain of Kerry Mills's immensely popular *At a Georgia Campmeeting* (Ex. 4-2). Note that the sixteen-measure strain consists of four four-measure phrases. Melodically, the first and third phrases were usually similar, with the second phrase sometimes also being in the same mold. The final phrase might have a changed texture, such as bare octaves. As illustrated by the Mills piece, the thematic pattern could be diagrammed as a-a'-a''-b. Other common patterns are a-b-a'-b' and a-b-a'-c. Harmonically, the second phrase frequently cadenced on the dominant.[8]

Brun Campbell relates that Joplin told him of completing his first rag, *Original Rags,* in 1897.[9] Joplin tried to place *Original Rags, Maple Leaf Rag,* and possibly *Sunflower Slow Drag* in 1898 with the Sedalia publisher A. W. Perry & Son, but was turned down.[10] He had not yet approached John Stark, the Sedalia publisher with whom he would later form a fruitful relationship, possibly because in the 1898–99 directory Stark listed himself as a retailer of music and instruments, not as a publisher.[11]

Ex. 4-2. Kerry Mills, At a Georgia Campmeeting *(New York: F. A. Mills, 1897),* A1-16.

Joplin then brought his music to Kansas City publisher Carl Hoffman, who accepted only *Original Rags,* issuing it early in 1899.[12] Apparently there was an important stipulation regarding the publication, for it bears the legend "Arranged by Chas. N. Daniels." On the Claimant Card in the Copyright Office, Daniels is listed as composer. In Hoffman's advertisements for the piece (selling for 14 cents), Daniels is sometimes named as arranger, whereas Joplin's name never appears.[13]

Daniels was a competent composer who, under both his own name and the pseudonym Neil Moret, wrote many notable works during the ragtime years. His *Hiawatha* (1901) was a particularly popular (non-rag) piece within the "Indianist" phase of popular music.[14]

But I doubt that Daniels arranged Joplin's *Original Rags.* I do not see any part of the score that is closer to Daniels's style than to Joplin's. It is possible that Daniels's name was added simply as a requirement of publication, a common practice, or perhaps even without Joplin's prior knowledge.

Even if Daniels was not the arranger, there is one part of the composition on which he may have had some influence. This is the fourth strain, which bears a marked resemblance to a work published by Hoffman the previous year: H. O. Wheeler's *A Virginny Frolic* (Ex. 4-3).[15] The keys and harmonic

Ex. 4-3. (a) H. O. Wheeler, A Virginny Frolic *(Kansas City, MO: Hoffman, 1898), A1-8; (b) Scott Joplin,* Original Rags *(1899), D1-8.*

designs are identical; the melodic contours are similar, and the unusual off-beat right-hand pattern that occurs in measures 3-4 and 7-8 is the clincher. It would certainly appear that *A Virginny Frolic* is the source of Joplin's strain. But what a difference! The reworking marvelously demonstrates what separates a hack from a genius. We do not know whether it was at Daniels's insistence that Joplin used this strain, but his transformation of it into his own idiom is masterful.

It is not just Joplin's transformation of the fourth strain that makes *Original Rags* notable. With this piece Joplin reached his stride. He finally published a work in the genre in which he was to excel, and it is already a superior sample of the style. Of all the rags written up to this time, perhaps only Tom Turpin's are of comparable quality.

We can point, for example, to the assuredness with which he directed his harmonies toward a tonal center (Ex. 4-4a); how he perceived the richness of contrary motion in places where his contemporaries brought in bare octaves

Ex. 4-4. Scott Joplin, Original Rags: *(a) E13-16; (b) D13; (c) B1-4.*

(Ex. 4-4b; cf. mm. 13–14 in Mills's *At a Georgia Campmeeting,* Ex. 4-2). Of particular interest from the view of Joplin as a black composer are his blues-suggesting appogiaturas (Exx. 4-4c, 4-5a). This was more than a decade before the blues emerged as a recognizable genre in printed sheet music.[16] With *Original Rags,* Joplin found his idiom.

It is likely that Joplin sold the rights outright to the publisher and was unhappy for having done so. In listings of his works on sheet music, he never mentioned *Original Rags* even though it remained in print for a considerable time, being reissued first by Whitney-Warner, and then by Remick. Ultimately, it may have been to Joplin's benefit that he struck such a poor deal for *Original Rags,* allowing credit to be taken by Daniels. It may have taught him a valuable lesson about the pitfalls of publishing. Perhaps it was this experience that led Joplin to hesitate before publishing another rag, and finally to retain a lawyer to assist him in negotiating a contract.

Echoes of Scott Joplin, Part 1: Original Rags

 In using Wheeler's *A Virginny Frolic,* Joplin apparently borrowed from the music of another composer. More often, echoes of Joplin are heard in the works of others. In 1903, the Sedalia publisher A. W. Perry issued L. Edgar Settle's *X. L. Rag,* dedicated to the X. L. Club of New Franklin, Missouri. In the third theme of this work we hear what could have come from

Ex. 4-5. (a) Scott Joplin, Original Rags, *E1-8. (b) L. Edgar Settle,* X. L. Rag *(Sedalia: Perry, 1903), C1-8.*

the fifth theme of *Original Rags* (Ex. 4-5). The resemblance is not so close that it could not have been accidental, but given Settle's connection with a Sedalia publisher and his other borrowings from Joplin (to be discussed), we must take notice.

Will H. Etter's *Whoa! Maude,* from 1905, looks more suspicious. The C theme of this work must certainly have come from the opening of *Original Rags* (Ex. 4-6).[17]

Joplin's next publication was the *Maple Leaf Rag,* his most important composition and the best-known instrumental rag of the period. It was important to Joplin, first, because it made his reputation; thereafter he was always known as the composer of the *Maple Leaf Rag.* Second, because the

Ex. 4-6. *(a) Scott Joplin,* Original Rags, *A1-8; (b) Will H. Etter,* Whoa!
Maud *(Galveston: Goggan, 1905), C1-8.*

publication included a royalty contract, the income from this one work
changed the conditions of his life.

Whether *Maple Leaf* was composed in 1897 or 1898 is uncertain, but it
seems to have been known in Sedalia prior to its publication in August or
September of 1899.[18] Brun Campbell told of reading from a manuscript copy
around 1898.[19]

Joplin recognized prior to its publication that the *Maple Leaf* was a
special piece of music, unlike any other rag ever published. He said to his
protégé Arthur Marshall, "Arthur, the *Maple Leaf* will make me King of
Ragtime Composers."[20]

Given that the publication of the *Maple Leaf Rag* was one of the momentous incidents in Joplin's life, as well as a signal event in ragtime history, it is
not surprising that it has become subject to myths. At this point, we are not
sure precisely how the publication came about. One of the most frequently

repeated stories is one that Blesh and Janis characterize as a "legend" that "may also be fact." In this version, Sedalia music and instrument dealer John Stark stopped on a hot summer afternoon at the Maple Leaf Club for a beer. That was when he heard Joplin play the *Maple Leaf Rag*. Impressed, he asked Joplin to bring the music to his store the next day and the deal was consummated.[21]

Because of Blesh and Janis's initial characterization of this story, we might hesitate to accept it, but Mildred Steward, Arthur Marshall's daughter, gave the same account:

> Poppa said Joplin took it to Kansas City, but didn't make it there. He was sitting up here at the Maple Leaf Club on Main Street and playing it and John Stark walked in and heard him playing it and that's where the *Maple Leaf Rag* was bought right then.[22]

Still, this account leaves unresolved issues. Sedalia was a town overflowing with saloons; there was a saloon two doors away from the Maple Leaf Club. If Stark wanted a beer, why would he have gone to a black social club for it, one that did not even have a liquor license? On a hot summer day in a town known for uncomfortable humidity, why would he have gone to an establishment on the second floor, where the Maple Leaf was situated, rather than one on the cooler street floor? Was the Maple Leaf Club even open during the summer? The Black 400 Club, across the street, closed down during the hot months.

The biggest problem with accepting this account is that there are other versions of what happened, from equally authoritative sources. William P. Stark, John Stark's son and manager of the music store in Sedalia in 1899, told a story set entirely in the Stark office. He makes no mention of a meeting at the Maple Leaf Club.

> Mr. [John] Stark, when Scott first played *Maple Leaf Rag* for him, shook his head and said: "It's too difficult. Nobody will be able to play it." So Scott said: "If I find someone right outside your house in the street who can play it, will you then publish it?" Mr. Stark said he would. Scott went out and came back with a little Negro boy of about fourteen or so who settled down at the piano and played Maple Leaf Rag right off the sheet and without a flaw. Mr. Stark slapped his thigh and said: "I'll publish it." What Mr. Stark did not know at the time was that the little boy could not even read music and that Scott had brought him all the way from Kansas City to Sedalia after coaching him for months in the intricacies of the *Maple Leaf Rag*.[23]

It would seem from this account that Joplin had prepared for his sale of the music, ruling out the chance encounter at the Maple Leaf Club. But this story, too, has its problems. Why would Joplin have prepared so carefully before presenting the music to an insignificant publisher? Though Stark advertised his sales of instruments, he did not even list himself as a publisher in the town directory or in newspaper advertisements.

The William Stark account also has its variations. In a version reported by William's wife, there is again a boy accompanying Joplin into the office, but he does not play the music; he dances to it. It is also William Stark, rather than John Stark, who makes the decision to publish the music.

> Reminiscing about the firm of John Stark & Son and how it came to print this classic [the *Maple Leaf Rag*], Mrs. Stark [William Stark's widow] recalled that it was back in 1899 that the meeting took place between Stark and the now legendary Joplin. "A good many stories have been published about this encounter," she emphasized, "but most of them have been fanciful. In fact until the day Will died in 1949 he never tired of laughing at some of the highly embroidered versions that he read, and of telling me exactly how he happened to meet Joplin.
>
> "According to Will," Mrs. Stark explained, "Joplin wandered into the Stark store in Sedalia one day holding the *Maple Leaf Rag* manuscript in one hand, and a little boy's hand with the other. Sitting down at the piano, Joplin began to play the now-famous tune while the youngster stepped it off. Grandpa [John Stark] thought nobody would play it because it was too difficult," she recalled, "But Will was so taken with the lad's dance, that he decided to buy it."[24]

When John Stark himself told of the event, there was nothing about the Maple Leaf Club, a boy dancing or playing piano, or of Will convincing him to publish:

> When he first came into the office of the Stark Music Company some years ago, with the manuscripts of Maple-Leaf Rag and the Sunflower Slow Drag, he had tried other publishers, but had failed to sell them. Stark quickly discerned their quality, bought them and made a five-year contract with Joplin to write only for his firm, which firm has all of his great compositions.[25]

Of the frequently mentioned five-year contract little is known, but the original royalty contract came to light in 1975, leading to yet another version of the Joplin-Stark meeting.[26]

A signatory witness to the contract was R. A. Higdon, who we now know was Joplin's attorney. According to Higdon's daughter, Lucile Higdon Lloyd, her father and Joplin became acquainted at dances where Joplin worked. These were held in a club room over the St. Louis Clothing Store, 201–203 Ohio (corner of 2nd and Ohio). Robert Higdon, a man of about 29, had just finished his law degree in 1898, had been admitted to the Missouri bar, and had opened his practice in Sedalia. At this time, he also advertised

Contract for publication of the Maple Leaf Rag.

that he would lend money at 5 percent interest, indicating he was already fairly well off.

He attended the dances regularly and frequently stood by the piano while Joplin played. On one occasion, Joplin demonstrated the new rag he had just completed—the *Maple Leaf Rag.* Impressed, Higdon suggested that the rag be published. When Joplin responded that he did not know how to go about having it published, Higdon offered to act on Joplin's behalf, and did so without charge. Higdon then spoke to Stark about it and drew up the contract. Higdon and Will Stark, who was running the store at this time, were already acquainted and may have been friends. They attended the same church, the Epworth Methodist Episcopal, and served on several church committees together.[27]

The contract, signed on August 10, 1900, provided Joplin with a one cent per-copy royalty, ten free copies, and the right to purchase additional copies at five cents each. The music was not to retail, by either party, for less than twenty-five cents. There is no mention here of a five-year agreement, as indicated above by Stark, but there may have been a later contract.

While provision for a royalty was not unique for the time, it was unusual.[28] Most often, publishers purchased music outright for $25 or $50, this being especially true when the composer was black. That Joplin had a royalty contract for what became the best-known instrumental rag of the period, gave him sufficient income to change the conditions and course of his life.

There is no question as to the popularity of the *Maple Leaf Rag*. Everyone interested in piano ragtime played it, or at least tried to play it. Pianist and composer J. Russel Robinson (1892–1963), in recalling the period around 1908, said:

> One of the tunes I played a lot while touring the South was Scott Joplin's
> "Maple Leaf Rag." . . . I think it is one of the finest tunes ever written, . . . the
> King of Rags, and in my way of thinking, nothing that Joplin or any other rag
> writers wrote ever came close to it.

Jelly Roll Morton was sufficiently egotistical to refer to himself as "the inventor of jazz," but he nevertheless deferred to Joplin, calling him "the greatest ragtime writer who ever lived and the composer of *Maple Leaf Rag*." Stride pianist pioneer James P. Johnson, in speaking of the rags he played in 1912, named only three Joplin rags, but of the *Maple Leaf* he said, "Everybody knew that by then." Ragtime pedagogue Edward R. Winn referred to "Scott Joplin, the world's greatest composer of Ragtime, who wrote the celebrated 'Maple Leaf Rag.'" Axel Christensen, who operated ragtime music schools throughout the country, used the *Maple Leaf* to identify John Stark: "It was he who discovered Scott Joplin, who put on paper for the first time that wonderful composer of classic ragtime. It was he who gave to the public the famous and never dying 'Maple Leaf Rag.'" John N. Burk, who in later years was to become a biographer of classical music composers and program annotator for the Boston Symphony Orchestra, wrote in 1914 a defense of ragtime and used an example from the *Maple Leaf Rag* to support his case. Alfred Ernst, director of the St. Louis Choral Symphony Society, said in 1901 he would be introducing the *Maple Leaf* to Germany on his next trip there.[29]

The *Maple Leaf Rag* was the king of piano rags, and it made Scott Joplin the "King of Ragtime Writers."

Much has been written about the prodigious sales of this famous rag, but we must be judicious in evaluating the claims. Songwriter and journalist Monroe Rosenfeld wrote in 1903 of the *Maple Leaf Rag*: "Almost within a month from the date of its issue, this quaint creation became a byword with musicians, and within another half a twelve-month circulated itself throughout the Union in vast numbers." John Stark wrote of the sales in his advertising copy: "One million copies have been sold and no abatement of demand. There will be a temporary stop to its sale when every family in the civilized world has a copy."[30]

As popular as the *Maple Leaf* was, it did not sell as much as the most popular vocal music. Charles K. Harris's *After the Ball* (1892), for example, probably sold far more than *Maple Leaf*. As for *Maple Leaf* selling a million, we should also ask by what year that figure was reached. It is not likely that it was during Joplin's lifetime.

In the May 1905 issue of *The Intermezzo,* a monthly magazine that Stark published, he stated that it took a year to sell the initial printing of four hundred, and that by 1905 sales were 3,000 a month. William Stark also cited four hundred as the first year's sales, claiming that sales were initially hurt by the music's difficulty.[31] At the later rate of 3,000 per month, or 36,000 per year, it would have taken more than twenty-seven years to reach a million.

This may have been an unusual piece in that sales increased as time passed. William Stark said in 1909 that sales of *Maple Leaf* had reached a half-million copies.[32] Since only a negligible amount was sold in the first year, the half-million was sold in eight-and-a-half years, averaging to 60,000 per year. This was the same figure cited by John Stark's grandson, who recalled that in the late teens and early 1920s they were selling 5,000 a month.[33] Although we would expect that by the 1920s sales would have slackened considerably, *Maple Leaf Rag* had unusual staying power: it was recorded as often in the decade of the 1920s (six times) as it had been in its first two decades of existence.[34]

Can we credit the figures given in interviews? John Stark's ledger books from April 1907 to November 1908 show sales even more modest than anything so far mentioned: about 15,500 per year. However, the books, which list sales of even a single copy as well as of several thousands, are incomplete: an entry for November 10, 1908, is crossed out and has an added comment: "Paid cash." The ledgers, then, reflect only credit sales. It is also evident, since the books are in John Stark's handwriting, that they were from his New York office, where he lived during those years. Perhaps there were separate accounts kept in the St. Louis office. The actual sales amounts revealed in the books therefore tell only part of the story.[35]

Regardless of specific amounts, the ledgers do indicate proportional sales. While the *Maple Leaf* was selling 15,500 per year, James Scott's *Frog Legs* (1906), which was second in sales in Stark's catalogue, sold about 5,000 annually. The Joplin-Hayden collaboration *Sunflower Slow Drag* sold 1,600.

Stark's wholesale prices are also revealing. This was a period of price wars in the New York sheet music business, retail prices at one point dropping to one cent for best-selling songs. Most of Stark's works wholesaled for between three and seven cents, but because of the demand for *Maple Leaf,* he maintained a higher price for this one work: an average of eight-and-a-half cents during the worst years of the wars, and sometimes as high as twelve-and-a-half cents.

Our interest is to learn how much Joplin earned from *Maple Leaf.* At the contractual rate of one cent per copy, he would have netted four dollars in the first year. As sales picked up, perhaps going to a half-million by 1909, his royalties would have averaged about $600 annually. This was not a princely sum, but neither was it insignificant, for it equaled the average industrial wage during the same years. Letter carriers in 1899 had salaries ranging between $600 and $850. In turn-of-the-century Sedalia, five- to eight-room houses were renting for between $7 and $35 per month.[36] From this one work, therefore, Joplin was probably able to meet most of his basic expenses.

There have been assertions that Joplin was not the sole composer of *Maple Leaf,* that others had a hand in it. Otis Saunders was one to stake a claim. Saunders and Joplin met at the Chicago Fair in 1893, became fast friends, and on leaving Chicago traveled together to St. Louis and then to Sedalia. For a while the two were inseparable. Campbell reported that where you saw one, you saw the other. But by the time Joplin left Sedalia in 1901, he and Saunders no longer seemed to be associates. Campbell reported learning from Saunders some years later that the two friends had quarreled over authorship of *Maple Leaf Rag* and Joplin's later *The Favorite* (1904).[37]

Campbell seemed to believe there was something to Saunders's claim. We note, however, that Saunders claimed also to have composed Tom Turpin's *St. Louis Rag*—suggesting that he may have been prone to such assertions—and that he is not known to have published any music under his own name.

Mildred Steward, daughter of Arthur Marshall, said her father had co-composed the *Maple Leaf.* However, Arthur Marshall lived until 1968, was interviewed many times, and is never reported to have made a claim for the *Maple Leaf.* He wrote to Blesh and Janis that he alone was responsible for *Lily Queen* (1907), a work that lists Joplin as co-composer, but made no mention of the *Maple Leaf.*[38]

If Marshall had been cheated out of credit for the pre-eminent rag of the period, one would think he would have expressed some bitterness toward Joplin. Instead, he had only unqualified praise for the older composer: ". . . one of the most pleasant men you'd ever want to meet. . . . He was kind to all of us musicians . . . he was an inspiration to us all. . . . there wasn't a better man known in my whole lifetime."[39] Unlike the rupture with Saunders, Joplin's relationship with Marshall remained close, and the two corresponded after their travels to different parts of the country separated them physically.

I believe that Mrs. Steward was mistaken, and she gave a clue to the source of her error. She referred to an early, pre-Stark publication of the *Maple Leaf,* crediting both Joplin and her father as co-composers. The cover, she said, displayed the pictures of her father and Joplin in the lower corners. What she apparently had in mind was the first printing of the Joplin-Marshall collaboration *Swipesy Cake Walk,* which has such a cover.[40]

There is also the issue of Marie Walker, the woman from Hannibal whom Campbell said had aided Joplin, possibly including the *Maple Leaf.* However, as with the earlier pieces, we have no information on the extent or nature of Mrs. Walker's input.

The most credible testimony of another person having something to do with *Maple Leaf* is Joe Jordan's, and by his own word his involvement was minimal. Jordan, an important arranger, conductor, and songwriter during the ragtime years, reported that Joplin originally composed and played the *Maple Leaf* on the piano in the key of A-major, the key in which he also played the work with his band. (Though the band was composed mostly of B-flat instruments, slides pitching the instruments to A were in common use.) According to Jordan's story, he suggested that the key of A-flat would be more appropriate for piano publication, and that Joplin thereupon transposed the work, but never played it as well in the new key.[41] Since the orchestration published by Stark is in A-major, Jordan's report is believable.

The *Maple Leaf Rag* was published between August 10, 1899—the date Joplin and Stark signed their contract—and September 20, the date the Copyright Office received two copies of the music. The first printing, of 400 copies, was on a rough paper and had for a cover a crude black-and-white drawing of two cakewalking black couples. This drawing was copied from one of a series of cakewalk pictures distributed by the American Tobacco Company for cigarette coupons. The pictures portray the famous black stage stars Bert Williams (the taller man) and George Walker. The women were the future spouses of Williams and Walker, respectively Lottie Thompson and Aida Overton. The music was printed in St. Louis, where Stark was living at the time, but lists the publication city as Sedalia, where Stark's son Will was still running the family music store.

Stark reissued the music in 1900 or 1901 from his St. Louis store, using the same music plates but a new cover, displaying simply a green maple leaf, Scott Joplin's photograph, and a dedication "To the Maple Leaf Club." Later printings omitted the photograph and the dedication.[42]

The title of this rag has also been the subject of speculation. Did the title precede the naming of the Maple Leaf Club, or vice-versa? If the former, does the name have any significance?

Original cover of The Maple Leaf Rag, *and the print on which it was based.*

A 1903 article in the *Sedalia Times,* the newspaper edited by Joplin's colleague W. H. Carter, states that the rag was named after the club. This was also what Arthur Marshall had said. Tom Ireland, though, wrote the opposite in 1947: "A young colored man opened up a club room . . . after Joplin arranged his 'Maple Leaf Rag' and named it the 'Maple Leaf Club'."[43] Testimony is therefore inconclusive.

The earliest documented notice I have of the club is November 18,

THE CAKE WALK N°1

1898, when it was referred to in all three Sedalia daily newspapers, but it may have existed before then. The rag almost certainly existed by then, as well, but we do not know when it acquired its title. Documentary evidence on which came first, the rag title or the club name, is, then, also inconclusive.

Whether referring to the rag or the club, why "maple leaf"? One attractive idea links the name to Canada, which has the maple leaf as a national symbol. Since Canada was the goal of many black slaves escaping to freedom, this hypothesis suggests that the *Maple Leaf Rag* and the Maple Leaf Club both commemorate the black hope for true freedom and equality. The hypothesis, though, has no historical support. There is no evidence that the "underground railroad" to Canada was associated with the maple leaf, and there is no indication in the black newspapers of turn-of-the-century Sedalia that Canada was any longer thought of as a beacon of equality.

Another idea links the name to the Chicago Great Western Railway, called "The Maple Leaf Route" because the route map resembled the outline of a maple leaf. Advertisements for this railway, displaying the maple leaf, were printed in Sedalia newspapers in 1899 and were probably seen by

Joplin. But there is no reason to believe that this was the inspiration for Joplin's title or for the club.

The truth is likely to have been much simpler. Maple trees were abundant in Sedalia. A postcard from the period showing a tree-lined path at Liberty Park bears the legend "Under the Maples." The town's most majestic residence was called Maple Square; on the corner of Ohio and Broadway, it was nine blocks from the Maple Leaf Club.[44] There was also a nearby community with a "maple" name: Maplewood, an area in the nearby town of Smithton.[45]

"Leaf" club names were similarly known in Sedalia, there being a Clover Leaf Club and an Autumn Leaf Club, both being white organizations. Inasmuch as "maple" and "leaf" names were common in Sedalia, there is no reason to read special significance into the maple leaf title used by Joplin and by the social club.

The reason for the enormous popularity of the *Maple Leaf Rag* is another unanswered, and probably unanswerable, issue. True, when it was first brought out there was no other rag that could compare with it in terms of rhythmic vitality, imagination, and originality. But innumerable other superior works remain ignored while, too often, the public makes hits out of music that has little to offer in a purely musical sense.

Public taste is a mystery. We cannot by analyzing the *Maple Leaf* show why it was so loved. All we can do is discuss the craft and suggest highlights and features that may further enhance our own appreciation of what Joplin achieved.

We have already established the musical context with Mills's *At a Georgia Campmeeting* and Wheeler's *A Virginny Frolic* (Exx. 4-2 and 4-3a). On the most cursory examination, it is evident from the opening strain that *Maple Leaf* is far more varied and complex (Ex. 4-7). It is not just the rhythms, the mid-measure and bar-crossing syncopations. Though these were still uncommon in ragtime in 1899, they were not unique. What catches our attention are how single notes may play rhythmic rather than melodic functions, such as the last note in the first measure springboarding to the following E-flat octave; how accented syncopated notes are emphasized with octave reinforcement (such as the E-flats in measures A1-4); how the bass in this strain completely abolishes the stereotypical pattern of octave-chord alternations.

Most of all, Joplin recognized that patterns of durations were not the only source of rhythmic vitality; he had discovered the propulsive power of harmonic motion, of resolving dissonances, and of goal-directed voice-leading. These principles are also apparent in the very first measures. He clearly establishes A-flat tonality with tonic and dominant chords, but introduces in the first measure a dissonant A-natural as a lower-neighbor to B-flat, these two notes at the same time forming part of a three-note ascending chromatic motive (A-flat, A, B-flat) used throughout the piece.

Ex. 4-7. Scott Joplin, Maple Leaf Rag *(1899), A1-16.*

The real shocker in the piece's opening moments is the slide away from the dominant E-flat to the flatted sixth (spelled correctly as F-flat, mm. A4-6), using this "bluesy" relationship, along with a chromatic three-note bass pattern, through the rest of the A strain.

The structure of this first (A) strain retains the standard pattern of four 4-measure phrases, but is atypical in its details. The thematic pattern, rather than have the third phrase repeat the opening material (such as the common a-b-a-c or a-b-a'-b'), follows a more progressive and dynamic pattern whereby the second half of the strain is unrelated to the first half: a-b-c-c.

This is the remarkable opening of the work that was to change ragtime. The rest of the composition is equally marvelous. The second (B) strain begins with the right hand playing high and descending against a bass that makes judicious use of the three-note chromatic ascending motif (A-flat, A, B-flat in B4-5; B-flat, B, C in B6-7). Joplin also introduces a device he was to use throughout his career: a reiterated right-hand pattern with changing harmonies (B3-8, Ex. 4-8).

Ex. 4-8. Scott Joplin, Maple Leaf Rag: *(a) B1-16; (b) analytic reduction of B1-8.*

After a reprise of the A strain, Joplin presents a C strain that has a treble part suggesting Latin-American rhythms (C1-2), but with little melodic interest. The bass, in contrast, is remarkably melodic while, at the same time, providing a strong tonal direction (Ex. 4-9).

Ex. 4-9. Scott Joplin, Maple Leaf Rag: *(a) C1-16; (b) bass line reduction,*
C1-9.

Most rags, both before and after the *Maple Leaf,* had only three strains.
In *Maple Leaf,* Joplin used four strains, and this became the general pattern
for Joplin rags and for rags modeled after his. This final (D) strain again
makes use of the lower neighbor motion and ends with a phrase that becomes
one of Joplin's characteristic designs: an inner voice that descends chromatically the interval of a fourth (Ex. 4-10). The counterpoint between this voice
and the bass line is evidence of an unusual musical thinking for a composer
working in what was essentially a homophonic form.

Where did Joplin learn such voice-leading? From his early teacher?
From his instruction at the college? There is no direct evidence on this

Ex. 4-10. Scott Joplin, Maple Leaf Rag: *(a) D1-16; (b) analytic reduction showing inner voice against bass, D13-16.*

question, but I tend to think he derived his contrapuntal perception from his experience as a quartet singer. Even today, "untutored" black quartet groups associated with churches and other traditional organizations demonstrate an extraordinary linearity in their musical thinking. This was probably true during Joplin's time as well.

With the *Maple Leaf Rag,* Joplin's career took a giant leap forward, and ragtime entered a new age.

Echoes of Scott Joplin, Part 2: The Maple Leaf Rag

 It is not surprising that the *Maple Leaf,* the most popular of all instrumental rags, was imitated by other composers. As Scott DeVeaux put it in his edition of James Scott's music, the *Maple Leaf* "cast a long shadow."[46] James Scott's use of this masterpiece, as well as other Joplin works, was subtle and are properly viewed more as influence than as theft. Similarly, Joe Lamb's suggestion of the opening gesture of *Maple Leaf* in his *Sensation,* published on Joplin's recommendation (see Chapter Ten), can be seen as tribute. Many other composers, though, committed blatant and clumsy plagiarisms.

Using Joplin's A strain, we note an obvious copy in the A strain of Frank S. Butler's *The Tantalizer* (Ex. 4-11a). Leon Donaldson's

Ex. 4-11. Frank S. Butler, The Tantalizer *(St. Louis: Mellinger, 1916): (a) A1-8; (b) B1-4.*

Latonia Rag has the gesture of *Maple Leaf,* but is melodically and harmonically more like the A of *Original Rags* (Ex. 4-12; compare also with Ex. 4-6). *That's Goin' Some,* subtitled "A Syncopated Reminiscence" and described on the music as "The Crookedest Rag

Ex. 4-12. Leon Donaldson, Latonia Rag *(St. Louis: American Music Syndicate, 1903), A1-4.*

That Ever Happened," is a special case since the composer James Nonnahs (Shannon backwards) acknowledges his theft (Ex. 4-13).

For a sample imitation of Joplin's B strain, we can return to Butler's *The Tantalizer* (Ex. 4-11b).

Ex. 4-13. James Nonnahs [Shannon], That's Goin' Some *(Detroit: Grinnell, 1909), C1-4.*

We have seen how L. Edgar Settle used the E strain of *Original Rags* for the third strain of his own *X. L. Rag.* In the same work, Settle clearly based his B strain on the C of *Maple Leaf,* taking some of the right-hand gestures but missing entirely the melodic bass (Ex. 4-14).

Ex. 4-14. L. Edgar Settle, X. L. Rag, *B1-4.*

Will Etter, who had borrowed from *Original Rags* for his *Whoa! Maud,* based the D strain of that work on the D of *Maple Leaf* (Ex. 4-15).

Ex. 4-15. Will H. Etter, Whoa! Maud, *D1-4.*

These are only a few of the *Maple Leaf Rag* imitations. As we will see, Joplin himself was to become the major imitator of *Maple Leaf Rag*, and some of his reworkings were plagiarized by other composers, making them third-generation copies of the *Maple Leaf.*[47]

CHAPTER FIVE

The Ragtime Dance, 1899–1901

Let me see you do the "rag time dance"
Turn left and do the "Cake walk prance"
Turn the other way and do the "Slow drag"
Now take your lady to the worlds fair. . . .
And do the "rag time dance."
Words and music by Scott Joplin.

MUSIC PUBLISHER JOHN STARK played an important part in the Scott Joplin story. Stark believed in ragtime. He recognized an excellence in the best piano rags and proclaimed them the equal of classical music. This was especially true for the publications of his "House of Classic Rags."

"Maple Leaf Rag" marks an era in music composition. It has throttled and silenced those who oppose syncopations. It is played by the cultured of all nations and is welcomed in the drawing rooms and boudoirs of good taste.

"The Cascades" . . . is as high-class as Chopin. . . .

If there ever was a song without words this [*Sunflower Slow Drag*] is that article; hold your ear to the ground while someone plays it, and you can hear Scott Joplin's heart beat.

Tell me ye winged winds that 'round my pathway roar,
—We know one house of classic rags—pray are there any more?
The answer filtered through the leaves and whispered 'long the shore:
"There's only one classic Rag House."

We mean just what we say when we call these instrumental rags classic.

They are the perfection of type. "The glass of fashion and the mold of form." They have the genius of melody and the scholarship of harmonization. They are used in the drawing rooms and the parlors of culture.

Stark's position was bold and ran counter to his era's most widely held truism about music: that music—"the Divine Art"—possessed distinct hierarchical qualities of class and virtue. Within this scheme, ragtime ranked near the bottom in both categories.

One of the reasons for the low regard of ragtime was its acknowledged source: African-American music. In this respect, too, Stark was a maverick. Rather than conceal the race of his black composers, as many publishers did, he took pride in it:

A Fierce Tragedy in One Act

SCENE: A Fashionable Theatre. Enter Mrs. Van Clausenberg and party—late, of course.

MRS. VAN C: "What is the orchestra playing? It is the grandest thing I have ever heard. It is positively inspiring."

YOUNG AMERICA (*in the seat behind*): "Why that is the *Cascades* by Joplin."

MRS. VAN C: "Well, that is one on me. I thought I had heard all of the great music, but that is the most thrilling piece I have ever heard. I suppose Joplin is a Pole who was educated in Paris."

YOUNG AM.: "Not so you could notice it. He's a young Negro from Texarkana, and the piece they are playing is a rag."

Sensations—Perturbation—Trepidation—and Seven Other Kinds of Emotion.

MRS VAN C: "****** The idea. The very word ragtime rasps my finer sensibilities. *(Rising)* I'm going home and I'll never come to this theatre again. I just can't stand trashy music."[1]

Note that Stark used a capital "N" in "Negro." This usage was rare except in African-American publications and reveals Stark's awareness of what black intellectuals wanted, and his respect for that desire. The elaborate advertisement also shows Stark's resentment of the segment of the musical public that, without discerning qualitative distinctions, automatically rejected any music that had the ragtime label. He frequently addressed that point: "There is nothing in common with these inventions [his publications] and the flood of Rags, Drags, and Jags on the market."

Stark never became a major publisher, but he holds a special position in the history of ragtime because he issued so many high quality rags. This quality was recognized at the time by aficionados. Axel Christensen, who operated a chain of ragtime music schools, was an admitted devotee of Stark publications. Joe Lamb, before becoming a composer for Stark, was an avid purchaser of his ragtime output. Jelly Roll Morton said that St. Louis was a great town for black ragtime partially because of Stark's activity in that city.[2]

Stark is reported to have been fair in his dealings with his composers. His royalty agreement with Joplin for *Maple Leaf* was one of the important financial events in the composer's life. Arthur Marshall said he received $50 and 3-cent royalties for his own compositions *Ham and !, Kinklets,* and *The Peach.* He also claimed that James Scott received royalties, and added, "That's one thing about old man Stark. He was pretty fair with us people that he bought rags from. He'd give 'em a royalty on it—and that's the only way it should be."[3] Yet Marshall seemed to have forgotten about the difficult dealings he had with Stark in 1906—possibly for his rag *The Pippin* (1907)—in which he finally agreed to Stark's non-royalty offer of $10 and 200 copies of the music.[4]

Stark, for all his love of ragtime, was a businessman. He was born in Kentucky in 1841, grew up on a farm in Indiana, and was a bugler in the Union army during the Civil War. He married, tried his luck for a few years as a farmer, then went into the ice cream business, selling from a wagon. After a few years, he opened a piano and organ dealership briefly in Chillicothe, Missouri, moving his store to Sedalia around 1882.[5]

Even in so musical a town as Sedalia, business must have been difficult, for there were usually between two and four competing music stores. Nevertheless, Stark remained in business, changing his address almost yearly. Around 1895, he bought out J. W. Truxel Music Co., in the process acquiring seven music copyrights. Stark thereby became a music publisher, but it was a small part of his business. Though he added a few compositions in following years, he did not advertise himself as a publisher.

His fanciful and exuberant manner with words was already in evidence during his pre-ragtime years. Typically, he would advertise as follows:

> Of all the arts and sciences Music is the most worthy of cultivation. Since David charmed into peace the evil nature of Saul millions have been led from sadness of soul to sunshine and gladness by its gentle power. Sad indeed is the home where not one can bring optimistic rosy views of life to the circle with some musical instrument. Happy will be the memories of the old home where music made evenings glad.[6]

In December 1898, W. Sharp Music announced a "going-out-of-business" sale, seriously cutting into Stark's sales during the Christmas season. Since Sharp had a similar sale the previous year, Stark was justifiably annoyed. He countered with his own sale, poking fun at his competitor with terms that were impossibly extravagant: "We will sell during this sale 100 percent below cost. We intend to quit business twice a month for the next six months. . . ."[7] So clever was Stark's advertising that a trade journal in New York, *Music Trades,* commented upon it.[8]

Music was a way of life in Stark's family. He was very familiar with the classical music he wrote of. His son Etilmon, known as "Till," taught violin in several Missouri colleges near his home in Lexington and gave frequent recitals throughout the state. Under the names "E. J. Stark" and "Bud Manchester," Till composed highly competent rags which his father published. He probably played an advisory part in the family business.

Stark's daughter Eleanor, known as "Nellie," was the outstanding musical talent in the family, as well as in Sedalia. With her father providing the finances, she studied piano in Berlin from 1895 to 1897 with the celebrated virtuoso and composer Moritz Moszkowski. After a short return home in the

Eleanor Stark, 1898.

John Stark, ca. 1909.

summer of 1897, she went to Paris, where Moszkowski had relocated, and studied with him for one more year. On her return in the summer of 1898 (responding to her parents' fears for her safety during the Spanish-American War), she gave local recitals both as a pianist and as a singer. Her piano recitals were rigorous, a typical program consisting of Beethoven's Variations in C Minor, Grieg's *March of the Dwarfs,* Moszkowski's *Caprice Espagnol* and *Etincelles,* and Chopin's Scherzo in B-flat Minor. In January 1899, as soloist with Carl Busch's orchestra in Kansas City, the city's leading orchestra, she played Weber's *Konzertstück.*

After Eleanor had completed her studies in Europe, it quickly became clear to her and her family that she had outgrown the musical and cultural life of Sedalia. Consequently, in February 1899 she opened a music studio in St. Louis at 3210 Lucas Avenue. Her mother and father joined her in St. Louis the following month and the three lived at 3848 Washington Boulevard. John Stark opened a furniture moving business in St. Louis, and his son William remained in Sedalia to run the music store.[9]

Eleanor also played ragtime. Ted Browne (a.k.a. Ted Brownold), composer of *Manhattan Rag* (Stark, 1901), recalled: "She played a fine popular and classic piano and could do a great job on Scott's tunes."[10] She was also a publications adviser for her father. It was at her insistence, in 1902, that Stark published Joplin's *Ragtime Dance,* which was first performed in 1899.[11] We also have Stark's testimony that he relied upon his daughter's advice. In a letter to Arthur Marshall on July 28, 1906, he stated that he was awaiting his daughter's return before deciding upon music Marshall had submitted for publication.[12]

We have no record of Eleanor performing ragtime at recitals, but this is not surprising; no reputable concert artist of the time would have done that. All we know is that she played ragtime well and advised her father on ragtime publications. We are left wondering to what extent she truly understood and appreciated the special qualities of Joplin's music, how well they knew each other, and whether she influenced Joplin in any way.

The evidence in newspapers that the Starks moved to St. Louis early in 1899 demonstrates the dangers of relying on testimony from even the most authoritative sources. Decades later, the Stark family consistently reported that John Stark had moved to St. Louis *after* the September 1899 publication of *Maple Leaf Rag* in order to promote and market it more effectively from the larger city.[13] The newspaper reports prove that the family legends were not entirely correct.

But they may have been partially correct. Although Stark was living in St. Louis, had the *Maple Leaf* printed there, and was planning on additional publications,[14] he probably had not yet opened a music store in the larger city. That is why he used the Sedalia address on the first printing of *Maple Leaf.* I would guess that it was only after sales showed promise that he put aside his furniture-moving business and opened his music company and printing plant in St. Louis.

Stark would have been satisfied to have Joplin compose only ragtime, and to compose it only for him. This was not what Joplin envisioned. Joplin wanted to branch out beyond ragtime; he wanted to compose for the lyric theater.

Piano ragtime was widely regarded as reflecting a lower order of the art, a music of saloons, brothels, and lower-class blacks. While the theater was also suspect in the minds of many, it nevertheless carried with it an aura of sophistication and almost of intellectuality, especially for the African-American performers and composers. As the 19th century was coming to a close, it became clear that the musical theater was one venue in which the talented African American could make his mark. With the visits to Sedalia of black minstrel troupes that played at Wood's Opera House, Joplin had doubtlessly heard of, and perhaps read of in the *Indianapolis Freeman,* the success of such black theatrical works as Cole and Johnson's *A Trip to Coontown* (1897), of Will Marion Cook's *Clorindy* (1898), and of the performers Williams and Walker, portrayed so elegantly in the American Tobacco Company prints.

In his earlier days, Joplin had worked as a minstrel, but by the turn of the century minstrelsy was considered a low type of musical theater and a bit old-fashioned. Joplin wanted to make his mark in musical theater of a higher rank. His first effort toward this end was *The Ragtime Dance,* a tableau for singing narrator and dancers.

Blesh and Janis, based on information that Arthur Marshall recalled with some difficulty, tell us that in the fall of 1899, after publication of *Maple Leaf,* Joplin formed a drama company to put on his "ballet" of African-American dances. As Marshall reported it:

> I had to concentrate to be as correct as I could remember about the facts. The Drama Company was formed and rehearsed very strenuously. Mr. Will Joplin was a lead Character in the featuring of Joplin's Rag Time Dance. . . . I did ragtime specialities on the piano and some numbers with other members of the company. Scott Hayden did some of the same. Joplin played piano when we were performing other than quartette and specialities.[15]

Joplin, according to Blesh and Janis, then rented Wood's Opera House for a performance. The previous month Stark had already announced he was going to publish "A Black American Cake Walk," which may have been an earlier title of *The Ragtime Dance.*[16] However, Stark did not publish a piece by either title at this time. Perhaps he was not sufficiently impressed with the performance, or was concerned about the commercial potential of sheet music that had nine pages of music rather than the usual four.[17]

Eventually, Stark did publish *The Ragtime Dance* (copyright December 23, 1902). He and his daughter Eleanor attended a performance in St. Louis, this time with an orchestra consisting of Lij Cross and the Vassar Boys. Eleanor was enthusiastic about the music and urged her father to publish it.[18]

The Ragtime Dance is a bold idea, and Joplin tells us with his lyric what the idea is. He had provided music for black dances held at the Black 400 Club, the Maple Leaf Club, and other places as well. He saw that there were many whites eager to attend these dances as viewers. Why not re-create in a theater piece the events of such an affair? He sets the scene in the opening strain:

> I attended a ball last thursday night
> Given by the dark town swells.
> Ev'ry coon came out in full dress alright
> and the girls were society belles.
> The hall was illuminated by electric lights
> it certainly was a sight to see.
> So many colored folks there without a razor fight
> twas a great surprise to me.[19]

The second strain tells how the participants prepare for the dance. The dance begins in the third strain, and from there on, for eight strains, the narrator calls the steps. The dances mentioned by the narrator or written as brief comments in the score are the rag time dance, cake walk prance, slow drag, World's Fair dance, clean up dance, Jennie Cooler dance, rag two step, back step prance, dude walk, town talk, stop time, and Sedidus walk. There is also a country dance component, with dancers circling around and forming lines. (The steps were advertised as being available for sale, but no copy is known to exist.)

According to Marshall's report of the 1899 performance, the show consisted of more than what we today know as *The Ragtime Dance*. Scott Hayden and Marshall both played "specialty numbers," and "Joplin played piano when we were performing other than quartette and specialities." The production was probably only one component in an entire evening of vaudeville entertainment.

Marshall said that Will Joplin, Scott's brother, was the lead, that is, the narrator. A newspaper report confirms that Will was in Sedalia during the fall of 1899. He had been singing baritone with the Kentucky Rosebud Quartette in Omaha and, on the way home, the group stopped in Sedalia to visit Scott Joplin. In Sedalia they were heard by Albert Kahn, owner of the St. Louis Clothing Store, who immediately hired them to give evening concerts at his store.[20]

It is possible that Joplin's brother Robert was also in town, since his presence was indicated in several newspapers two weeks later. He may have arrived with brother Will, for two papers specified he was from Omaha (one said he was from St. Louis). If he was in Sedalia at the time of the performance, he probably would have taken part since he was a noted dancer.

> The social club, composed of colored people, will give a cake-walk and entertainment at the D. O. H. hall tomorrow night. Rob't Joplin, of Omaha, Neb., a professional cake-walker, will be among those who take part.

> R. B. Joplin, of St. Louis, a brother of the celebrated Scott Joplin, assisted by Virgil Bradley, Jesse Holland, Adah Burress and other colored folks, of Sedalia, gave a cake walk last night at the Social club, 106 Main.

> A number of colored young people, including R. B. Joplin, Richard Smith, James Ellis and their ladies, went to Houstonia today to give a cake walk and entertainment in the hall there tonight. . . . Henry Jackson will furnish the music.[21]

About eight years later, by which time Robert was a celebrated vaudevillian, he directed and narrated a performance of *The Ragtime Dance*.[22]

Blesh and Janis do not specify when in the fall of 1899 that *The Ragtime Dance* was performed at Wood's, and the newspapers never mention the piece by title. But I think we can deduce when the performance occurred. The newspapers note only a single occasion during the fall of 1899 on which local blacks put on a show at Wood's Opera House.

> The "Home Talent Cake Walkers and Minstrel company," composed of colored people, have arranged a fine programme to be rendered at Wood's opera house next Friday evening.
>
> One of the features will be a cake walk for prizes between professional cake walkers from Kansas City and Sedalia.
>
> Besides the cake walk, there will be buck and wing dancing and music by the "Pork Chop Greasy quartette," composed of John and Len Williams, Richard Smith and John Nelson. This quartette is one of the best in the city and all the latest songs will be sung.[23]

It is my guess that it was at this event, which featured some of Joplin's close associates, that *The Ragtime Dance* was first staged. A newspaper item the next day reported that the attendance was poor, but that the "dancing and singing were very creditable." A few days later, the paper reported that "the Dramatic company of Sedalia young colored people will give an entertainment at Marshall."[24] This "Dramatic company" was probably the company that Arthur Marshall said Joplin had formed to put on his work. The performance in Marshall, Missouri, would have been on December 2. Unfortunately, no Marshall newspapers mention the performance.

From these newspaper notices we surmise that *The Ragtime Dance* was performed as part of an evening's entertainment that included also a cakewalk competition, quartet singing, and piano exhibitions; that Joplin's brother Will was narrator, and his brother Robert may have been one of the dancers. The date of the event causes us again to reflect on the date that Lottie Joplin many years later cited as Scott Joplin's birthday: November 24.

Was this another coincidence? The Maple Leaf Club held its first dance on November 24, 1898. *The Ragtime Dance,* Joplin's first effort in serious musical theater, was premiered on November 24, 1899. Did Joplin in later years speak of the importance of this date in his life, leading to Lottie's mistake?

Joplin and Coon Songs

 The text of the *Ragtime Dance* may seem shockingly offensive. How could Joplin have written such a lyric? Understanding this issue requires historical perspective. Though such texts are clearly offensive, they were an accepted part of show business. This was a period that had little concern with ethnic sensibilities; blacks bore the brunt of the insults, but "comic" impersonations of Chinese, Italians, Germans, and Jews were also commonplace, and black performers included such impersonations in their own repertories.

The performance of coon songs by African Americans was not merely a reluctant adherence to a convention expected by the dominant culture. Rather, it was felt that a bit of mockery, even self-mockery, was "good fun." This attitude was true both in major cities and in towns. In Sedalia, for example, the black Queen City Cornet Band performed for a black audience the song *Coon! Coon! Coon!*, which has as its text:

> Coon! Coon! Coon!
> I wish my color would fade;
> Coon! Coon! Coon!
> I'd like to be a different shade. . .[25]

Theatrical pages of black newspapers referred to "coon songs" without any apparent embarrassment, and traveling black shows assumed such titles as "Hottest Coon in Dixie." Bert Williams and George Walker, long after achieving international success, continued billing themselves as "The Two Real Coons." Even Will Marion Cook, who was well educated, proud, and haughty, wrote such texts as:

> Warm coons a-prancin', Swell coons a-dancin',
> Tough coons who'll want to fight;
> So bring 'long yo' blazahs, Fetch out yo' razahs,
> Darktown is out to-night![26]

Not all African Americans were happy about the "coon" conventions, and Joplin himself would eventually speak out against them. But the conventions continued throughout the ragtime years and were not viewed with the same discomfort that they cause today.[27]

Early the next year, in January 1900, Joplin began work on a piece for the local Augustain Club, a white organization that had just been formed the previous month.

> Scott Joplin, well known as a composer, is at work on a piece of music to be dedicated to the Augustain club. The piece will be a waltz and will be quite a "hit" in this city.[28]

This piece was *The Augustan Club Waltzes* (misspelling "Augustain"), composed between January 2 and 31, 1900, about a year before it was published by Stark (copyright notice on the music is 1901; copyright registration was March 25, 1901). It is a pleasant waltz, showing Joplin's usual professional assuredness, but is not particularly distinctive. It is surprising that Joplin worked on it for an entire month.

An announcement indicated the waltz would be performed by the Second Regiment Orchestra, a white organization, at a masquerade ball given by the Augustain Club on February 19. This means an orchestration would have been needed. Whether the performance occurred is not known. The report of that affair indicated that the orchestra played sixteen numbers, but none were named.[29]

The work was apparently a commission, and that a black man was asked to compose music for a white club is significant. White organizations frequently had black musicians provide music for their social functions, but Joplin was not called upon to perform for the Augustain Club. From Joplin they wanted the more intellectual and creative task of composing a piece of music that would represent their club. This suggests that Joplin, months, or even more than a year before he was to achieve nationwide renown as composer of the *Maple Leaf Rag,* already had significant fame in Sedalia.

There are other indications of this local renown. In December 1898, Joplin was referred to as "one of the best pianists in the country." A year later, the *Sentinel,* a paper that frequently expressed an extremely disdainful attitude toward blacks, described Robert Joplin as "a brother of the *celebrated* Scott Joplin." In January 1900, Scott Joplin was cited as "the well known piano player" and "a well known composer."[30] Although we lack the detailed information to account for Joplin's reputation in Sedalia, it is clear that he had both fame and—unusual for a black man—*respect.*

Among Joplin's publications are several on which his name appears together with that of a student or a younger colleague. These works offer both problems and opportunities for the student of ragtime. Which works were really collaborations, as opposed to pieces to which Joplin simply attached his name to promote sales? In cases of authentic collaborations, is it possible to distinguish Joplin's contribution from his collaborator's? Since testimony has it that Joplin and his collaborators worked on separate strains, is there any evidence that Joplin touched up the work of his co-composer?

The earliest collaboration according to copyright dates is *Swipesy. Cake Walk,* co-composed by Joplin's student Arthur Marshall. This work bears a 1900 copyright notice, Joplin's only work with that year, and was registered on July 21, 1900. Dates can be deceptive, though. We have seen that *The Augustan Club* was composed more than a year prior to its copyright registration, and we shall see further that Stark frequently delayed submitting copyright applications. Blesh and Janis imply that *Swipesy* was composed while Joplin was still living in the Marshall household,[31] but this would place the composition in 1894 or 1895, which is much too early.

The original cover has small photographs of Joplin and Marshall in the lower corners and a large photograph of a black boy—"Swipesy"—in the center. Later printings eliminated the Joplin and Marshall photos.

Blesh and Janis report that the center photograph was of a local shoe-shine boy. Based on the boy's sheepish appearance, which Stark thought made him look as if he had just swiped some cookies, Stark suggested they name the piece "Swipesy."[32]

In a 1960 interview, Marshall gave another version of the title's origin. He and Joplin had just delivered the manuscript and were standing in Stark's office when they noticed two newspaper boys squabbling outside. Stark observed that one boy swiped a newspaper from the other and proposed that they name the piece "Swipesy." In a variant of this story, told by Marshall's daughter Mildred Steward, it was Joplin and Marshall, before entering Stark's store, who witnessed the scene and came up with the title.[33]

That Joplin did not turn his students' works into something closely resembling his own suggests that he permitted them latitude in expressing their own musical personalities. However, there is evidence that he also touched up the music of his collaborators, smoothing out rough spots and adding improvements. Joe Lamb told how Joplin, on first hearing his *Dynamite,* immediately suggested that a two-octave chromatic ascent in both hands be changed to a passage with contrary motion, with the left hand descending chromatically.[34] Similarly, Joplin told John Stark that a work of Arthur Marshall needed some rewriting before being engraved.[35] It was part of Joplin's musical nature to perceive alternatives and make changes. He did this even with his own music. Patterson reported that Joplin, while playing, "would not finish one piece without stopping to make changes and work things out."[36] He was always composing and recomposing.

According to reports, Marshall composed the A, B, and D strains of *Swipesy* and Joplin added the C strain.[37] The musical evidence supports this contention. The A, B, and D strains, all in B-flat, are unlike anything composed by Joplin. For example, the prominent fifth and fourths in the right hand in D3-4 would be uncharacteristic of Joplin (Ex. 5-1a). It is not possible to know whether Joplin suggested changes in Marshall's strains, but I would propose that a likely detail would be the final measure of the introduction, which has voice-leading and contrary motion characteristic of Joplin (Ex. 5-1b).

Joplin's section, strain C, is clearly his and shares with another collaborative work from this period a trait that is apparently commented upon by Campbell. It is a seemingly peculiar statement and, on the surface, seems to have little significance: "Joplin became very fond of the 7th chord."[38] But considered in the context of this strain, we see that with the upper-neighbor motif in measures 2 and 3, he forms not seventh *chords,* as measured from the bass, but prominent seventh intervals in the right hand (Ex. 5-1c). This may be what Campbell had in mind.

The other collaborative work from this period was *Sun Flower Slow*

Ex. 5-1. Scott Joplin and Arthur Marshall, Swipesy. Cake Walk *(1900): (a) D1-4; (b) Introduction 4, A1; (c) C1-4.*

Drag, co-composed with his student Scott Hayden. It bears a 1901 copyright notice and was registered on March 18, 1901, but Stark wrote that it was presented to him at the same time as the *Maple Leaf Rag* (in the summer of 1899). Support for Stark's assertion comes from W. H. Carter, editor of the *Sedalia Times,* who said it came right after the *Maple Leaf.*[39]

Carter said *Sun Flower Slow Drag* was "a great favorite among Sedalians." It is listed in a performance of the Queen City Cornet Band in 1901, and Tom Ireland said the band was still playing it in the 1920s.[40] P. G. Lowery, one of the leading black band leaders of the time and the dedicatee of Joplin's *A Breeze from Alabama* (1902), also performed *Sun Flower:* "P. G. Lowery is featuring "The Sunflower Slow Drag," by Scott Joplin of St. Louis. If you want the finest slow drag published get it. . . . Everybody wants to hear the 'Sun Flower Slow Drag,' by Scott Joplin, played by P. G. Lowery's band."[41]

Blesh and Janis suggest that the first two strains were composed by Hayden, and I concur.[42] As with *Swipesy,* the A and B strains have gestures uncharacteristic of Joplin. In the first strain, measures 2-3, the upward leap of a ninth after a descending line lacks the melodic continuity of a Joplin melody (Ex. 5-2a).

Joplin's contribution begins with the trio, highlighted once again by

Ex. 5-2. Scott Joplin and Scott Hayden, Sun Flower Slow Drag *(1901): (a) A1-4; (b) A15-16; (c) C1-4.*

right-hand sevenths (Ex. 5-2c), as in *Swipesy.* In addition, the first two measures of C echo the cadential motif of Hayden's A and B strains, just heard (Ex. 5-2b). Though the key is different, the melody notes are retained on the same pitches. Joplin further adds variety with a changed and fuller harmony, rhythmic displacement, and sequential repetition. He then, in C3-4, picks up the opening idea of Hayden's A strain and creates a variant, which is itself spun into new metamorphoses.[43] In staying so close to his student's original conception, Joplin paid tribute to the younger man, at the same time unifying the entire work in a most convincing manner.

Echoes of Scott Joplin, Part 3: Sun Flower Slow Drag

Roscoe Carter in 1909 made an obvious imitation of the opening of *Sun Flower* (Ex. 5-3). This is actually an echo of Scott Hayden rather than of Scott Joplin.[44]

Stark wrote in advertising copy for *Sun Flower:* "This is Joplin's favorite, and there are many who think it superior to the world famous Maple Leaf. It was written during Joplin's courtship and is immensely sentimental."

Ex. 5-3. Roscoe Carter, Crazy Horse Rag *(Chicago: Kremer, 1909), A1-4.*

There is much of the usual Stark hyperbole in this copy, and we have no evidence that *Sun Flower* was Joplin's favorite. We must also wonder why Stark invariably ignores Scott Hayden in the advertising. But was it written during Joplin's "courtship"?

The reference probably was to Joplin's marriage to Belle, who was the widow of Scott Hayden's brother Joe. Belle, referred to as "Belle Hayden" in earlier biographies, had a son named Alonzo who lived with his grandparents, Marion and Louise Hayden (Scott Hayden's parents), at 133 West Cooper. The 1900 census of Sedalia, taken on June 6, 1900, shows that Joplin was living in an unnumbered house on Washington Street owned by Susannah Hawkins.[45] The only other lodger was a Belle Jones, who had one child, living elsewhere. Belle Jones was probably the same person as Belle Hayden.

The listing in the census is the first reference to Belle, and her marriage to Joplin lasted until 1903, at the latest. There is no direct evidence on whether the marriage was contractual or common law, but a document we will examine in Chapter Eight *suggests* the former.

This was the last reference that we have for Joplin in Sedalia before his move to St. Louis with Belle. His years in Sedalia had been good to him. He had arrived in town in 1894 as an unknown musician and an unpublished

United States Census, Sedalia, 214 A, SD 7, ED 111, Sheet 1. Taken on June 6, 1900.

Scott Joplin is on line 8. In the columns following his name are "Lodger," "B" (black), "M" (male), "Oct. 1872" (date of birth), "27" (age at last birthday), "S" (single). Belle Jones is on the next line. The columns following her name are: "Lodger," "B," "F," "Mar. 1875," "25," "S," "3" (Mother of how many children), "1" (Number of these children living).

composer. He found Sedalia a nurturing place in which he could develop his skills and gain respect. On his departure in 1901, he was well known locally as both a pianist and composer, was quickly gaining nationwide fame as the composer of the *Maple Leaf Rag,* and was soon to be known as "the king of ragtime writers."

CHAPTER SIX

The King of Ragtime Writers, 1901–1902

Director Alfred Ernst of the St. Louis Choral Symphony Society believes that he has discovered, in Scott Joplin of Sedalia, a negro, an extraordinary genius as a composer of ragtime music.

St. Louis Post-Dispatch, Feb 28, 1901.

THROUGH MOST OF THE 1890s, ragtime was still considered a novelty and made only tentative inroads on public taste. By 1900, though, the music was clearly winning over the American public, or at least the youthful portion of that public. The music had vigor and excitement, its syncopated rhythms were marvelous to dance to, and its lyrics were a welcome relief from the sentiments of the prevailing "tearjerker" ballad. From that point on, the music's racial associations were weakened, and by 1905 or 1906 ragtime was the music not just of black Americans, but of all Americans. And unlike so much music whose models were clearly European, ragtime was distinctively American. This was one of the strongest arguments in the music's favor.

> Of the character of "rag-time" there can be no doubt—it is absolutely characteristic of its inventors—from nowhere but the United States could such music have sprung.
>
> As you walk up and down the streets of an American city you feel in its jerk and rattle a personality different from that of any European capital. . . . No European music can or possibly could express this American personality. Ragtime I believe does express it. It is to-day the one true American music.[1]

At a time when many American composers were struggling to develop a distinctive national style, one not beholden to European prototypes, ragtime was suggested as a valuable source material and point of departure. Just as Eastern European countries had based national styles on indigenous folk musics, the argument ran, American composers could build upon ragtime. Antonín Dvořák, the great national composer of Bohemia, had already advised American composers in 1895 to develop a national style based on the musics of Native and African Americans,[2] and had demonstrated the feasibility of this approach with his "New World" Symphony in 1893.

The idea took hold in many circles and was argued forcibly:

> Here, perhaps, then, for those who have ears to hear are the seeds from which a national art may ultimately spring. . . . We look to the future for the American composer, not, indeed, to the Parkers and MacDowells of the present who are taking over a foreign art and are imitating it with more or less success and with a complete absence of vital force, but to someone as yet unknown, perhaps unborn, who will sing the songs of his own nation, his own time, his own character.
>
> But here and nowhere else are the beginnings of American music, if American music is to be anything but a pleasing reflection of Europe. Here is the only original and characteristic music America has produced this far.[3]

Ragtime also had its detractors. Its success, its potency, made it a danger. Had it been, as frequently asserted, just a passing fad that would soon blow off in the wind like any piece of fluff, it could have been ignored. But it did not pass from the scene so quickly. It developed from a curiosity in the black subculture to the major popular musical expression of the American mainstream. Above all, as a black-originated music that was embraced by the youth of America, it was perceived as a negative influence that had to be confronted and eliminated.

How would American youth ever develop tastes for "good" music, when they were subjected to the constant and seductive syncopations of ragtime? Ragtime, it was argued, was lowering musical tastes. Soon there would be no public for good music, no public for symphony concerts.

> Ragtime has dulled their taste for pure music just as intoxicants dull a drunkard's taste for pure water. Ragtime becomes a habit, and like all other habits, it is very difficult if not impossible for its victims to break away from it.
>
> Especially with young people ragtime takes up so much time and thought that they lose in higher musical cultivation. This is the harm of ragtime.[4]

But more than the nation's musical health was at stake; its moral, spiritual, mental, and even its physical well being were under attack by ragtime.

The counters of the music stores are loaded with this virulent poison which, in the form of a malarious epidemic, is finding its way into the homes and brains of the youth to such an extent as to arouse one's suspicions of their sanity.

These "crochety" accents, these deliberate interferences with the natural logic and rhythm, this lengthening of something here and shortening of something there, must all have *some* influence on the brain.[5]

There are several underlying issues here. One is the tendency, particularly conspicuous in the present century, of older generations to reject the music of youth. We have witnessed similar condemnations and warnings accompanying the advent of many successive styles of popular music, jazz, and rock-and-roll. Another is the fear that a style that evolved outside conventional musical circles was taking the place of "legitimate" music. In this regard, in New York in 1915 the complaint arose that black musicians "who haven't the slightest conception of music" were in greater demand for dance music than their "more properly trained" white counterparts.[6]

Probably the greatest source of disturbance was that American youth had so eagerly accepted an African-American expression. This was the first time a black art had such impact on American cultural life. Given the generally low esteem in which blacks were held, the enthusiasm for ragtime was viewed as both confusing and tragic.

Closely allied to the racial aspect was the sexual. Ragtime amounted to a cultural miscegenation, an issue that presents a host of complex emotional ramifications. The argument ran that adoption of ragtime would lead to adoption of other African-American traits. The fear of this linkage was made explicit in the following:

Can it be said that America is falling prey to the collective soul of the negro through the influence of what is popularly known as "rag time" music? . . . If there is any tendency toward such a psychological amalgamation, toward such a national disaster, it should be definitely pointed out and extreme measures taken to inhibit the influence and avert the increasing danger—if it has not already gone too far. . . .

The American "rag time" or "rag time" evolved music is symbolic of the primitive morality and perceptible moral limitations of the negro type. With the latter sexual restraint is almost unknown, and the widest latitude of moral uncertainty is conceded.[7]

The linkage was present also in the adoption of dances that were perceived to have black roots. These dances of uninhibited abandon led, it was argued, to loosening morals and the corruption of many young ladies.

Ragtime's racial identity prompted other arguments against the music and against the thought of developing it into an American art music. The

freshness of ragtime suggested that an "inferior" race had developed something that was innovative. Opponents to this idea denied that ragtime *was*
new. They argued that Haydn, Mozart, Beethoven, and Brahms all used
syncopations similar to those found in ragtime and that the same rhythms
could be found in the Spanish, Gypsy, and Hungarian dances. This reasoning
led to the conclusion, recognized as absurd by some, that ragtime originated
in Europe.

Even if ragtime's genealogy was acknowledged as black, a view that
many saw as inescapable, the thought of basing a national school on black
music was considered unacceptable. One answer was to deny the America-
nism of African Americans: "The so-called negro melodies, even if they be
original with the colored race, cannot be considered as American, for the
negro is a product of Africa, and not of America."[8]

But there was also black opposition to ragtime. As we have seen, the
black preachers of Sedalia, in their opposition to the Maple Leaf and Black
400 clubs, protested against ragtime, and their attitude was typical of many
in the African-American community. Churchgoing blacks, those who were
educated, and those who sought respectability tended to accept the stated
values of white America. In doing so, they became vocal critics of ragtime.
How did they reconcile this attitude with the blatant antiblack rhetoric of
some ragtime critics? One approach was to deny ragtime's black heritage:

> White men also perpetuate so-called music under the name "rag-time," repre-
> senting it to be characteristic of Negro music. This is also a libelous insult. The
> typical Negro would blush to own acquaintance with the vicious trash put forth
> under Ethiopian titles. If the Negro Music Journal can only do a little mission-
> ary work among us, and help to banish this epidemic it will go down in history
> as one of the greatest musical benefactors of the age.[9]

Scott Joplin sought acceptance as an educated, cultured, and respectable
individual. In this regard, he could be expected to ally himself with others in
the black community who had similar aspirations. But his art, ragtime, was
rejected by these very same circles. This was one of the conflicts of his life, a
conflict that took on mythic proportions.[10]

With his success, Joplin had outgrown Sedalia. It was time to return to
St. Louis, where he had probably lived at various times in the late 1880s and
early 1890s, and where he frequently visited while living in Sedalia. Though
almost 200 miles apart, St. Louis and Sedalia were linked by direct lines on
the M. K. & T. and the Missouri, Pacific railways, with roundtrips costing
from $2.50 to $7.55.[11]

The area of St. Louis that Joplin moved to was a black neighborhood,
the city's red-light district, and a major center for the early development of
black Midwestern ragtime. It was situated in the easternmost bulge of the

Scott Joplin's St. Louis. (1) Scott Joplin's residence, 2658A Morgan (1901); (2) Sam Patterson's residence, 2215 Morgan; (3) Scott Joplin's residence, 2117 Lucas (1902–03); (4) Daniel Davenport, 2339 Chestnut; (5) Tom Turpin's Eureka Club, 2208 Chestnut; (6) Booker T. Washington Theater, Market & 23rd; (7) Scott Joplin's residence, 2221 Market (c. 1905–07); (8) Tom Turpin's Rosebud Bar, 2220–2222 Market; (9) Louis Chauvin's residence, 1613 Linden; (10) Stark Music, 1516 Locust; (11) T. Bahnsen Piano Mfg. Co., 1522 Olive; (12) Crawford Theatre, 14th & Locust; (13) The Alcove Cafe, 1300 Market; (14) Val A. Reis Music Co., 1210 Olive

city, beginning about a mile from the Mississippi River. While no firm boundaries can be drawn, the district could be said to have extended from 12th Street on the east, Beaumont on the west, Clark on the south, and Morgan on the north (see map). The main thoroughfare was Market, on which the Union Railroad Station was situated. Blues pioneer W. C. Handy was there around 1892 and gave this description:

> I have tried to forget that first sojourn in St. Louis, but I wouldn't want to forget Targee Street as it was then. I don't think I'd want to forget the high-roller Stetson hats of the men or the diamonds the girls wore in their ears. Then there were those who sat for company in little plush parlors under gaslights. The prettiest woman I've ever seen I saw while I was down and out in St. Louis.[12]

It is unfortunate that Handy did not discuss the music he heard in the St. Louis district in 1892, for he could have cast some light on ragtime's earliest days. The term may not have been in use yet, and the style was to remain unknown to the public at large until unveiled at the Chicago World's Fair the following year. But in the St. Louis district in 1892, ragtime was already jangling on pianos in the saloons, dance halls, and brothels.

This is where Tom Turpin (1873–1922) was to make his mark. In 1890, and perhaps earlier, Tom and his brothers Charles and Robert were working at their father's saloon, the Silver Dollar, at 425 South 12th Street. Tom also worked as a pianist and listed himself as a musician in the 1891–92 city directory.[13] He knew the ragtime style, for by 1892 he had composed his first rag, the *Harlem Rag*.[14] A young black man of that time and place was not about to have a piece of music published, especially a music associated with disreputable aspects of society, so he had to wait five years to see it in print. When this event finally occurred, in 1897, other piano rags had already been issued, but Turpin's was the first rag publication by an African American.

Harlem Rag has more than historical interest as a "first." It is also a good piece of music, and it preserves a record of improvisation by having three of its four themes followed by variations—in effect, written-out improvisa-

Tom Turpin.

tions.[15] After being published in St. Louis by Robt. De Yong, *Harlem Rag* was reissued in a simplified arrangement in 1899 by Jos. W. Stern & Co., a major New York firm. Turpin did not publish many rags, but each one is a quality work. They include *Bowery Buck* (1899), *A Ragtime Nightmare* (1900), *St. Louis Rag* (1903), and *The Buffalo Rag* (1904).[16]

Whatever the impact of Turpin's publications, it was as a performing musician in St. Louis that he had his major influence. He was an imposing, massive man whose hands looked too large to fit a piano keyboard. But he was a skilled musician, musically literate, at ease playing in any key, and a master at "cutting contests," events in which pianists would pit their virtuosic skills against each other. Musicians coming to town were interested in hearing him, and he was equally interested in learning what they could do. One report describes how Turpin would test out newly arrived musicians on the piano in the parlor of "Mother" Johnson's brothel, on Market and 22nd. After the pianist displayed his talents, other pianists would join in and for the next few hours a contest would ensue.[17]

Turpin was a leader in ragtime circles in St. Louis and became the mentor to many younger musicians. Of these, three stand out: Louis Chauvin (1881–1908), Sam Patterson (c. 1881–1955), and Joe Jordan (1882–1971).

Louis Chauvin was from a family of musicians: others were Sylvester, Peter, and Abraham.[18] Louis, a small man of about 5'5", was an outstanding pianist with a technical facility that outstripped the abilities of everyone else in his circle, including Turpin. Patterson recalled that, to warm up, Chauvin would play a Sousa march in his own arrangement, covering the keyboard with double-time octaves in contrary motion. He could play anything he heard, but would improve it with his own harmonies and put it in different keys. Though unable to read, he once played an entire show after having the music hummed to him. He was also a brilliant composer and used harmonies so original that his authorship could not be mistaken. But unless someone else wrote the music out for him (as Joplin would do in one case), it was lost. His talents did not stop at piano playing. He was also a superb singer and dancer, and did both professionally.[19]

Sam Patterson, like Chauvin, was a small man, about 5'3". He and Chauvin were childhood friends and by their mid-teens had teamed up as dual pianists, singers, dancers, and black-face minstrel comedians. Among the groups they formed was the Mozart Comedy Four, a vocal quartet consisting of Chauvin as first tenor, Patterson as second tenor, Charles Young, baritone, and Baley Alexander, bass. They performed selections from light operas and many spirituals. Patterson never published any compositions, but in addition to piano he was proficient on the trumpet, saxophone, and xylophone and was to become very successful in the black vaudeville world as an all-around musician.[20] He was also to become one of Scott Joplin's closest friends.

Louis Chauvin.

Joe Jordan was born in Cincinnati but spent his teen years in St. Louis. He attended the Smith College in Sedalia.[21] He was a proficient pianist and was declared the winner in at least one ragtime contest sponsored by Turpin. Early in his career he gained respect as a major musical talent and he wrote and directed a number of black shows in New York and Chicago.

The St. Louis district was a virtual school of music for the young ragtimers who frequented it. The brothels, saloons, and dance halls provided ample employment opportunities for black musicians. As one ragtimer described it:

> The players would get together in the clubs and wait for calls from the houses. They were not competitive, tried to help each other. They would learn a song and try to teach the ones who didn't know it. There was plenty of work. They could pick up money any time they wanted to, were satisfied with $2 or $3 at a time. You could always get free lunches.[22]

Scott Joplin's move to the area was to enrich it, and at the same time put him into closer contact with people who were to influence his life.

The accolades Joplin had received in Sedalia pale besides the notice he received in St. Louis early in 1901. There, shortly before moving to that city, he met with Alfred Ernst, conductor of the St. Louis Choral Symphony Society, the major music organization of that city. Ernst, like the leaders of almost every significant orchestra in the United States, was German. As leader of the St. Louis Symphony from 1894 to 1907, he was one of the most important musical figures and opinion makers in the city.

America's cultural leaders generally scorned ragtime, and this was especially true of the Germans who were imported to lead the musical life. Karl Muck, for example, conductor of the Boston Symphony Orchestra, reacted predictably:

> I think what you call here your ragtime, is poison. It poisons the very source of your musical growth, for it poisons the taste of the young. You cannot poison the spring of art and hope for a fresh clear stream to flow out and enrich life.[23]

It must have been unusual for Ernst even to meet with Joplin. However the meeting came about, the result was extraordinary, for Ernst went counter to expected form. Putting aside the usual biases of race and musical style, he recognized something special in Joplin's music and in Joplin the man. The encounter was reported, accompanied by a photograph of Joplin, in the *St. Louis Globe-Democrat* of February 28, 1901:

> Director Alfred Ernst of the St. Louis Choral Symphony Society believes that he has discovered, in Scott Joplin of Sedalia, a negro, an extraordinary genius as a composer of ragtime music.
>
> So deeply is Mr. Ernst impressed with the ability of the Sedalian that he intends to take with him to Germany next summer copies of Joplin's work with a view to educating the dignified disciples of Wagner, Liszt, Mendelssohn and other European masters of music into an appreciation of the real American ragtime melodies. It is possible that the colored man may accompany the distinguished conductor.
>
> When he returns from the storied Rhine Mr. Ernst will take Joplin under his care and instruct him in the theory and harmony of music.
>
> Joplin has published two ragtime pieces, "Maple Leaf Rag" and "Swipesey Cake Walk," which will be introduced in Germany by the St. Louis musician.
>
> "I am deeply interested in this man," said Mr. Ernst to the Post-Dispatch. "He is young and undoubtedly has a fine future. With proper cultivation, I believe, his talent will develop into positive genius. Being of African blood himself, Joplin has a keener insight into that peculiar branch of melody than white composers. His ear is particularly acute.
>
> "Recently I played for him portions of 'Tannhäuser.' He was enraptured. I could see that he comprehended and appreciated this class of music. It was the

TO PLAY RAGTIME IN EUROPE

Joplin photo from St. Louis Globe-Democrat, *Feb. 28, 1901. The microfilm copy of this newspaper issue (no paper copy is known to exist) is too dark, rendering Joplin's portrait virtually in silhouette.*

opening of a new world to him, and I believe he felt as Keats felt when he first read Chapman's Homer.

"The work Joplin has done in ragtime is so original, so distinctly individual, and so melodious withal, that I am led to believe he can do something fine in composition of a higher class when he shall have been instructed in theory and harmony.

"Joplin's work, as yet, has a certain crudeness, due to his lack of musical education, but it shows that the soul of the composer is there and needs but to be set free by knowledge of technique. He is an unusually intelligent young man and fairly well educated."

Joplin is known in Sedalia as "The Ragtime King." A trip to Europe in company with Prof. Ernst is the dream of this life. It may be realized.[24]

The account raises many questions, most of which we can not answer. How did the meeting come about? Did Ernst have prior familiarity with Joplin's music? Perhaps Eleanor Stark, well known in St. Louis musical circles, arranged the meeting? Was it the meeting, and the hearing of Wagner's music, that inspired Joplin to write his own operas? Did Joplin and Ernst meet again, perhaps for the instruction mentioned? Was it to be near Ernst as a mentor that Joplin moved to St. Louis? If Ernst introduced Joplin's music to colleagues in Germany, what was the reaction?

The account in the Ernst interview also provides information and suggests issues to explore. The most explicit point is about Joplin's musical ear. Ted Browne had reported that Joplin had perfect pitch and composed away from the piano.[25] An interview in 1907 mentioned that Joplin always had music paper with him so that he could jot down ideas.[26] Ernst's statement that "His ear is particularly acute" does more than just support the above observations. Ernst was a major conductor who had worked with hundreds of talented and finely trained musicians. For him to make particular note of Joplin's hearing suggests that it must have been of an unusually special order.

The newspaper notices in Sedalia showed that Joplin was admired and respected, but provide no clues as to how he had earned such respect. Certainly his musicianship was a significant factor, but as a member of a disdained group there had to be more. The Ernst interview gives the first clues. He described Joplin as intelligent, fairly well educated, eager to learn, and responsive to the "higher class" of music.

Joplin's intelligence could have been judged by what he had to say. His comments might also have suggested his degree of education, but another significant factor could have been the manner in which he spoke. We know that Joplin equated manner of speech with education, for he made that an important part of the characterizations in his opera *Treemonisha* a decade later. For this reason, we suspect that Joplin cultivated speech patterns that avoided the stereotypes of African-American dialect.

Blacks were not expected to be well educated or well spoken, so Joplin's manner called attention to itself. Monroe Rosenfeld, a songwriter and journalist, after noting Joplin's accomplishments in the formal, "educated" aspects of music—"a tutored student of harmony and an adept at bass and counterpoint"—said that, despite Joplin's race, "he is attractive socially because of the refinement of his speech and demeanor."[27] It is clear that Joplin had a manner that allowed whites to be comfortable with him. This appraisal is confirmed by several reports from white associates. His lawyer described him as "a fine young man"; the widow of one of his late students remembered him as "a charming man with a pleasant personality"; John Stark's grandson said, "Scott Joplin was a real gentleman."28

But another issue is raised by the last quotation: Joplin's "demeanor." Joplin was consistently described as being extremely quiet and modest. Rag-timer Charley Thompson, who was associated with Joplin in St. Louis, described Joplin as "a fine chap . . . who never talked above a whisper." Rosenfeld said Joplin had "a retired disposition," and then, more expansively: "One of the interesting characteristics of Scott Joplin's personality is his conservatism. He rarely refers to his productions and does not boast of his ability." Another reporter, in 1907, wrote, "He is unassuming and never has much to say, and seldom speaks of his music." One black reporter noted in 1908, "he is very retiring in manner," and another, in 1909, that "Scott Joplin . . . hates notoriety." The most vivid description comes from Beatrice Martin, a woman who, when in her nineties, spoke of the Joplin she remem-bered from 1904: "He never talked very much. He was a very quiet man. He only answer what you say to him."29

We suspect this was more than just modesty. Joplin seemed to have had an extreme reticence. Sam Patterson said he was "not much socially." A student from around 1911 noted that Joplin rarely had anything to say in social circumstances and was almost "morose." The morose quality may have been brought on by his illness at the time, but the evidence suggests that he was an intense person totally absorbed with his music. In the middle of a conversation totally unrelated to music, Joplin might interrupt with a com-ment about how he would handle some musical problem. Beatrice Martin said he "always looked like he was in a deep dream and he was always studying." Arthur Marshall, while agreeing that Joplin's nature was sub-dued, saw some break in the bleakness of other descriptions: "one of the most pleasant men you'd ever want to meet," and "quiet, serious, jolly but not frivolous. He was always thinking of ideas, and said very little."30

Consistent with this serious nature, Joplin rarely smiled. Browne re-ported, "Scott had a reputation of never smiling, and the Starks spoke of it so often even in front of Scott." Sam Patterson agreed: "Joplin never smiled in his life. We would have lunch together and I used to tell him jokes, but couldn't get a smile out of him."31

Joplin clearly did not fit the pattern of a flamboyant showman. Unlike such ragtimers as Jelly Roll Morton and Willie "The Lion" Smith, who consciously developed attitudes of bravura and panache,32 Joplin was low-keyed, subdued, introspective, and non-offensive. It is in this perspective that we gain an understanding to his life.

The Ernst interview at the end of February 1901 indicated that Joplin was still "a Sedalian." The move to St. Louis probably took place in the spring, and by the summer of 1901 he was established there, for a theatrical notice in July referred to "Scott Joplin *of St. Louis.*"[33]

The earliest St. Louis address we have for Joplin is 1658A Morgan, which was between Beaumont and Jefferson and easy walking distance to the heart of the district.[34] He moved there with Belle; Marshall reported that they had married in St. Louis.[35]

Joplin was already well known and admired in St. Louis. Charles Thompson commented, "When Joplin returned for a visit 'there would be just like a parade down Market St.'"[36] In addition to the usual denizens of the district, there were other new arrivals with whom he was close. He and Belle were joined by Scott Hayden and his bride Nora. Will Joplin was five blocks away at 2117 Lucas, and Arthur Marshall's brother Lee was at 2939 Scott.[37] John Stark, who had moved to St. Louis in 1899, had opened a music printing plant nearby at 3615 Laclede, and his daughter Eleanor had a studio at 3210 Lucas. The brothers Tony and Charles Williams, formerly of the Black 400 clubs of Sedalia and Joplin, were also in town and, as usual, were being enterprising. Learning of the coming world's fair in St. Louis, they passed around a circular calling for a Negro pavilion at the fair. This proposal generated a debate. On one side, it was argued that such a pavilion would give blacks an opportunity to demonstrate what they could do for themselves, following the example of the Negro Pavilion at the Atlanta Exposition in 1895. The counter-argument was that a separate pavilion would give whites an excuse to exclude blacks from all other exhibits.[38] The counter-argument prevailed, but even without a separate Negro Pavilion as an excuse for African-American exclusion from fair events, black Americans suffered the usual discriminatory humiliations.

However Joplin felt about the racial situation in St. Louis, things were going well for him. He had the companionship of his new wife and some of his closest associates from both Sedalia and St. Louis. His reputation was growing, he had the endorsement of one of the city's leading cultural figures, and he had acclaim from a leading newspaper. He was also in the city with some of the best and most exciting ragtime players in the country. St. Louis in 1901 was the right place for Joplin.

Two of Joplin's publications in 1901 were actually from earlier years. *The Sun Flower Slow Drag,* co-composed with Scott Hayden, was from 1899 or 1900 but not registered for copyright until March 18, 1901. Since Ernst mentioned only *Maple Leaf* and *Swipesy* in the interview of the previous month, a copy of *Sun Flower* was probably not yet available. *The Augustan Club,* composed in January 1900, was registered for copyright on March 25, 1901. It was probably published at the same time as *Sun Flower,* since it is listed on the *Sun Flower* cover.

Another work in this same publishing batch was *Peacherine Rag,* registered on March 18. *Peacherine* has a number of unusual features. Compared with Joplin's previous rags, its opening strain is unexpectedly simple. Its melody has a restricted range, repeatedly circling within diatonic fourths, and its bass plays almost entirely on tonic-dominant relationships (Ex. 6-1a). The B strain, rather than continuing in the tonic key, is in the dominant. The probable reason for this is that the dominant key puts the strain into a more comfortable range; playing it in the expected tonic results in bass chords that are too low and somewhat muddy.

Ex. 6-1. Scott Joplin, Peacherine Rag *(1901): (a) A1-4; (b) D1-4; (c) Scott Joplin,* Combination March *(1896), D1-4; (d) Scott Joplin,* Peacherine Rag, *D12-16.*

The C strain returns to the more complex and busy textures of the earlier rags, also placing some emphasis on the right-hand sevenths, noted in *Swipesy* and *Sun Flower*. The D strain (Ex. 6-1b) echoes and develops one of the themes of Joplin's earlier *Combination March* (Ex. 6-1c). He ends the strain with an ascending chromatic bass line, driving to the final cadence (Ex. 6-1d).

Echoes of Scott Joplin, Part 4: Peacherine Rag

 A piece that bears a slight resemblance to a portion of *Peacherine*, strain A, is Al. Verge's *Whoa You Heiffer*, strain C (Ex. 6-2). Though the similarity is not so close as to be a conclusive case of plagiarism, our suspicions are heightened in recognizing that strain A of the piece suggests Joplin's 1903 *Palm Leaf Rag*.

Ex. 6-2. Al. Verges, Whoa You Heiffer *(New Orleans: Hakenjos, 1904), C1-4.*

A final rag from 1901 is *The Easy Winners,* which has an unusual publishing history. Joplin registered the piece for copyright in his own name on Oct. 10, 1901, and published it himself.

Why was *Easy Winners* not published by Stark? Did Stark reject it? Was he unwilling to meet Joplin's terms? Were they having difficulties with their professional relationship? Had Joplin, realizing he could earn more by publishing music himself, decided to strike out on his own? These questions remain unanswered.

Whatever Joplin's plans, he apparently learned that he could not handle the distribution, for by 1903 *Easy Winners* was republished by Shattinger Music Co., a St. Louis instrument dealer that published piano music primarily for teaching purposes.[39] Possibly with pedagogy in mind, Shattinger simplified the music slightly and put it in F rather than A-flat. Aside from the change in key, among the significant details distinguishing the two publications is one occurring in D2 and 10: in the Shattinger version, the sixteenth notes in the bass are excised (Ex. 6-3a, b). A puzzling detail is in B15. The

Ex. 6-3. Scott Joplin, Easy Winners *(1901): (a) D1-4; (b) Shattinger version, D1-4.*

(a)

reprint edition by the New York Public Library has a tie at mid-measure that is lacking in the Copyright Office copy. In the Shattinger version, however, the tie is present. Did Joplin add the tie? By 1908 *Easy Winners* was in the hands of Stark, who published it using a new cover, but Joplin's original plates, without the tie.[40]

Easy Winners must be judged one of Joplin's great works. It has a classical balance between its strains, its moods, and in its progression from the smooth calm of strain A to the sporadic agitation of strain D. On the technical level, it reveals musical thinking unusual for ragtime. We referred earlier to the inner voice descending chromatic fourth that Joplin used in the final strain of *Maple Leaf.* In *Easy Winners,* he built the entire C strain on that concept. The strain has four similar phrases, the first three showing an inner voice chromatic descent, from B-flat to A-flat, that is interrupted before spelling out a full fourth (Ex. 6-4a). The last phrase, though, starting an

Ex. 6-4. Scott Joplin, Easy Winners: *(a) C1-4; (b) C13-16; (c) analytic reduction showing chromatic descent in inner voice, C13-16.*

extra semitone higher (C-flat), completes the descent (Exs. 6-4b, c). Joplin indicates the importance of this descending motif by double-stemming the notes, suggesting that they should be brought out in performance.

One other piece, the song *I Am Thinking of My Pickaninny Days,* bears a copyright notice of 1901, although it was not received by the Library of Congress until April 9, 1902. It was published by another St. Louis firm, Thiebes-Stierlin Music Co., again raising the question whether something had disturbed the relationship between Joplin and Stark.

This was the first published Joplin song in which lyrics were by some-one else, in this case Henry Jackson. Jackson was an associate from Sedalia, a musician, and a porter on the M. K. & T. He was also a member of the Joplin Dramatic Company when it performed *The Ragtime Dance* in Sedalia in 1899.[41]

As suggested by the title, the song is in the 19th-century tradition of sentimental "darkey" ballads, of which Stephen Foster's "Old Folks at Home" (1851) is a classic example. It follows the expected lyric and music clichés and has nothing to commend it; it is, however, handled with the utmost professionalism. I see no reason to doubt Joplin's authorship of the music. Since Joplin wrote the lyrics for his earlier songs, as well as for his later opera, the idea for the collaboration was probably Jackson's. The likelihood is that Jackson had a lyric, or possibly a lyric and a melody, and asked Joplin for his assistance. It may therefore have been Jackson who arranged for publication.

In the latter part of 1901, the theatrical pages of *Indianapolis Freeman* began to contain notices extolling Joplin's compositions—*Maple Leaf, Sun Flower, Easy Winners,* and *Peacherine.* Most of these notices were from P. G. Lowery, a leading black cornetist and bandleader. A notice by Lowery on November 16, 1901, mentions another work: *A Blizzard.* No piece by that name has survived and this is the only known mention of it.

It is evident that Joplin and Lowery were friends. Aside from the good notices, Lowery visited with Joplin in November. The following year, Joplin was to dedicate a work to him.

In April 1902, W. H. Carter, trombonist for the Queen City Concert Band and editor of the *Sedalia Times,* visited Joplin and wrote of the compos-er's fame and his life in St. Louis:

> Mr. Scott Joplin . . . is gaining a world's reputation as the Rag Time King. Mr. Joplin is only writing, composing and collecting his money from the different music houses in St. Louis, Chicago, New York and a number of other cities. Among his numbers that are largely in demand in the above cities are the Maple Leaf Club [*sic*], Easy Winner, Rag Time Dances and Peacherine, all of which are used by the leading piano players and orchestras.[42]

Actually, in 1902 Joplin had not yet published with Chicago or New York firms. But in other respects, the report describes the essential change in his life. Unlike most ragtime pianists in St. Louis, for whom the saloons were both a second home and a place to find employment, Joplin, in the words of

Artie Matthews, "did not hang around joints."[43] He may have walked over to the district occasionally to socialize, but he earned his living at home composing, collecting royalties, and teaching.[44]

An additional reason why Joplin did not work at the Market and Chestnut Street resorts was that, as a performer, he was probably outclassed by the competition. He had worked in Sedalia and elsewhere as a pianist with apparent success. A Sedalia newspaper in 1898 referred to him as "one of the best pianists in the world";[45] his Sedalian lawyer Robert A. Higdon liked the way he played,[46] as did a student from in New York in 1915–16[47]. In 1911, a columnist for a New York-based nationwide music magazine wrote: "It takes Scott Joplin to play ragtime on the piano. There are ragtime players, but when it comes to playing in a musicianly way, Joplin is there with the goods every time."[48]

But Chestnut Valley had a different standard. This was the scene of hot piano playing, of seething cutting contests, and there Joplin could not compete. Among the St. Louis pianists, few had anything favorable to say about Joplin's playing. Only his good friend and student Arthur Marshall could come up with anything favorable to say. Marshall reported Joplin played slowly, "but exceedingly good. Played more piano than you would want to see. . . . had an execution that you would stand back and listen and wonder how he got to do that stuff." Charlie Thompson spoke of Joplin only reluctantly and expressed total disdain for his piano playing, saying he never played anything except the *Maple Leaf Rag*. Joe Jordan agreed that he never played anything other than his own pieces, but that he did play these well, more or less as written. In another interview, Jordan was more critical, describing Joplin's playing like that of a "stationary Indian." Patterson said Joplin "never played well," and Artie Matthews reported that the St. Louis pianists delighted in outplaying Joplin with his own music.[49]

A comment by Stark's son highlights the extent of Joplin's deficiencies as a pianist:

> Mr. W. P. Stark . . . remembers that Scott . . . was a rather mediocre pianist and that he composed "on paper" rather than "at the piano" as all the real ragtime virtuosos did. This became a real problem when Scott had to play one of his own compositions and found that he had to rehearse it carefully before he could play it convincingly.[50]

Ted Browne amplified upon this theme: Joplin not only had to practice his own pieces before performing them, he required instruction. "'Sunflower Slow Drag' was quite difficult and John Stark said that Scott had to take some lessons on that tune before he could execute it correctly."[51]

It may seem strange that such a poor pianist should have favored the piano as his major medium of expression and that he should have developed a style that is so eminently pianistic. Is it possible that Joplin was not always a mediocre pianist? Could he, by 1901, have already been experiencing the debilitating effects of the disease that was to take his life sixteen years later? His death certificate lists the diagnosis of dementia paralytica—cerebral form, with syphilis as a contributory factor. Dementia paralytica normally sets in between twenty and thirty years after contraction of syphilis, and early symptoms include discoordination of the fingers.[52] By this time, Joplin may have already had the disease for some years. Whereas it might have been a little early in the course of the disease for a victim to notice a deterioration of digital coordination, the fine finger control required by a pianist could well have been affected. It is possible, then, that Joplin's pianistic skills were in decline.[53]

Despite any liabilities Joplin may have had as a performer of ragtime, as a composer of the music he was widely respected. As Patterson put it, "Everyone had a lot of respect for him, . . . he had something to offer. His music was something to talk about."[54]

Arthur Marshall and others left valuable testimony regarding Joplin's life in St. Louis, but except in cases where we can cite documentary evidence, the chronology of these events remains vague. For example, when W. H. Carter wrote of his visit with Joplin in April 1902, he wrote of other Sedalians also in St. Louis at the time; among them Robert O. Henderson (the Sedalia Comedian and Queen City Cornet Band member), Tony and Charles Williams, and Lee Marshall. That he made no mention of Scott Hayden and his wife, who were reportedly living with the Joplins, suggests that they were no longer there in April. We are able to locate the Haydens only in the summer of 1902, when they were back in Sedalia. There, in July, Nora Hayden gave birth to a girl; the following month, Nora died.[55]

Again from testimony, we are told that the Joplins left their residence on Morgan and purchased a thirteen-room house a few blocks away at 2117 Lucas. There they rented some of the rooms to boarders, including Arthur Marshall and Scott Hayden. Scott Joplin's brother Will may also have been one of the lodgers, since the 1902 directory placed him at that address. However, by 1903 Will had died, for the 1904 directory has "Sopharina Joplin, widow of William."

That Scott Joplin and Belle lived at that address is confirmed by the 1903 city directory as well as by other sources, but we cannot determine precisely when this move occurred. A letter Joplin wrote to the Copyright Office shows that he was on Lucas in February 1903, but he may have moved there as early as the spring of 1902.

One of Arthur Marshall's anecdotes has Marshall and Hayden both living with the Joplins. This had to have been in late 1902 or 1903, for Marshall had been on a two-year tour with the Dan McCabe's Coontown 400 until the fall of 1902.[56] It was probably at the end of the tour that he moved in with the Joplins, for Dan McCabe in December was in St. Louis and visited with Joplin.[57] Hayden could have returned to St. Louis at any time after his wife's death; it is likely that he was there by the autumn, for he and Joplin issued a new collaborative rag early in 1903. Although I could be off by a few months, I place the time of Marshall's anecdote in 1903.

A work W. H. Carter mentioned in April 1902 as being among Joplin's publications was *The Ragtime Dance.* This piece was not registered for copyright until December 29, 1902, but apparently was already in print; the list of works displayed on the cover includes only pre-1902 pieces.

As recounted in Chapter 5, *The Ragtime Dance* was performed at Wood's Opera House in the fall of 1899, but rejected by Stark for publication, possibly because of its length of nine pages. Now, however, Joplin was famous and his works were selling, so Stark could reconsider publication. The Starks attended a performance in St. Louis with an orchestra consisting of Lij Cross and the Vasser Boys. Eleanor was enthusiastic about the results and convinced her father to publish the piano-vocal score.[58] Once published, it attracted some attention and was programmed by McCabe & Young's Company (also known as Dan McCabe's Coontown 400):

> D. W. McCabe called on Scott Joplin, the king of all rag-time composers, while in St. Louis and Mr. Joplin gave him quite a reception, also, three of his latest compositions, the latest being "A Ragtime Ball," [Ragtime Dance] which is without a doubt, the best piece of music of its kind ever written and which we are now producing to great success.[59]

A publication from the spring was *Cleopha,* issued by S. Simon, a small St. Louis firm with nine copyrights between 1902 and 1905. Why this work was not published by Stark is not known. Perhaps Stark had rejected it.

Cleopha is a difficult piece to classify and is subtitled simply "March and Two-Step." The first strain is very square, on the beat, almost stodgy, although still somewhat attractive. Strains two and three loosen up, but are still unsyncopated. The final strain is a rollicking rag. Despite this closing, three-quarters of the piece is unsyncopated, and Joplin did not call it a rag; we therefore agree to the classification stated in the subtitle.[60]

The Strenuous Life. A Ragtime Two Step bears a 1902 copyright notice but was never registered. Since the publication lists no other works and there is no Copyright Office copy to examine, we have little evidence on which to judge when in 1902 it appeared. The one clue is that it is advertised on the back cover of the copyright copy of *A Breeze from Alabama,* registered on December 29 but issued by early September (see below).

The title *The Strenuous Life* refers to a speech, and then a published collection of speeches, by Theodore Roosevelt. The original address was given in Chicago on April 10, 1899, when Roosevelt was governor of New York.

> I wish to preach, not the doctrine of ignoble ease, but the doctrine of the strenuous life, the life of toil and effort, of labor and strife; to preach that highest form of success which comes not to the man who desires mere easy peace, but to the man who does not shrink from danger, from hardship, or from bitter toil, and who out of these wins the splendid ultimate triumph.

Roosevelt proceeds to extend this concept to the actions of a nation, stressing the need to accept responsibilities in world affairs, emphasizing military preparedness, commercial adventurism, and willingness to intervene in world events. He thereby justifies the Spanish-American War and its territorial results, and the building of a Panama Canal. The speech was reported and quoted extensively.[61] In 1900, it was printed as the title essay in the collection of thirteen in *The Strenuous Life: Essays and Addresses*.[62] The volume was reprinted and enlarged in 1902 and reprinted again in 1905.

Had Joplin read Roosevelt's address, or was he simply attracted by the phrase, which itself had become famous?[63] In the same year, William J. Short, an otherwise unknown composer, published *Strenuous Life. March & Two Step,* and in following years there were five more marches registered for copyright with that title.[64]

Whether or not Joplin was inspired by the message of the address, he was probably inclined to pay tribute to Roosevelt. In 1901 Roosevelt shocked much of the nation with the politically dangerous act of inviting African-American leader Booker T. Washington to dine at the White House. Newspapers in the South condemned the invitation as an unwarranted attempt to place the black man on the same social plane as the white man; Roosevelt's act put him in a category with Ulysses S. Grant, and he would never be forgiven. The *Sedalia Sentinel* printed a poem on page one entitled "Niggers in the White House," which concludes with a black man marrying the President's daughter.[65]

Major newspapers in the North had a more charitable view. They recognized Washington's unique achievements and suggested that the invitation was Roosevelt's way of demonstrating he was President of all the people.

Scott Joplin did not specify a dedication for *The Strenuous Life,* nor is Roosevelt's picture on the cover. But the phrase "strenuous life" was so closely associated with the President that the audience of 1902 would have recognized the reference. In the midst of Southern denunciations that were to continue for years, this ragtime tribute to Theodore Roosevelt was Scott Joplin's response.

The Entertainer has a copyright registration date of December 29, 1902, but, like *The Strenuous Life,* is mentioned on the copyright copy of *A Breeze from Alabama.* We can therefore assume that it was published by September.

The Entertainer is most famous today for having spearheaded the Joplin and ragtime revival of the 1970s, but even during the ragtime period it received attention. In an article in 1903, writer Monroe Rosenfeld reproduces the first page of the music and cites the piece for singular praise:

> Probably the best and most euphonious of his latter-day compositions is "The Entertainer," . . . . It is jingling work of a very original character embracing various strains of a retentive character, which set the foot in spontaneous action and leave an indelible imprint upon the tympanum.[66]

In the same year, a report of one of the leading black minstrel companies elevates *The Entertainer* above *Maple Leaf:* "Prof. H. S. McDade's orchestra opened with Scott Joplin's masterpiece, 'The Entertainer,' and closed with 'Maple Leaf.'"[67] John Stark also found the two works to be comparable. In one advertisement he wrote:

> When we flashed the immortal "Maple Leaf" onto the dark moods of mankind our confidence was expressed in language supposed by a few people of lymphatic cast to be extravagant.
>
> All will now agree, however, that we but feebly stated the facts. It is not only ingenious and scholarly above all others of its class, but it has actually throttled and silenced the senseless knocker of ragtime—so called—and forced its way into the halls of the highest culture and refinement.
>
> All that we have ever said of "Maple Leaf Rag" is true of "The Entertainer" and more may be said of that subtle, soulful quantity not seen in the notes and never found by the marble heart. Hear "The Entertainer" well played and if the harp of your affections don't sound an aeolian chord, then we don't want to know you. . . .[68]

The simplicity of the A strain recalls the approach used in the opening of *Peacherine*. Both demonstrate remarkable diversity within musical economy. The essence of *The Entertainer* A strain is the E-C alternation over changing harmonies, preceded by the D-D-sharp pickup. After the initial statement, in single notes, the motive is repeated in filled octaves (the C now going to a higher E) and embellished (Ex. 6-5). Texturally, the first three phrases, with

Ex. 6-5. Scott Joplin, The Entertainer *(1902).*

bare, single-note beginnings followed by fuller, filled octaves, suggest the *solo-ripieno* alternation of a concerto. The final *tutti* phrase summarizes the E-C dominated strain.

Nothing is known of "James Brown and his Mandolin Club," but the music reflects the dedication since it lends itself very nicely to mandolin performance. Oddly, the mandolin arrangement published by Stark is in D, which is a more difficult key for that instrument.[69]

Echoes of Scott Joplin, Part 5: The Entertainer

Once again we return to L. Edgar Settle's *X. L. Rag* (1903). This time, it is the introduction, which bears a suspicious resemblance to B13-16 of *The Entertainer* (Ex. 6-6).

Ex. 6-6. (a) Scott Joplin, The Entertainer, *B13-16; (b) L. Edgar Settle,* X-L Rag, *Introduction.*

In 1920, John Stark added his own lyric to Joplin's first strain (acknowledging Joplin as the originator), transforming it into a song: *Oh You Tommy. Reminiscent of Tom Moore* (Ex. 6-7). The reference is the Irish poet Sir Thomas Moore (1779–1852), whose image was used to advertise cigars in Stark's day. The poor quality of the transformation from Joplin's original is evidence that Joplin had nothing to do with it.[70]

Jazz impresario and author Al Rose recalled that, during his childhood in New Orleans in the late teens and early 1920s, street hawkers selling watermelon would sing, to the opening half-phrase of *The Entertainer,* "Watermelons, they're wet, they're cold."[71]

A Breeze from Alabama was published in the last week of August or the first week in September. This information comes from accounts appearing in the September 6 issue of the *Indianapolis Freeman:*

Ex. 6-7. John Stark (words), music adapted from Scott Joplin, Oh You Tommy, Reminiscent of Tom Moore *(St. Louis: Stark, 1920), Verse 1-4.*

The title page of Scott Joplin's latest ragtime two step has a large picture of P. G. Lowery. . . .

Piano players should not forget to write to John Stark and Son, Music Publishers, St. Louis, Mo., and secure a copy of Scott Joplin's latest ragtime two-step, "A Breeze from Alabama," dedicated to P. G. Lowery. . . .

P. G. Lowery writes:—It is a great pleasure for me to state that the late composition by Scott Joplin, "A Breeze from Alabama" is a hit everywhere. It is fast growing popular. Get it while it is fresh from the press. Any one that has played his Maple Leaf Rag, knows that the name of Scott Joplin is a guarantee of the merits of "A Breeze from Alabama." Mr. Joplin is justly termed the king of rag-time writers.[72]

That it was not registered for copyright until four mouths later (entry date of December 29, 1902; copies received by the Copyright Office on January 8, 1903) is further evidence of Stark's procrastination in these legal matters. He registered *The Ragtime Dance, The Entertainer* and *Elite Syncopations* on the same date, and neglected to register *The Strenuous Life* and *March Majestic.*

It is on the basis of the newspaper accounts about *A Breeze* and mention of other Joplin rags on that publication that I have approximated the publication order of several other Joplin works of 1902. This ordering rests also on the assumption that the copies of *A Breeze* filed with the Copyright Office the following January were from the original printing. If the copyright copies were from a new printing, the advertisements could have been changed. It is less likely, but possible, that the list of works on the cover could also have been changed.

A Breeze is dedicated to "P. G. Lowery. World's Challenging Colored Cornetist and Band Master." Cornet virtuoso Perry G. Lowery (c.1870– c.1930) started with minstrel bands, became prominent as a soloist, and then organized his own vaudeville and circus bands. Major black newspapers of the period, such as the *Indianapolis Freeman* and *New York Age,* contain regular notices of his engagements during the entire ragtime period. In 1902, at the time of these announcements, he had his own organization: the P. G. Lowery Band and Vaudeville Company.

A Breeze is not one of Joplin's better rags, but it reveals an interesting facet of his musical thinking. He had long made dramatic use of flatted-sixth chords, examples being in his song *Please Say You Will* (on the word "gold"), or in the fifth and sixth measures of the *Maple Leaf*. In *A Breeze* he expanded the concept from a momentary harmonic surprise to a principle of tonal organization. John Stark recognized there was something different here; in his advertisements, he called the piece "a story in transitions." The "transitions" or tonal design can best be shown graphically:

$$\text{Intro} \parallel: A :\parallel: B :\parallel \quad C \quad :\parallel \text{trans} \parallel: D :\parallel \text{interl} \parallel: B :\parallel$$

$$\underbrace{\text{I-}\flat\text{VI-I}}$$

$$\text{C:} \quad \text{I} \qquad\qquad \flat\text{VI} \qquad \text{IV} \qquad\qquad \text{I}$$

The crux of the experiment is the middle strain, C; instead of being in the expected subdominant (F-major, or IV), it is in the key of the flatted-sixth, A-flat. Halfway through this strain there is another surprise: a further move by an interval of a flatted-sixth to F-flat, notated enharmonically as E. Joplin finally goes to the subdominant, after a somewhat clumsy four-measure transition, and then ends in the tonic. The tonic, however, stands as a flatted sixth from F-flat, and, to follow the logic, should be notated as D-double-flat. The overall design of this study in flatted-sixths is:

$$\text{C - A-flat - F-flat - D-double-flat}$$

This is Joplin experimenting. If the result fails to produce an outstanding rag, it nevertheless provides a rare glimpse into his musical mind.

Echoes of Scott Joplin, Part Five: A Breeze from Alabama

The introduction to James Scott's *A Summer Breeze* from 1903 closely resembles Joplin's introduction. Since the word "breeze" appears in both titles, Scott may have purposely mimicked Joplin's introduction (Ex. 6-8a, b). In Scott's *The Fascinator*, his A strain matches up with Joplin's A strain (EX. 6-8C).

Ex. 6-8. (a) Scott Joplin, A Breeze from Alabama *(1902), Introduction, A1-4; (b) James Scott,* A Summer Breeze *(Carthage, Mo.: Dumars), 1903, Introduction; (c) James Scott,* The Fascinator *(Carthage, Mo.: Dumars), 1903, A1-4.*

March Majestic was another work from 1902 dedicated to a cornet-playing bandmaster, this time James Lacy. Lacy led the band in Billy Kersand's Georgia Minstrels, perhaps the greatest black minstrel company of the time.

It is a good march, in 6/8 meter, but not one to call attention to itself. It would appear to have been designed for concert performance, with the B section as a showcase for the cornet. The copyright was never registered and there are no clues as to when in 1902 it was published.

Elite Syncopations is another rag registered for copyright on December 29, but it may have been composed much earlier: William P. Stark said it was the first work Joplin composed after arriving in St. Louis in 1901.[73] If it was composed in 1901, it was not published until 1902, for that is the date Stark printed on the music. Since none of the other publications of 1902 advertise it, it was probably issued late in the year.

Right from the first phrase of the A strain, *Elite Syncopations* stands out as a new approach to ragtime. It forgoes the stereotypical ragtime left hand of bass-note/chord alternations in favor of a chromatic bass line in octaves, first ascending and then descending. The upper notes of the right hand are a static D-C-D-C, but gain interest in the context of changing harmonies. The third measure, which begins the chromatic descent (again, filling in the interval of a fourth), is unsyncopated, but is articulated with alternating left- and right-hands, recalling the similar articulation of a diatonic ascent in the upbeat (Ex. 6-9a).

The bass line of the final strain is also notable, being far more melodic than the treble. It supports a right-hand part that is little more than a rhyth-

Ex. 6-9. Scott Joplin, Elite Syncopations *(1902): (a) A1-4; (b) B1-8; (c) D1-8.*

mic riff, but a riff of a particularly obsessive character (Ex. 6-9c). (Stark, in trying to depict the excitement of this obsessive riff, wrote in his later advertising: "Syncopations in the last part are actually frenzied.") The final phrase, while continuing the persistent momentum, shifts to a fresh tonal color with a flatted-sixth chord in measure 13. It is a stunning conclusion to an outstanding work.

Echoes of Scott Joplin, Part Six: Elite Syncopations

 The A strain of *Elite Syncopations* is clearly, if clumsily, echoed in the opening of A. H. Tournade's *Easy Money*, of 1904 (Ex. 6-10a).[74] The B strain is barely

Ex. 6-10. (a) A. H. Tournade, Easy Money, Ragtime Sonata *(New Orleans: Hackenjos, 1904), A1-4; (b) James Scott,* A Summer Breeze, *D1-8; (c) Al. Morton,* Fuzzy Wuzzy Rag *(Memphis: Pace & Handy, 1915), C1-4.*

disguised in the D strain of James Scott's *A Summer Breeze* (Exx. 6-10b, 6-9b). Al. Morton's 1915 imitation of the final strain is a shameless theft (Ex. 6-10c).

As difficult as 1902 was for Joplin on the personal level, with the deaths of his brother Will and Scott Hayden's wife Nora, his professional life continued to thrive. He issued eight pieces of music in that year, with at least half being significant works. And his music was being heard. P. G. Lowery was performing Joplin on his extensive nationwide tours. A notice from The Old Plantation and Southern Carnival Co., then in Arizona, indicates that Joplin had reached the Southwest: "Mrs. Douglas Banks Jones, our pianist, is 'cleaning up' with Scott Joplin's rags."[75] In St. Louis, John Eason, one of the city's popular dance band leaders, was featuring and applauding Joplin's music: "John Eason is making a big hit in St. Louis, playing Scott Joplin's ragtime two-steps. Eason says, Joplin is certainly the king of ragtime writers."[76]

Joplin ended the year with an announcement that he was to move on to an entirely new level of music composition: "Mr. Joplin is writing an opera in ragtime to be staged next season."[77]

CHAPTER SEVEN

A Guest of Honor, 1903

The day is fast approaching when a great colored composer will be recognized in this country, especially if he advances from being a ragtime idol . . .

Sylvester Russell,
Indianapolis Freeman,
May 2, 1903.

THE TERM "RAGTIME OPERA" was used frequently in the first years of the century, but more often than not its use was facetious. The very idea of "ragtime opera" was viewed by many as a self-contradiction. Opera was regarded as the highest form of musical art, the marriage of music and theater by such supreme masters as Wagner, Verdi, Beethoven, and Mozart. Ragtime was at the opposite pole. To many, it was the expression of an inferior race—the fidgety, syncopated music of saloons and brothels, the coon songs of the minstrel and vaudeville stages.

Many traveling minstrel and vaudeville companies chose names juxtaposing "opera" with the more popular entertainment forms: Foy Elliott's Rag-time Opera Co., Lew Hall's Ragtime Opera Company, Well's Operatic Minstrels, Johnson Operatic Cake Walkers, Mahara Vaudeville, Operatic and Polite Minstrel Carnival, and the like.[1] Usually, such companies made not the slightest gesture toward opera. Occasionally a company, such as Mahara's, would close with a comic "operatic" skit, but little precise information is available about these, and the "operatic" content is doubtful.

A few companies with singers of operatic quality would devote a portion of the show to concert performances of operatic music. An outstanding exponent of this practice was Black Patti's Troubadours, directed by Sissieretta Jones, a black concert artist who was dubbed "the black Patti" in reference to diva Adelina Patti. Whereas most of the show would be typical minstrelsy or vaudeville, time was always reserved for more ambitious vocal display, as illustrated by the following review:

> Particular attention should be given the ensemble number before the final curtain. It is called the "Operatic Kaleidoscope" and in it Mme. Jones and the greater number of her assistants appear to musical advantage, singing selections from various serious and light operas. . . . But perhaps the most effective selection in this ensemble number is during the finale when a quintette and the chorus carry one to the third act of "Martha."[2]

A black composer who was ideally qualified to write a true ragtime opera was Will Marion Cook. He had the talent, background, and training, having studied violin with Joseph Joachim and composition with Antonín Dvořák. Finding opportunities in the concert world blocked to him, he went into black musical theater. There, he and librettist Paul Laurence Dunbar made a decisive break with the minstrel–vaudeville format of black theater. They created book shows with strong ragtime flavoring, such as *Clorindy, or the Origin of the Cakewalk* (1898) and *In Dahomey* (1902). Cook referred to these extremely popular and successful works as "operas," but black columnist Sylvester Russell decried the claims as exaggerations: "'Clorinda' was no opera. It was simply a good, first-class, rag-time musical cake-walk. There is no such thing as rag-time opera; no opera can be written all in rag-time."[3] Despite Russell's opinion, the idea of ragtime opera was in the air, people were talking about it, and the stage was set for Scott Joplin.

By the beginning of 1903, Joplin was living in the large house at 2117 Lucas Avenue in St. Louis. It is likely that Arthur Marshall, who had returned to St. Louis the previous fall, and Scott Hayden were living there too. Tom Turpin was operating his newly opened Rosebud Bar (also spelled as two words, "Rose Bud") a few blocks away at 2220 and 2222 Market. It had a connecting pool room, two dining rooms, and "furnished apartments for gentlemen" upstairs. He generally advertised his bar as "Headquarters for Colored Professionals and Sports," referring to those in entertainment and sporting circles, and cited himself as the "Ragtime King."[4] Turpin was also managing a vaudeville company, The Dandy Coon Co., that was to play on April 9 at the nearby 14th Street Theatre. The cast included the team of Patterson and Chauvin and listed "Jas. Jordan" as music director. This last was possibly a misprint that should have read "*Joe* Jordan."[5]

Toward the end of 1902 Joplin and Hayden had collaborated on a second piece, *Something Doing. A Ragtime Two Step,* which was received by the Copyright Office on January 10, 1903. (The application was not received until the end of the following month.) However, it was published not by

Stark but by Val. A. Reis Music Co. at 1210 Olive Street, a few blocks from
Joplin's home. Since the Joplin–Hayden collaboration *Sun Flower Slow Drag*
was very successful, we must wonder why Stark did not publish this second
work. Was Stark unwilling to publish it, or was Joplin simply offered better
terms by Val Reis?

The cover of this piece is reminiscent of the first *Maple Leaf* cover, with a
black cakewalking couple in the same pose. In the background is an audience
of white people in society dress, showing again how black cakewalks had
become spectator events for whites.

There is no testimony allocating authorship of the specific sections in
Something Doing, but I propose that, as in the previous collaboration, Hayden wrote the first two strains and Joplin the last two. Strain B does have
right-hand sevenths that might be characteristic of Joplin, but measures 9-12
are too static harmonically to be his.

Joplin's first strain, section C, takes its point of departure from a frequently repeated melodic countour in strain B (Ex. 7-1). We see again how
Joplin builds upon, and pays tribute to, the work of a younger colleague.

Ex. 7-1. Scott Joplin and Scott Hayden, Something Doing *(1903): (a) B1-2;
(b) C1-4.*

Around this same time Joplin had a second rag published by Val Reis:
Weeping Willow. Though it was not registered until June, it was advertised
for sale (12 cents) by a St. Louis department store in late March. The same
advertisement offers, also for 12 cents, *Something Doing* and *Elite Syncopations.* The more popular Cole and Johnson song hit *Under the Bamboo Tree*
sold for 19 cents.

Joplin dedicated *Weeping Willow* to the Pawnee Club, about which I
have no information. The music is typical of Joplin in its contrapuntal implications, starting right off in the A strain with an inner-voice descending

fourth of g-d, lacking only the f-sharp to make it his familiar chromatic descent. (See measures 1-4; the descending motif falls most conveniently to the thumb of the right hand.)

Some details of Joplin's personal life, as related by Arthur Marshall, probably stem from this period, when both Marshall and Hayden were in town. A story tells of difficulties Joplin was having in his marriage:

> Mrs. Joplin wasn't so interested in music and her taking violin lessons from Scott was a perfect failure. Mr. Joplin was seriously humiliated. Of course unpleasant attitudes and lack of home interests occurred between them.
>
> They finally separated. He told me his wife had no interest in his music career. Otherwise Mrs. Joplin was very pleasant to his friends and especially to we home boys. But the other side was strictly theirs. To other acquaintances of the family other than I and Hayden and also my brother Lee who knew the facts, Scott was towards her in their presence very pleasing.
>
> A shield of honor toward her existed and for the child. As my brother, Lee Marshall, Hayden and I were like his brothers, Joplin often asked us to console Mrs. Joplin—perhaps she would reconsider. But she remained neutral. She never was harsh with us, but we just couldn't get her to see the point. So a separation finally resulted.[6]

The child was a girl born to the Joplins, who died in infancy. Marshall also reported that after the breakup, Belle "went in poor health, and passed," her death being about two years after the child's.[7]

Marshall was wrong about Belle's death. According to the Reverend Alonzo Hayden, Belle's son from her prior union with Joe Hayden, Belle moved to Chicago after her separation from Joplin and lived there until she died in 1930.[8] We cannot account for Marshall's error, but he may have confused Belle with the next woman in Joplin's life.

Another anecdote from this period tells of a Joplin performance and his customary pay. The information comes from a letter written in 1980.

> When I was a little girl, aged 10 in 1903, my mother gave a dance for my 16-year-old sister. She got a young black man named Scott Joplin to play the piano. This young man played a number called Maple Leaf Rag, which he had just written.
>
> The teenage crowd just loved it, and when the party was over the boys asked my mother for his name so they could give a dance and have him play.
>
> But they were all very sad to hear that he charged $5.00, and of course they could not possibly pay that.
>
> For that measly sum, Scott Joplin rode all the way out from St. Louis to Webster Groves on the street car and walked 6 blocks to our home down on Big Bend.[9]

Despite the death of his child and the breakup of his marriage, Joplin continued working. On February 16, 1903, he applied for a copyright on his new opera. The application is lost, but an accompanying letter exists.

The letter, the only letter from Joplin known to have survived, is interesting as much for its appearance as for its content. This was a time when educated people took pride in penmanship, and Joplin's efforts are evident: some of his characters, especially at the beginning, are ornate and almost

Scott Joplin's letter to the Copyright Office, accompanying his registration application for A Guest of Honor.

elegant. But the polish is lacking, and there are incorrect and inconsistent spellings, suggesting a rudimentary early education. A notable change occurs toward the end of the letter: decreased neatness, scratching out of a word, and a disproportionately sized "s" in the word "yours." Was writing a difficult process for him? Did he lack the patience to complete the letter with the same care with which he started? It appears he might have had a problem with coordination, a condition that would certainly have affected his piano playing.

The content is interesting because it makes clear—and this is important in view of the opera's history—that Joplin at this time included only the application and payment; he did not enclose the score. He was probably waiting for publication, for there is an entry in the Copyright Register, dated two days later, listing Stark as publisher,

By April, Joplin had formed a company to perform the opera, and had enlisted both Marshall and Hayden. This was the report Hayden gave on a visit to Sedalia:

> Scott Hayden has been in the city all [week] visiting parents and friends. He has signed a contract with the Scott Joplin Drama Company at St. Louis in which Latisha Howell and Arthur Marshall are performers.[10]

The next mention of the opera was two months later, when Joplin was treated with a glowing report in a major white newspaper, the *St. Louis Globe-Democrat*. The article, by Monroe Rosenfeld, both a journalist and one of the leading songwriters of the period, was accompanied by a reprint of the first page of *The Entertainer* and a new photograph of Joplin.

> St. Louis boasts of a composer of music who, despite the ebony hue of this features and a retired disposition, has written possibly more instrumental successes in the line of popular music than any other local composer. His name is Scott Joplin, and he is better known as "The King of Rag-Time Writers," because of the many famous works in syncopated melodies which he has written. He has, however, also penned other classes of music and various vocal numbers of note.
>
> One of the interesting characteristics of Scott Joplin's personality is his conservatism. He rarely refers to his productions and does not boast of his ability despite the fact that he is possibly one out of three score of composers who arranges his own compositions. This negro is a tutored student of harmony and an adept at bass and counterpoint; and, although his appearance would not indicate it, he is attractive socially because of the refinement of his speech and demeanor.
>
> Scott Joplin was reared and educated in St. Louis. His first notable success in instrumental music was "The Maple Leaf Rag," of which thousands upon thousands of copies have been sold. A year or two ago, Mr. John Stark, a publisher of the city, and father of Miss Eleanor Stark, the well-known piano

Photograph of Joplin as it appears on the cover of The Cascades. *The same photograph accompanied the article by Monroe Rosenfeld the previous year, on June 7, 1903.*

virtuoso, bought the manuscript of "The Maple Leaf" from Joplin for a nominal sum. Almost within a month from the date of its issue, this quaint creation became a byword with musicians, and within another half a twelve-month circulated itself throughout the Union in vast numbers. This composition was speedily followed by others of a like character, until now the Stark list embraces nearly a score of the Joplin effusions. Following is a list of some of the more pronounced pieces by this writer, embodying these oddly titled works:

"Elite Syncopations."
"The Strenuous Life."
"The Rag-Time Dance" (song).
"Sunflower Slow Drag."
"Swipsey Cake Walk."
"Peacherine Rag."
"Maple Leaf Rag."

Probably the best and most euphonious of his latter-day compositions is "The Entertainer," a few bars of which are herewith given. It is a jingling work of a very original character embracing various strains of a retentive character, which set the foot in spontaneous action and leave an indelible imprint upon the tympanum.

Joplin's ambition is to shine in other spheres. He affirms that it is only a pastime for him to compose syncopated music and he longs for more arduous work. To this end he is assiduously toiling upon an opera, nearly a score of the numbers of which he has already composed and which he hopes to give an early production in this city.

Monroe H. Rosenfeld.[11]

That the list of works is exclusively of Stark publications—not even Joplin's self-published *Easy Winners* is included—strongly suggests that Stark had arranged the interview. However the interview came about, Rosenfeld seems to have had a sincere appreciation of Joplin, both as a composer and as an individual. His description of Joplin's personality supports the picture we have of his dignified manner, modesty, and extreme reticence. Rosenfeld's surprise that a black man should have advanced compositional skills and social graces should, generously, be attributed to attitudes commonly held during that time.

The final paragraph states that Joplin had ambitions to grow beyond ragtime and alludes to the opera. It suggests that the opera was not yet complete, which raises the question as to how he could have applied for a copyright five months earlier. Perhaps his "assiduous toils" were on orchestrations. The "early production in this city" was probably what Marshall described:

As for the Rag Time Opera, *A Guest of Honor* was performed once in St. Louis. In a large hall where they often gave dances. It was a test-out or dress rehearsal to get the idea of the public sentiment. It was taken quite well and I think he was about to get Haviland or Majestic Producers to handle or finance the play, also book it. I can't say just how far it got—as I was very eager for greater money, I left St. Louis for Chicago.[12]

This dance-hall tryout would have been in August, for that is when Joplin's company of thirty-two went into rehearsal at the Crawford Theatre, at Fourteenth and Locust. As the time neared for the show to go on the road, he wrote to his friend W. H. Huston, publisher of the *Sedalia Conservator,* who put the following in the paper:

> Scott Joplin's opera is rehearsing daily at Crawford's theatre. Their present number is about 32 people; he has just received the book of the play from the publisher's hand, the title of the book and play is "A Guest of Honor." Joplin is backed by a strong capitalist who for many years has been manager and proprietor of several well known high class operas (white), this being his first adventure into Negro Opera. They open the season at East St. Louis Aug. 30; then five engagements at Sedalia. His Opera is entirely his original composition including songs and drills.[13]

The item supports Marshall's statement that Joplin had the interest, and possible support, of a theatrical producer, but no more is known of that connection. The final sentence requires interpretation. That Joplin wrote the "songs" means he wrote the lyrics as well as the music. The "drills" referred to would have been military routines, staples of the minstrel stage and, as we shall see, integral to the opera's setting. Finally, in pointing out that he was responsible for everything in the opera—music, text, and staging—Joplin showed that he modeled himself after Richard Wagner. This may have reflected the influence of Alfred Ernst, who played Wagner's music for him.

On the Road with a Black Company

For performers in most touring shows, conditions could be difficult and unpleasant. For black performers, conditions were usually many times worse, even precarious. On one occasion Billy McClain, a famous and extremely successful minstrel, was arrested in Kansas City for wearing too much jewelry. The charge was not bad taste. Rather, it was the opinion that a black man could not have obtained such expensive ornaments honestly.[14]

Travel for touring shows was usually by rail, and since accommodations in a town might be uncertain, the most successful companies owned their own railroad cars, using them both for travel and sleeping quarters. Other companies might take their chances in finding a rooming house that would accept black patrons, but in traveling through areas known for racial hostility, would rent a railroad car.

Some towns prohibited blacks from entering except under special circumstances. One of the circumstances was to put on a theatrical performance. On arriving in such towns, the performers would try to defuse racial tensions and win over the local population by marching through the main street playing the tune *Dixie*.[15]

Racial situations were not always avoidable. The story of Louis F. Wright, a 22-year-old trombonist with Richards and Pringles Famous Georgia Minstrels, illustrates the horrors that could develop.

Arriving Saturday afternoon, February 15, 1902, in New Madrid, Missouri, the troupe gave the traditional street parade to attract an audience for their evening show. Afterwards, while walking to the opera house, Wright and a companion were bombarded with icy snowballs thrown by young white men. Wright cursed at them and continued on his way.

The attackers decided that Wright's "vile insult" could not be overlooked and resolved to whip him. That night, after the show, the gang ran backstage and tried to drag Wright away. The rest of the company came to his aid, and a fight ensued in which there was shooting. Several of the performers were wounded, and one white man claimed to have been grazed by a bullet. The minstrels were all arrested and spent the night in jail.

The next day, each one was taken to the court and asked who among them had fired the gun. They all denied that any of their members owned a gun. They were returned to jail for another night. At 11 on Sunday night, Wright was taken from the jail. Two young boys had testified they had seen the handle of a revolver sticking out of his pocket.

The following morning the minstrels were released and found Wright hanging from a tree near the train depot. He was later cut down, put in a box, and shipped C.O.D. to his mother.[16]

Wright was among eighty-six African Americans lynched in 1902. Throughout the United States there were places in which citizens bragged that a black man had no rights in their town. Too often, this was a statement of fact.

Despite the assertion in the *Conservator,* the tour did not go directly to Sedalia after the opening in East St. Louis, and probably never reached there at all (see tour schedule).[17] While the company was usually called "Scott Joplin's Ragtime Opera Company," or some close variant (occasionally with a misspelling of Joplin's name), it was at times referred to as a minstrel compa-

Tour Schedule of A Guest of Honor

   1 August 30: East St. Louis, Illinois. Theater not identified.

   2 September 2: Springfield, Illinois. Opera House. Scott Joplin's Rag-Time Opera Co.

   3 September 3: Galesburg, Illinois. Auditorium. Scott Joplin's Ragtime Opera Co.

   4 September 12: Webb City, Missouri. New Blake Theatre. Scott Joplin Ragtime Opera Co.

   5 September 17: Pittsburg, Kansas. Opera House. Scott-Joplin Minstrels; also Scott-Johnson [sic] Minstrels.

   6 September 18: Parsons, Kansas. Edward's Opera House. Scott Joplin's Rag Time Opera.

   7 September 29: Ottumwa, Iowa. Grand Opera House. Joplin Opera Co.

   8 September 30: Cedar Rapids, Iowa. Greene's Opera House. Scott-Joplin Opera Co.

   9 October 6: Beatrice, Nebraska. Paddock Opera House. Scott's [sic] Joplin Ragtime Minstrels.

 10 October 7: Fremont, Nebraska. New Larson Theatre (or Love's Theatre). Joplin Ragtime Opera; also Joplin Ragtime Minstrels; also Joplin Minstrels.

 11 October 12: Mason City, Iowa. Wilson Theatre. Scott Joggins [sic] Rag Time Opera Co.; also Scott Joplin Opera Co.

ny, suggesting that some theaters would book a black show only for minstrelsy or vaudeville.

On August 31, the *Illinois State Journal,* a morning Republican paper covering Springfield, printed an advertisement for the opera, repeated in the afternoon Democratic paper, the *Illinois State Register.* An advertisement for the Galesburg performance the next day is similar.[18] No other advertisements have been found and no reviews located. Since reviews were not the rule for minor productions, this absence is not significant. Only one item about the tour has been found, and this, from the September 26 issue of the *Freeman,* contains devastating news:

> We are sorry to note the misfortune Mr. Scott Joplin met with his Ragtime Opera company while filling an engagement in Springfield, Ill. He has been doing big business, but his Bufay representative embarks with the receipts, leaving them in a hole. They are in Chicago for the present.[19]

What happened after the Springfield misfortune is open to conjecture. The company may have folded at that point, for the tour must certainly have been over by the time Joplin arrived in Chicago.[20] But did anything significant occur between September 2 and September 26, the date the item appeared in the *Freeman?* For the notice to be included on that date, it could have been offered to the paper as late as September 21, or maybe even a day or two later.[21] And there is a possibility that the tour did not end in Springfield, that Joplin tried to recoup his losses with the next few bookings.

This hypothesis is based on something Lottie Joplin had said in 1949. She told Blesh and Janis that Joplin had lost a trunk with unpublished music manuscripts (possibly including *A Guest of Honor*), personal photographs, and letters. He had left the trunk against an unpaid bill in a theatrical boarding house in Pittsburgh, and never retrieved it.[22]

Had Lottie, whose testimony we have found to be so unreliable, garbled her facts on this incident, as well? Perhaps she misunderstood the location of the boarding house. She said it was in Pittsburgh, Pennsylvania, but it may actually have been *Pittsburg, Kansas,* where the troupe was scheduled to appear on September 17. The hypothetical scenario we then draw has the troupe continuing its tour, perhaps making other stops along the way (but not Sedalia, where an appearance would have attracted notice in the newspapers), finally reaching Pittsburg on September 16 or 17. The show in Pittsburg did not produce sufficient income to pay the players' wages, and the troupe disbanded. Unable to pay the boarding bill for a company of thirty, Joplin was forced to leave his trunk.

This is a possible scenario, but one for which I cannot offer evidence. A search of newspapers in Webb City, Pittsburg, and Parsons produced no sign of Joplin.

The tour must have been a financially devastating failure, one on which

OPERA HOUSE

One Night Only

WEDNESDAY, SEPT. 2

SCOTT JOPLIN'S

RAG-TIME OPERA CO.

Management of Meiser and Amier, presenting

A Guest of Honor

A rag-time opera in three acts by Scott Joplin.

Pretty Girls---Sweet Singers

Elaborate Wardrobe

30---People---30

The only genuine rag-time opera ever produced.

Prices: 25c, 35c, 50c, and 75c. Seats at Chatterton's.

Advertisement in the Illinois State Journal *for the Springfield performance of* A Guest of Honor. *The listing of three acts was an error that was corrected to two acts in the following advertisements, in the* Illinois State Register *(Aug. 31, 1903) and the* Galesburg Daily Republican Register *(Sept. 2, 1903).*

Joplin lost all the money he had so recently earned from sales of the *Maple Leaf Rag*. If, in addition, it was at this time that he lost his trunk, it might explain why Joplin apparently tried to forget the opera. I have not discovered a single reference to the opera after 1903. Whereas from 1907 on, Joplin rarely missed an opportunity to mention he was working on an opera (which was to be *Treemonisha*), or, from 1911, that he had completed it, he never cites it as his *second* opera. It is as if he wanted to pretend *A Guest of Honor* had never existed.

In 1906, a clerk at the Copyright Office stamped the claimant card for *A Guest of Honor* "CARD FILED MAR 27 1906 WITHOUT CREDIT." This stamp, along with the unfilled blank on the line "2 copies received . . . ," indicates the score was never received. *A Guest of Honor* is lost.[23]

In 1944, two researchers came upon this card and inquired of William P. Stark about it. He wrote back: "While *A Guest of Honor* had some pretty good music, the story and lyrics were weak, and it was never published."[24]

Aside from the information already examined, one other item about the opera appeared in a newspaper in 1903. Just before beginning the tour, or possibly, right after the beginning, Joplin sent the following notice to the *Freeman*.

> Scott Joplin, who is termed "the king of rag-time writers," has written a rag-time opera entitled "A Guest of Honor," which is a most complete and unique collection of words and music produced by any Negro writer. The opera is in two acts, something on the order of grand opera, with not a piece of music in the whole opera other than from the pen of Scott Joplin, in which he introduces a lot of big numbers, some of which are "the Dudes' Parade," "Patriotic Patrol" and many others which go to make it grand.[25]

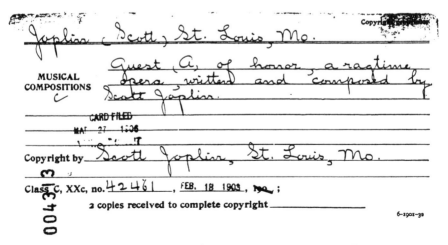

Claimant card for A Guest of Honor, *Copyright Office.*

The notice indicates that *A Guest of Honor* was not an operetta or musical comedy, but a grand opera. The two numbers he mentions, *Dudes' Parade* and *Patriotic Patrol,* are the only authenticated titles from the opera. Curiously, the title *Patriotic Patrol* was listed in a flyer distributed by publisher John Stark around 1906.[26] Since no music with this title has ever surfaced, knowledgeable collectors have assumed that the title was a discarded name for one of Joplin's marches, perhaps for *Antoinette* (1906).

This is not a bad hypothesis, for composers frequently recycle music that is unsuccessful the first time around. And *Antoinette,* for musical reasons, seems to fit the bill. Though it is a march, the 20-measure interlude within the trio suggests a dramatic *misterioso* that is more suggestive of opera than a march (Ex. 7-2).

As for the libretto, the story of *A Guest of Honor,* we can only speculate. Here is my guess.

Ex. 7-2. Scott Joplin, Antoinette *(1906), Interlude.*

I believe the title is itself a clue. The most famous "guest of honor" of the period was Booker T. Washington, on the occasion of his dinner at Theodore Roosevelt's White House. An editorial in the *Freeman,* a newspaper that Joplin read, used that very term in this context.[27] As we have seen, Joplin was sufficiently impressed by the event to honor Roosevelt with the rag *The Strenuous Life.* Would he have done less for the black leader? I propose that he commemorated the episode with an opera. The titles of the two known selections, *Patriotic Patrol* and *Dudes' Parade,* are consistent with this hypothesis. If we are correct about the libretto, *The Strenuous Life* could have been re-used to represent Roosevelt.

We have little information on Joplin's activities in Chicago after the failed opera tour. We speculate that he arrived there around September 20, 1903. In need of money, he arranged the song *Little Black Baby* for Louise Armstrong Bristol, registered for copyright October 7. The cover does not mention Joplin at all, but he is listed on the music as composer, along with Bristol as lyricist. However, it hardly seems credible that Joplin could have been responsible for the simplistic, monotonous melody; more likely, he was given the words and melody and he provided the piano arrangement—unexciting, but faultless as far as it goes.

A second publication was of *Palm Leaf Rag,* copyrighted by Victor Kremer Co. of Chicago on November 14. This rag is pure Joplin. At the end of the B strain he again uses the descending chromatic fourth motive (G-D, mm. 13-16), this time in an embellished form. The most notable feature of the rag is the single trio theme. It is characterized by arabesques that make it the most ornate melody Joplin had yet written (Ex. 7-3).

Ex. 7-3. Scott Joplin, Palm Leaf Rag *(1903), C1-4.*

Echoes of Scott Joplin, Part 6: Palm Leaf Rag

 Al. Verge, who lifted a section of Joplin's *Peacherine Rag* for his *Whoa You Heiffer* (see Chapter Six), took the trio of Joplin's *Palm Leaf Rag* for his A strain (Ex. 7-4).

Ex. 7-4. Al. Verge, Whoa You Heiffer, *A1-4.*

There was one other Joplin work issued by Stark in 1903: a song version of the *Maple Leaf Rag*. It was not copyrighted until August 22, 1904, but the music displays a copyright notice of 1903. The lyricist was Sydney Brown, an office boy for the Stark Music Co.,[28] and I believe the musical evidence indicates that Joplin had no part in the transformation of the rag into a song (Ex. 7-5).

Ex. 7-5. (a) Scott Joplin (music) and Sydney Brown (words), Maple Leaf Rag. Song *(1904), Chorus 1-4; (b)* Maple Leaf Rag, A9-12, *harmonic reduction and transposition to E-flat.*

The song, entirely in E-flat, consists of a verse and chorus, both taken from the A strain of the original (the verse uses measures A1-8 of the rag; the chorus uses A9-16), and a "Dance" section, consisting of the B and D strains of the original. What in the original was a prominent diminished-seventh chord with a tritone between the bottom and top notes, becomes in the song an ordinary triad (spelled awkwardly) with the bottom and top notes forming a major third. This one change eliminates the harmonic tension of the original and seriously reduces its harmonic variety and contrapuntal motion, duplicating a chord that appears again only two measures later. I doubt that Joplin, with his finely developed sense of harmony and voice leading (and knowledge of proper musical spelling), would have weakened his piece as this arrangement does. Furthermore, in the Dance section, the transposition of the B and C sections places the left-hand chords in an unusually low register, producing a thick sound. This arrangement is uncharacteristic of Joplin and was probably made by someone else.

It may have been around this time that the *Maple Leaf Rag* had its first recording. The chances are that Joplin did not know about it. The recording was made on cylinder in Minneapolis by a young clarinetist named Wilbur Sweatman, leading a six-piece band.[29] Sweatman was to become a significant ragtime and jazz musician and would one day play a larger role in the Scott Joplin saga.

In December of 1903, Joplin placed a notice in the paper mentioning the popularity of *Palm Leaf* in Chicago and indicating he was about to return to Arkansas to visit relatives. The visit would be momentous.[30]

CHAPTER EIGHT

Freddie, 1904

Meet me in St. Louis, Louis,
Meet me at the fair
Don't tell me the lights are shining
any place but there;
We will dance the Hoochee Koochee,
I will be your tootsie wootsie;
If you will meet me in St. Louis, Louis,
Meet me at the fair.

> Kerry Mills and Andrew B. Sterling,
> "Meet Me in St. Louis," 1904.

WITH THE ARRIVAL OF 1904, the St. Louis rag-timers, along with virtually everyone else in and near the city, were preparing for the St. Louis World's Fair—officially, the Louisiana Purchase Exposition. Tony Williams, the cakewalker and onetime proprietor of Black 400 clubs in Sedalia and Joplin, was now managing The Alcove, a cafe on Market Street that featured music and dance.[1] A new 400 Social Club opened at 717 North Thirteenth Street (corner of Morgan) in late February. It featured dance instruction and music by solo pianists and bands, but neither Tony nor Charles Williams seem to have been involved.[2]

On February 22, Tom Turpin's Rosebud Club sponsored its third annual ball and piano contest, holding the event at the New Douglass Hall, at the corner of Beaumont and Lawton. The advance publicity highlighted a contest between Louis Chauvin and Tom Turpin, confirming that these two were among the most formidable pianists in St. Louis. At the contest, Chauvin won the first-place gold medal and Joe Jordan and Charles Warfield tied for second.[3]

Sam Patterson had been in Chicago but returned to St. Louis in time to help run the ball. There is no indication that he competed in the contest, but once in town he again teamed up with Chauvin. In an event that was advertised for several weeks, he and Chauvin took part in a show sponsored by the Catholic Knights of America, held at the Masonic Temple (also known as Turner's Hall) on April 6.[4]

The Fair finally opened on April 30. Because of flagrant racial discrimination and insults, black newspapers urged its readers not to attend.[5] Nevertheless, blacks attended and participated. Arthur Marshall played piano for a while at the Spanish Cafe, earning about $12 a week plus tips until he was replaced by a band.[6] Sam Patterson and Louis Chauvin played piano and sang at a beer house called Old St. Louis.[7] There was another ragtime contest, although it is not certain whether it was held at the fairgrounds. Jelly Roll Morton told about it:

> It was about that time, in 1904, that they announced the piano-playing contest at the World's Fair in St. Louis. I was a half-hand bigshot on the piano around Mobile and the girls were willing to finance my trip. I had decided to go until I heard that Tony Jackson was going to appear at the contest. Of course, that kind of frightened me and so I stayed in Alabama. Later on I heard that Tony Jackson hadn't gone and that Alfred Wilson had won, which disgusted me, because I knew I could have taken Alfred Wilson.[8]

It was probably while Joplin was still in Chicago, at the end of 1903 or beginning of 1904, that he arranged with Will Rossiter for the publication of *The Sycamore* (copyrighted July 18, 1904). Though it is subtitled "A Concert Rag," it bears no musical intimation of a move toward a new, "classical," direction. Rather, it steps back from the ornate lines of *Palm Leaf Rag,* suggesting that it may have been an earlier piece, although in the summer of 1904 he cited *The Sycamore* as a recent work.[9] In *The Sycamore,* we see the usual craftsmanship of Joplin: significant, expressive harmonies (such as the augmented sixth chord at B16) and important bass motifs, double-stemmed to call them to the attention of a performer. The last four measures of the final strain are unusual. Though the strain is dominated by eighth- and sixteenth-notes, in the closing phrase (D13-16) it suddenly slows to half speed with quarter- and eighth-notes, apparently to have us dwell upon the chromaticism of the ending.

Early in 1904, Joplin went to Arkansas to visit relatives in Texarkana and Hot Springs. He also visited Little Rock, where he met a young woman of nineteen years named Freddie Alexander. He was very much taken with her, and it is my guess, for reasons that will become clear in this and later chapters, that she helped him clarify and crystalize his ideas concerning racial pride and African-American heritage.

He responded in his most eloquent manner. He composed a rag. This

Dedication to Freddie Alexander on the original cover of The Chrysanthemum. *The words are barely distinguishable because of the poor choice of colors, yellow and green. When the cover was redone, the dedication was omitted.*

was *The Chrysanthemum,* described significantly as "An Afro-American Intermezzo," and inscribed: "Respectfully Dedicated to Miss Freddie Alexander, Little Rock Ark."[10] Stark, in his advertising for the piece, wrote that the composition had been inspired by a dream after Joplin read *Alice's Adventures in Wonderland.* We have no way of judging the truth of this information, but we must wonder if Miss Alexander was in any way responsible for Joplin's reading of the book.

The Chrysanthemum follows the direction initiated with *Palm Leaf Rag,* though is less ornate. It is certainly one of Joplin's most graceful and lyrical rags (Ex. 8-1).

Joplin was in St. Louis when *The Chrysanthemum* was published in late March. From there he went to Sedalia, arriving around April 2 or 3, where he distributed the music, including a copy to the editor of the *Democrat.*[11]

Ex. 8-1. Scott Joplin, The Chrysanthemum *(1904), A1-4.*

That he made no mention of *Cascades,* copyrighted on August 22, the same day as *Chrysanthemum,* suggests that it had not yet been published. (*The Maple Leaf Rag. Song* was also copyrighted on that date, but, as indicated in Chapter Five, it was probably published in 1903 without Joplin's participation.)

In Sedalia he initially stayed at 113 East Main Street, which was a black-owned restaurant run by Felix Warfield. After a few days, he moved to the home of Solomon Dixon, 124 West Cooper. Dixon, who was a highly respected black citizen of Sedalia, had his granddaughter Lucile living with him. Many years later, she recalled that Joplin would play for hours on the family's Emerson Square Grand.[12] Joplin apparently intended to remain at that residence for a while, for he had himself listed there in the 1904 Sedalia directory, indicating "music rms."

In Sedalia, Joplin lost no time in finding work, playing for a dance held at D. O. H. Hall on April 11. Later that week, on April 14, he again performed at the hall, inviting as guests the Billy Kersands' Minstrel Company, which was in town.[13] Among the performers with Kersands at this time were known friends: James Lacy, the cornet player and band leader to whom he had dedicated *March Majestic* in 1902, and Joe Jordan.

Joplin probably returned to St. Louis in time for the fair's opening on April 30. The focal point of the St. Louis fair was the Cascades Gardens, an extraordinary spectacle of artificial waterfalls, fountains, lagoons, and rapids extending into the midway. Inspired either by the view or by a perceived opportunity, Joplin entitled his next rag *The Cascades,* dedicating it to banjoists Kimball and Donovan, of whom nothing is known. We do not know precisely when the work was published (like *Chrysanthemum,* it was copyrighted August 22), but it was issued soon enough to be heard at the fair. Joplin reported to Mildred Steward, Arthur Marshall's daughter, that *Cascades* was played constantly at the fair.[14]

One might see in the almost continuous motion and undulations of the A section of *Cascades* a suggestion of tone painting, but it is not sufficiently explicit for us to know for sure that this was Joplin's intent. What is definite is the resemblance between the A and B sections of *Cascades* and *The Maple Leaf Rag.* The similarities in harmonic patterns (transposed from A-flat to C) are clear. In addition, there are striking parallels in melodic gestures and contours of A7-16. It would seem that Joplin was trying to recapture his earlier success by using *The Maple Leaf Rag* as a model and, in effect, recomposing the *Maple Leaf.*

The C and D sections of *Cascades* depart from the *Maple Leaf* model and are unusual in key relationships, section C being in B-flat rather than the expected subdominant F major, and section D being in E-flat. These key relationships suggest two possibilities: Joplin, in a hurry to finish *Cascades* in time for it to be played at the fair, may have drawn from some other, as yet incomplete, rag. The musical nature of the C section might also have suggested the key. The bass is unusually active, difficult on the piano and perhaps written with the trombone, baritone horn, or tuba in mind. Writing the part for these instruments in F major would have been possible, but it would have brought the instruments to the tops of their ranges, making performance less comfortable.

On *Cascades,* Stark had much to say. He wrote in an advertisement: "Hear it, and you can fairly feel the earth wave under your feet. It is as high-class as Chopin and is creating a great sensation among musicians." He also used the "earth waving" imagery in a letter to the *New York Sun* in September 1905. Stark does not reveal his identity as Joplin's publisher and offers the letter as a disinterested observation about ragtime. It is clear, though, that he used the letter to promote his own publications and to foster the idea of Scott Joplin's originality.

> With the troubles of "Old Man," in your edition of September 9, I can sympathize, but, since the St. Louis fair, I will have to dissent from his sweeping aspersion of ragtime. I heard there the "kilties band" of Scotland play a selection the intense rhythm of which made the ground wave under my feet. The weird and sentimental melody with unique harmonic treatment tempted me to reverie.
>
> I could look away into the dim misty distance and "see things." I was diligent to learn what this music was and found that it was one of Scott Joplin's rags, "The Maple Leaf," "Sunflower," "Cascades," or one of the others, I don't know which. Now, who do you think this Scott Joplin is? He is a young and untutored negro from the swamps of Arkansas.
>
> Is this not food for thought? Educated musicians have rehearsed the literature of all nations. Their compositions evolved from what they have absorbed. This negro had never heard what we call good music. His pieces bud out from his own consciousness and are real creations. They are not light and trashy. They are profound and difficult. I believe they are all published in the West and are little known in the East. There are a number of them, not all rags—and mark this prediction: They will find their way to all countries, be played by the cultured musicians everywhere and welcomed into the drawing rooms and boudoirs of good taste.
>
> John Stark.
>
> New York, September 11.[16]

Joplin's rag *The Favorite* was published in the summer, but not by Stark; it was issued by the long-time Sedalia publisher A. W. Perry & Sons, who copyrighted it on June 23 and announced it on July 24.[17] Why would Joplin have gone to Perry rather than Stark? According to Brun Campbell, Perry had actually purchased this rag in 1899.[18] It is not known whether Campbell's information was based on personal knowledge—he had been in Sedalia in 1899—but, as already indicated, his testimony is not always reliable. In this case I would tend to credit Campbell's report, for it is difficult to imagine why Joplin, at this point in his career, would have sold a work to such an insignificant publisher as Perry. Also, when Joplin announced a performance for July 28, he cited *The Sycamore* and *The Cascades* as "latest compositions," but not *The Favorite,* even though it was on the program.[19]

After spending time at the fair, Joplin went back to Arkansas. There, on June 14, in the Little Rock home of Freddie Alexander's parents, the 36-year-old Scott Joplin and the 19-year-old Miss Alexander were married. The marriage license is a five-part record that contains information that raises more questions. On the third part, "Marriage License," Joplin again gave an incorrect age, listing himself as 27. The probable reason for this was Freddie's much younger age. The third part and first part, "Bond for Marriage License," confirms that "Freddie" was the bride's true name, not merely a nickname. The first part also lists "Chas. Brooks" as surety, i.e., guarantor that Joplin would fulfill his commitment to marry Miss Alexander or forfeit one hundred dollars. We know nothing of Charles Brooks, but he was evidentally a friend.

The second part, "Affidavit," in which bride and groom both affirm that they are single, was left blank. This again raises the question of Joplin's relationship with Belle. Was there a legal marriage between him and Belle, and if so, was there a divorce? Or was Joplin a bigamist? If Joplin was indeed single at the time of his marriage, why did he not sign the affidavit?

Marriage certificate of Scott Joplin and Freddie Alexander, parts 1 and 3.

If we were to take the blank "Affidavit" as evidence that Joplin was already legally married, we would have to conclude that he had an inconsistent sense of values, one that permitted him to commit bigamy, but not perjury (while, at the same time, lying about his age). But why was the incomplete certificate of marriage even honored? I posed this question to the microfilm clerk at the Little Rock Court House and was told that during that period the "Affidavit" section was frequently left blank. We therefore can draw no firm conclusions from Joplin's failure to sign it.

Following the marriage, Joplin and his bride made their way to Sedalia, stopping along the way for him to give some performances. When they arrived, in the first or second week of July, Freddie was suffering from a bad cold and was confined to bed in the Dixon household.

There are reports that Joplin devoted much time to tending to his sick bride, but he also had to work, and he set about immediately to give performances in Sedalia. At his first performance, attendance was poor. The editor of the black *Sedalia Conservator* chided his readers for their failure to support black artists, and then focused on Joplin:

> Several months ago, a troupe of high classed Negro celebrities appeared at Wood's Opera house. Their talent had been recognized by the musical critics of the principal Metropolitan Journals, where they had appeared. Yet when they appeared here only a very few Negroes attended. The same conditions of affairs obtain in the recent concert given by Scott Joplin, recognized by the greatest musical critics of all nations, as the greatest of all "Rag Time" composers. He stands alone, without equal. Still, Sedalians of color fail to accord to him that encouragement and patronage that his efforts merit. Surely the musical critics of the *St. Louis Globe Democrat, Chicago American,* the *New York World,* and other Journals, know talent, when they come in contact with it. Let the Sedalia people hereafter see to it that established ability be not passed unhonored.[20]

Editor W. H. Huston was a graduate of Smith College and continually exhorted his readers toward intellectual and cultural improvement. Like many African Americans of this period who were striving for respectability and acceptance, he rejected aspects of black culture that were regarded as lacking in refinement and dignity. In this vein, he was an adamant foe of ragtime and the ragtime culture he saw along Main Street. Despite this attitude, it is obvious that he recognized special qualities in Scott Joplin, and perhaps in his music, for he gave Joplin unqualified support.

Huston was not alone in this admiration. The white newspapers also made favorable comments. The *Democrat* wrote, "Sedalia can proudly say that she has one of the most wonderful musicians in the world in the person of Scott Joplin, who is considered a wonder in the art of music by the leading musicians of this and foreign countries." Even the *Sentinel,* which routinely denigrated blacks, referred to Joplin as "the world's greatest composer of syncopated music."[21] The evidence mounts in showing how favorably Joplin impressed people of all circles.

Throughout the summer, Joplin performed for dances and in concerts, both as a piano soloist and as a singer with vocal groups. The above concert would have been prior to July 15. He held a successful musicale at Liberty Park Hall on July 28, at which he performed *Weeping Willow, The Cascades,* and *The Favorite;* singers Hortense Cook, Richard Smith, and the Sedalia Quartette presented other selections. Tickets were twenty-five cents, and nearly two hundred people attended. Unfortunately, an advertisement specified that the concert was "for whites only." He had a correction placed in the *Sedalia Conservator,* stating that the announcement should have read: "White friends invited." The correction, however, did not appear until after the concert, and only whites were in attendance.[22]

Other concerts announced in the newspapers were at Liberty Park Hall on August 2; with pianist and singer Arthur Channels, a benefit for the Taylor Street Chapel on August 18; a grand ball that Joplin organized on August 28th; and a vocal performance with the Sportie Boys at a picnic in the nearby town of Sweet Springs on September 8.

The major performance of the summer was at the annual Fourth of August Emancipation Day celebration held at Sedalia's Liberty Park, attended by African Americans from throughout the state. The program began with music (the Queen City Cornet Band played at the corner of Third and Ohio at 9:20 in the morning), ended with music (a 10:00 P.M. grand march led by band member Nathaniel Diggs and Sedalia Quartet member John Williams), and included music in most events during the day. The band gave several concerts and at 2:00 P.M. escorted the invited speakers to the park. The musical highlight was the "Scott Joplin Musicale" held in the evening, accompanied by the Sedalia Quartette—John Williams, Frank Bledsoe, Lynn Williams, and Arthur Channels—and the baritone soloist Richard Smith.[23] A report of the celebration the next day stated: "The night program and the Scott Joplin musicale was attended by many white people, and all pronounced the celebration complete and a marked departure from former celebrations."[24] In Sedalia, and as a quartet singer, Joplin could still succeed as a performer.

Emancipation Day in Sedalia

Emancipation Day, held annually on the Fourth of August, was a major celebration for blacks of Missouri. African Americans from all over the state would take advantage of the special railroad excursion rates to attend. Each year white Sedalians met the day with mixed feelings. On one side, it was felt that the convergence of thousands of blacks would bring serious disorders; on the other, storekeepers looked forward to the business that the visitors would generate.

After each event, the newspapers would comment on the good behavior of the celebrators, unlike "previous years." It seems not to have mattered that, at least from the mid-1890s on, all celebrations were orderly. White Sedalia, judging by the newspaper comments, still anticipated the worst. One year, the *Sentinel* summed up the opinion with a quotation from Will Marion Cook's popular song *Darktown Is Out To-Night:* "warms coons a-prancin', swell coons will be dancing, and tough coons a wantin' to fight."[25]

The celebrations were held mostly at Liberty Park and consisted of speeches by black politicians, clergymen, and educators; baseball and other sporting events; and much music. The Queen City Cornet band was usually (but not always) in attendance, opening the program with a march to the park and playing at various times during the day and evening. Scott Joplin may have performed in years other than 1904, but if so, he was not mentioned in the newspapers. By 1904 he was famous and was therefore a feature on the program.

But why was Emancipation Day celebrated on August 4? This date had nothing to do with the end of slavery in the United States. It was on September 22, 1862, that Abraham Lincoln issued the Emancipation Proclamation, providing that slavery be lawfully ended in the Confederate states the following January 1.

The event being celebrated on August 4 was actually the end of slavery in the British West Indies. The Abolition Act became law in 1834 on August 1 (not the 4th), providing that domestic slaves be freed on August 1, 1838, and agricultural workers two years later. Though it may seem curious that West Indies emancipation be commemorated in Missouri, that event was perceived as "freedom's first dawn to the colored race."[26]

There were some in the black community who argued for a more appropriate date for the celebration of Emancipation Day. January 1 was not a good choice because the season was too cold for outside events and there was an obvious conflict with New Year's Day. Reformers were successful in arranging celebrations for September 22, but these were not as popular or as well attended as the August 4 events. A major difficulty with the 22nd was that the best facilities were taken by another celebration that occurred on the same day: German-American Day. So for a while, black Missourians observed the ending of slavery in the British West Indies as their own day of emancipation.

Whatever professional success Joplin had that summer, he worked under a cloud. His young bride's illness worsened. An ominous sign appeared in the *Conservator* on August 12: "Mrs. Scott Joplin, who has been confined to her bed with pneumonia, suffered a relapse recently, but is somewhat improved at this date." On Thursday, September 8, Miss Lovie Alexander arrived from Little Rock to be at her sister Freddie's bedside.[28] Two days later, on September 10, Freddie died. The *Sentinel, Democrat,* and *Capital* all had brief items the next day:

Death of Mrs. Scott Joplin

Mrs. Scott Joplin, wife of the colored song writer, and whose maiden name was Freddie Alexander, died Saturday afternoon at the home of Solomon Dixon, 124 East Cooper street, of pneumonia, aged 20 years. The couple had been married only ten weeks. The funeral arrangements had not been perfected last night.[28]

A few days later, the black *Conservator* printed a longer obituary:

Mrs. Freddie, wife of Mr. Scott Joplin, and a bride of only two months, after an illness of seven weeks duration, died Saturday after noon at 3 o'clock p.m. Her death was not unexpected, as she had contracted cold which developed into a complication of complaints either of which might have resulted in death.

She was married to Mr. Joplin two months ago at the home of her parents in Little Rock, Ark., and had traveled some with Mr. Joplin who was billed to give piano recitals in western towns. They arrived here seven weeks hence, and from their first day here, Mrs. Joplin has been confined to her bed at their rooms at 124 E. Cooper St.

Thru-out her sickness, Mr. Joplin has administered to every want. Her sister, Miss Lovie Alexander, arrived here last Thursday and was constantly at her side until death separated them.

The interment was in the colored cemetery from the Morgan St. Baptist church, Rev. S. A. Norris preached the funeral which occurred at 2 o'clock in the afternoon.[29]

The Search for Freddie

 The very existence of Freddie was almost forgotten. None of Joplin's acquaintances who had been interviewed for earlier biographies referred to her. The first hint of her existence emerged in 1975 in an interview with Beatrice Martin, an elderly resident who had been born in Sedalia in 1890. Mrs. Martin told of her memories of Joplin and mentioned a wife who died in childbirth, with the child dying as well.[30]

In 1984, Lucile Martin, another aged Sedalian and a cousin by marriage to Beatrice Martin, related a slightly more elaborate story. When Lucile Martin was about four, which would have been in 1893 or 1894, she went to live with her grandparents Mr. and Mrs.

Solomon Dixon. Joplin was boarding in the home, and she remembered him as a young man in a brown suit and a slouch hat. In the home was an Emerson Square Grand that he frequently played. While living there, Joplin went to visit his family in Texarkana and returned with a wife. About two years later, the wife died of tuberculosis.[31]

While both Mrs. Martins remembered the essential elements of the marriage and the early death, they were wrong on many of the details. Once again we see the unreliability of testimony.

I decided the reports were sufficiently plausible to warrant a search. It at first appeared that 1896 might be a likely year for the marriage, for Joplin probably was in Texas then. Arguing against that year was the 1900 Census, in which Joplin listed himself not as a widower, but as single. The year 1904 also looked promising because of the directory listing putting Joplin in the Dixon household.

While at the Ragtime Festival in Sedalia in 1988, I went to the court house to look for death records, only to learn they had been sent to Jefferson City, the state capital. John Hancock, a Missouri state representative and a serious ragtime musician, checked into the matter and reported that the records had been destroyed in a Sedalia court house fire in 1920.

I then went to the historic black cemetery, which adjoined the white burial grounds. Looking through the record books for the 1890s and early 1900s, I found that records were kept only for white burials. A brief walk through the black section convinced me of the impracticality of trying to read all of the headstones myself. At the Festival the following year, I used a seminar session to explain the problem and requested volunteers to help in this historical search. Late that afternoon, the volunteers joined me and, in five or six groups of twos and threes, we walked through the cemetery. While we spotted several familiar names on stones, many of the stones had been eroded by weather and age, and we could find no sign of Joplin's wife or child. (Of course, there was no child, but we did not know that then.) The lesson to the volunteers was that good efforts do not always produce positive results. A search for record of her in funeral parlors was equally nonproductive.

The actual discovery of Freddie came about more simply. Because the town directory showed that Joplin lived in the Dixon household in 1904, I read through all of the available newspapers for that year. It was with the newspapers that I finally confirmed that Joplin had another wife. I then set about to learn more about her.

I had already determined that there was no sign of Freddie at the cemetery. The Morgan Street Baptist Church, where the funeral was held, has been succeeded by the Ward Memorial Church. Dr. William Singleton, current pastor of the church and president of Western Bible College in Kansas City, reported that no church records survive from that period.[32]

I succeeded in obtaining a copy of the marriage record in Little Rock, but could locate no listing in the 1900 Census for Arkansas or Texas that could refer to Freddie's family.[33] Despite the impasses, we have confirmed the existence of another wife and have obtained several facts about her. Slight as it is, this information allows us to reinterpret the rest of Joplin's life.[34]

It is strange that none of Joplin's associates who were interviewed in later years, including his widow Lottie and his friends Arthur Marshall and Sam Patterson, mentioned Freddie. Is it possible that he never spoke of her? Even if he had not, Arthur Marshall must certainly have known of her, for he returned to Sedalia to live with his own bride four months later, and someone would have told him of Joplin's tragedy. In reporting incorrectly on Belle's death, Marshall may actually have confused Belle with Freddie.

Perhaps the brevity of the marriage made it insignificant in the minds of his friends. But Freddie was not insignificant to Joplin, for some years later he was to memorialize her in his most ambitious work.

CHAPTER NINE

Final Days in the Midwest, 1905–1907

*B*Y 1905, JOPLIN COULD POINT to significant accomplishments. His *Maple Leaf Rag* was an unprecedented success, known nationwide. His subsequent rags did not summon the same degree of acclaim, but they did contribute to his popular and critical renown. With praise coming from the white press, there even were signs that he was overcoming the biases that normally blocked serious African-American artists.

He had reason to be pleased, but he also had reason to despair. His opera, an attempt to demonstrate that he was a serious artist who could transcend the limitations of ragtime, was a failure and probably had severe financial repercussions. On the personal side, his marriage to Belle had been a disaster, underscored by the death of their infant daughter. And if Freddie inspired him to new heights, her passing must have cast him to new depths.

We do not know where he went after Freddie's death in September 1904, but by early 1905 he was back in St. Louis. He may have roomed at 2221 Market, directly across the street from Tom Turpin's Rosebud, for he gave that address as his residence in the 1906 directory, listing himself as a composer.

Tom Turpin was attracting attention with his saloon. During the 1904 Christmas season he delighted many in the neighborhood by installing what was then an extraordinary novelty, an electric Christmas tree. On Christmas night, December 25th, he had gifts for all patrons, a gesture that cost him a reported three hundred dollars.[1]

Of Joplin's other known associates, a few were still nearby. Sam Patterson and Louis Chauvin were again a team, working around the area. In October 1905 they were hired for the First Grand Ball of the Lady Piano Players Club, Patterson presiding as the master of ceremonies and Chauvin billed as "King of Rag Time Players" (probably to distinguish him from Joplin, who was "King of Rag Time Writers").[2] Charles Williams worked in the area as a bartender, and he and his brother Tony were both organizing events for the Twilight Social Club.[3] Joe Jordan was in town only briefly, for by March he had moved to Chicago, where he became music director of the Pekin Theatre and co-composer (with black opera composer H. Lawrence Freeman) of *Rufus Rastus,* Ernest Hogan's new musical.[4] Arthur Marshall left St. Louis in January, returning to Sedalia with his bride Maude McManus. He remained there at least through May, working as a pianist for dances and other social functions.[5]

Blesh and Janis have written that James Scott, a young black ragtimer from Carthage, Missouri, went to St. Louis in search of Joplin in 1905, and that Joplin introduced him to Stark.[6] Stark published James Scott's *Frog Legs Rag* in 1906, and its sales were second only to those of *Maple Leaf* in Stark's catalogue.[7] Through the next sixteen years, Stark was to publish twenty-seven Scott rags, and for modern ragtimers James Scott is enshrined in the ragtime pantheon alongside Scott Joplin.

Scott Joplin and James Scott may have known each other, and they certainly knew each others' music. (Joplin orchestrated Scott's *Frog Legs Rag,* published by Stark in *Standard High-Class Rags*—known as "The Red Back Book"—around 1909.) Inasmuch as Joplin and Scott had similar temperaments, both being mild-mannered, quiet, and thoroughly engrossed in their music, they probably could have had a congenial friendship. However, no evidence has surfaced confirming that the two were personally acquainted.[8]

A friend of some importance, but one of whom we know virtually nothing, was Dan Davenport. Davenport is listed in several directories between 1901 and 1906 as a porter and a bartender, working on Chestnut and Market streets. We know his name today for one reason: Joplin dedicated *Bethena. A Concert Waltz* to "Mr. and Mrs. Dan E. Davenport of St. Louis Mo."

This is an unusual dedication. Davenport was not a prominent performer who could showcase Joplin's music. Nor was he a person of influence or wealth, one who could have helped Joplin's career or commissioned music. So why should Joplin dedicate this piece to Davenport and his wife?

The dedication must have had personal significance. Since this is Joplin's first copyrighted work (March 6, 1905) since Freddie's death, I think the Davenports might have helped Joplin through some difficult times, and the dedication was his way of showing appreciation. Supporting the idea that *Bethena* is connected with Freddie, or at least reflects Joplin's mood at this time, is that the music is sadly poignant. It is decidedly not a cheerful, merry waltz, and in this respect contrasts with all of Joplin's other waltzes.

But then, what is the meaning of the title? "Bethena" is an unusual name. Was there a Bethena? If so, is she the beautiful woman pictured on the cover? Or (and this thought is highly speculative) is this picture of Freddie? The woman's race is impossible to determine. She could be white, or she could be mulatto.

Musically, the opening strain of *Bethena* is most important. It appears

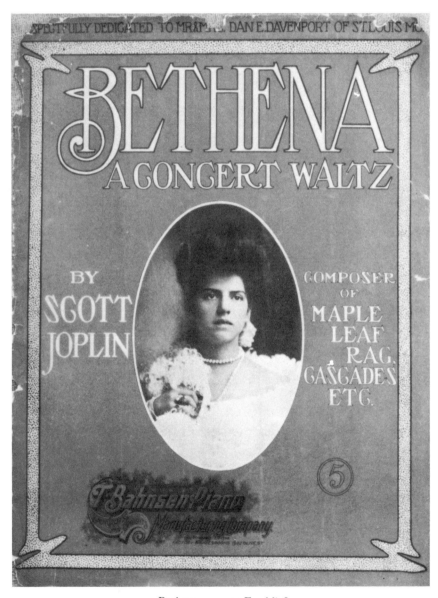

Bethena . . . or Freddie?

three times in the piece, and its four-note motif is used also for the introduction and coda (Ex. 9-1a). This motif is itself partially responsible for the music's mood, for it starts on a dissonance, the melody note A being harmonized against a G major chord. Joplin reharmonizes an exact repeat of the motif in the next measure before spinning it off into new guises. He then creates variants of the motif in the D and E strains.

Ex. 9-1. Scott Joplin, Bethena. A Concert Waltz *(1905): (a) A1-4; (b) B1-4; (c) C1-4; (d) analytic reduction showing inner and bass voices, C1-4.*

The piece also has effective contrapuntal gestures. The opening phrase presents an inner voice descending chromatically from D to B. In the B strain, the treble and bass parts that are in contrary motion in measures 1-2 are exchanged in measures 3-4 (Ex. 9-1b). The opening phrase of the C strain has an inner-voice descent from F to A (Ex. 9-1c). These all contribute to making *Bethena* an enchantingly beautiful piece that is among the greatest of ragtime waltzes.

Bethena was issued by T. Bahnsen Piano Mfg. Co., at 1522 Olive Street. This company was a thriving firm that advertised frequently in the *St. Louis Post-Dispatch,* but, except for three Joplin publications, does not seem to have been involved in music publishing. The two other works, also issued in 1905, were both copyrighted on August 11.

One of these works is *Binks' Waltz,* a nonsyncopated piece composed in honor of an infant boy named James Allen Morgens, nicknamed "Bing" by his older sister. The title should have been "Bing's Waltz."

The child's father, William A. Morgens, operated the Morgens-Scott Dry Cleaning store at 3400 Olive Street. According to Morgens family legend, Joplin frequently stopped by the store "looking for handouts." The stories are inconsistent on whether Joplin composed the waltz in gratitude for generous "handouts" or in response to a direct commission, but it is clear that it was, in one way or another, a composition for which Joplin received payment.

There is insufficient evidence to determine whether Joplin could have actually been begging. It would seem likely that he had lost a good deal of money on his *Guest of Honor* tour two years earlier. It is possible that he was actively seeking commissions or writing for hire. Perhaps he was also seeking a patron or a backer for another theatrical venture. To a small businessman, one not necessarily versed in the traditions of the arts, it could have appeared that this black man was simply "looking for handouts."[9]

What the family legends do not tell is that the Morgens had a connection with T. Bahnsen. The firm was incorporated in June of 1898 with a capitalization of twenty-five thousand dollars. Timotheus Bahnsen and Alvina Bahnsen owned ninety-nine percent of the company, but a single one-hundred dollar share was owned by Anna Morgens.[10] No one in the family today can identify Anna Morgens, but the unusual spelling of the family name would strongly suggest she was a relative. It would therefore seem likely that her connection with Bahnsen had something to do with the publication of *Binks' Waltz* and with Joplin's dealings with the company.

While *Binks' Waltz* is a pleasant, effective dance piece, one cannot think of it as a distinguished work. Yet, it is one in which Joplin apparently took some pride. He announced its composition in the *Freeman* on July 22, 1905, programmed it at a concert later that year, where it was "the hit of the bill,"[11] and used it again for a dance five years later.

The final Bahnsen publication was *Sarah Dear,* a song with words by Henry Jackson, Joplin's earlier collaborator on *I Am Thinking of My Pickaninny Days.* The title is apparently derived from the dedication subject, Miss Sarah Jarrett, of whom nothing is known. Though the title and non-dialect lyric might suggest a piece in a genteel manner, the cover depicts a minstrel scene with a woman scolding a black-faced clown. The scene was from an act by Charles Williams and Augustus Stevens entitled "A Darkey Scrimmage," with Stevens, a specialist in female impersonations, taking the part of the woman. The two were highly regarded and in 1905 were leading their own minstrel company.[12]

Announcements in the *Freeman* on August 5 and 19, 1905, indicate the song was written for Williams and Stevens. True to the nature of the comedy team, *Sarah Dear* is a rollicking ragtime song, with a chorus that hints at a phrase used in other ragtime classics. Specifically, the chromatically descending minor third of the chorus opening is similar to patterns in Ben Harney's *Cakewalk in the Sky,* C1-4 and Barney and Seymore's *St. Louis Tickle,* B1-2 (Ex. 9-2).[13]

Ex. 9-2. (a) Scott Joplin (music) and Henry Jackson (words), Sarah Dear *(1905), Chorus 1-4; (b) Ben Harney,* Cakewalk in the Sky, *(New York: Witmark, 1899), C1-4; (c) Barney and Seymore,* St. Louis Tickle *(Chicago: Kremer, 1904), B1-4.*

We have to wonder why Joplin published with Bahnsen, a company not really involved in music publishing. Was he, despite his fame, unable to arrange favorable terms with established publishers? This may be the case, since one piece he composed was not even published. He announced in July that he had completed the song *You Stand Good with Me, Babe,*[14] but no music by that title has ever surfaced.

Our last sign of Joplin in St. Louis in 1905 is on November 12, when he appeared in a Sunday afternoon concert at Douglass Hall, at the corner of Beaumont and Lawton. The Sunday afternoon concerts were weekly affairs that featured the 30-piece band of William Blue and a selection of speakers who lectured on literature and other edifying topics. On this occasion, the program included "Joplin's Concert and Piano Contest," in which "Bink's Waltz was the hit of the bill."[15] This scanty information leaves us guessing at what occurred. The term "Joplin's Concert" suggests that Joplin performed solo, without Blue's Band. We have no information on the nature of the piano contest or who competed. We might wonder how *Binks' Waltz* could have been "the hit of the bill," how its rendition could have been so outstanding that it edged out Joplin's other works. In all likelihood, his rags were not performed. The dignified setting of the concert would probably have ruled out performance of the *Maple Leaf* or any other piece that hinted at "sincopation."

By this time, John Stark had made major changes. In 1904 he had incorporated the business, and the legal papers raise some perplexing questions. The business was capitalized at two thousand dollars, with two hundred shares being issued at ten dollars apiece. The dominant shareholder was not Stark, who held only thirty shares, but Edward H. Clay, with ninety-nine shares. Eleanor had fifty, William one, and A. E. L. Gardner twenty. The directors were John Stark, William Stark, and Edward Clay. Clay, who has never been mentioned in the Stark–Joplin stories, is obviously a business associate, but he seems to have been more than that. He lived in St. Louis at 4509 Shenandoah Avenue, the same address as John, Eleanor, and William Stark. He may also have had a relationship with the Starks from earlier years, for from 1897 to 1899 two string players named Clay appeared on concert programs in Sedalia with Eleanor and Etilmon.[16] But if Edward Clay had a family connection with the Starks, it has been forgotten; no one in the Stark family today can identify him.

In 1905, Stark moved to New York, opening an office and store at 127 East 23rd Street.[17] His son William remained in St. Louis to take charge of the office and printing plant at 210 Olive. Stark published one Joplin piece in 1905, *The Rose-bud March,* but failed to register it for copyright, so we have no clue as to when in the year it was issued.

The title refers to Turpin's saloon, and the piece was dedicated to "my friend Tom Turpin." That Joplin composed a march in 6/8 meter—a meter that does not lend itself to ragtime syncopation—rather than a "Rose-bud Rag," suggests the continued importance of the march among black ragtime pianists. This importance is supported by testimony, as well, with the anecdote of how Louis Chauvin would always play Sousa marches as warming-up exercises.[18]

A final work from St. Louis in 1905 was *Leola,* published by American Music Syndicate of St. Louis. Little is known about this company, but it was apparently connected with Stark, for in his monthly magazine *The Intermezzo* he wrote that *Leola* is "the music publishing event of the entire season."[19] A copyright application was made on December 18, 1905, but copies of the music were never sent.[20] Ragtimer Charles Thompson reported that the title reflected another love interest: "Joplin's one love was a girl named Leola, who jilted him! For some time afterward, he was not interested in women."[21] We have no way of evaluating Thompson's remark, but we have to wonder if he was aware of Joplin's marriage to Freddie. And who was Miss Minnie Wade, to whom the work was dedicated? The city directories (1901 through 1906) tell us she was a teacher. Was she another love interest, or was she a music teacher who helped Joplin?

Leola is another rag in the mold of *Maple Leaf Rag:* it is in the same keys as the earlier masterpiece, and the first three strains trace the same harmonic patterns. Here, though, there is considerably more of a melodic emphasis, beginning with a florid, two-octave arabesque (Ex. 9-3). It is clear that Joplin was again trying to recreate a *Maple Leaf Rag*-type hit, but now in a more lyrical manner.

Ex. 9-3. Scott Joplin, Leola *(1905), A1-4.*

It was with this rag that Joplin, for the first time, included the tempo warning "Notice! Don't play this piece fast. It is never right to play 'rag-time' fast. Author." Perhaps with the *Maple Leaf* in mind, he wanted to head off the type of high velocity performances that rag inspired. A tradition was already emerging of using the *Maple Leaf* as a virtuoso showcase, with pianists competing to see who could play it fastest. The arranger Mayhew Lake (1879–1955), for example, referring to the year 1903 when he was in St. Joseph, Missouri, wrote: "In those days the piece de resistance for raz-ma-taz pianists was the *Maple Leaf Rag* by Scott Joplin and, with proper inspiration, I could probably play it faster, louder (and lousier) than anybody north of New Orleans."[22]

Ragtime in the White House

While many in the music establishment viewed ragtime with disdain, its quick acceptance into the White House was more attuned to its reception by the general public. At least two rags, Kerry Mills's *Whistling Rufus* and Abe Holzmann's *A Bunch o' Blackberries,* were played in the White House by the Marine Band for a Valentine's Day ball hosted by President McKinley on February 14, 1901.

Theodore Roosevelt continued this welcome as he cake-walked to the new music at a Christmas party in 1901: "The President chose a partner and led the cakewalk down the long East Room, executing fancy, buck and wing steps, while others in the party clapped 'Juba'." The music played for the cakewalk was, again, *Whistling Rufus,* and Theodore Metz's *A Hot Time in the Old Town To-Night,* associated with Roosevelt's "Rough Riders" victories in Cuba.

The Maple Leaf Rag was also performed early and frequently by the Marine Band, as revealed in a story about Alice Roosevelt. Alice was seventeen when her father became President, and she immediately charmed the public, prompted by an almost continuous flow of items in the press. She was a good-natured, fun-loving young woman who was as energetic and irrepressible as her father. There were accounts of her jumping into a swimming pool fully clothed, pulling a live snake from her pocket, and sliding down the White House bannister to meet a room full of dignitaries. Society matrons voiced their disapproval of her uninhibited behavior, which endeared her even more to the American public.

It was around 1905, at a diplomatic reception at the White House, that Alice approached the director of the Marine Band with a request. A bandsman present wrote of the incident:

> Miss Roosevelt came up and said, "Oh, Mr. Santelmann, do play the 'Maple Leaf Rag' for me. . . . "The 'Maple Leaf Rag?'" he gasped in astonishment. "Indeed, Miss Roosevelt, I've never heard of such a composition, and I'm sure it is not in our library." "Now, now, Mr. Santelmann," laughed Alice, "don't tell me that. The band boys have played it for me time and again when Mr. Smith or Mr. Vanpoucke was conducting, and I'll wager they all know it without the music."

Santelmann acceded to her request, and the *Maple Leaf Rag* was played.

Supporting Alice's contention that the *Maple Leaf* was in the Marine Band repertory, a researcher has found marching band parts dating from 1904 in the Marine Corps Historical Center.[23]

Sometime after the middle of 1905, Arthur Marshall and his wife moved to Chicago, taking an apartment at 2900 South State Street, above Beau Baum's Saloon and across the street from the Pekin Theatre, where Joe Jordan was music director. State Street in Chicago, running north-south, included a section that was the equivalent of Market Street in St. Louis. It was a center for theater and other entertainment, both legitimate and illegitimate, and for ragtime. Marshall worked as a pianist at several locations in the area—the Wintergarden at 3047 South State Street, Lewis's Saloon, and the Eureka Saloon.[24]

Toward the end of 1905, Joplin went to Chicago, staying first with Marshall for about three weeks, then getting his own apartment at 2840 Armour Avenue. He may also have spent time with his brother Robert B. Joplin, who is listed at 2635 Armitage in the 1905 directory.[25]

Joplin's purpose in going to Chicago was apparently to seek out additional publishers. He visited several in Chicago and reported to Marshall that they all welcomed him cordially. He also corresponded with Harry and Albert Von Tilzer, prominent songwriter-publishers in New York, but nothing came of these contacts.[26] He did, however, renew his relationship with Will Rossiter, a major Chicago publisher. Rossiter had previously issued Joplin's *The Sycamore,* and late in 1905 put out Joplin's newest rag, *Eugenia.* (The date 1905 appears on the music, but the copyright was not registered until February 26, 1906.) We have no information on whether the title refers to someone of Joplin's acquaintance.[27] There is no dedication.

On the music, Joplin repeats the admonition that appeared on *Leola* against playing ragtime fast, and then goes one step further. For the first time, he uses a metronome marking: ♩ = 72. This seems slow for most ragtime dances, and is certainly slow for the indicated tempo of "Slow March Time." But it may be suitable for *Eugenia,* which has a somewhat complex melody (Ex. 9-4).

Ex. 9-4. Scott Joplin, Eugenia *(1905), A1-4.*

"Don't play this piece fast. It is never right to play ragtime fast."

 That Joplin perceived a need for the above statement indicates that ragtime frequently *was* played fast, faster than he thought suitable. But how fast is fast, and what is the proper speed for ragtime?

There is no single answer to the last question. There were certainly ragtimers who played fast, too fast for Joplin's taste. He commented that white people seemed to think ragtime had to be played fast, and they were wrong in playing his music that way.[28] This protest of his illustrates that there was more than one performing practice.

Ragtime was succeeded in the 1920s by *novelty piano*, a style characterized by marvelously intricate figurations with virtuosic demands (Representative composers of this style were Zez Confrey and Billy Mayerl, the latter from Great Britain.) The few rags that survived in performance in the 1920s, now divorced from their function as dance music, were played according to the new esthetic: fast and intricately embellished. The tradition of fast ragtime performances prevailed through the next half-century.

In the late 1960s and early 1970s, several classically oriented pianists took notice of Scott Joplin's music and decided it should be performed as the composer had put it down. The result was Joplin's ragtime played as written rather than as a springboard for improvisation and virtuoso display.

This new attitude, discussed in Chapter Thirteen, was a radical departure from what had been considered "the ragtime tradition." The new respect for the score was frequently accompanied by tempos that some considered agonizingly slow. Is this really what Joplin had intended?

Many of Joplin's rags are marked "Slow March Tempo" or "Slow March Time." *Eugenia* has such a tempo marking as well as the metronome specification of $\downarrow$ = 72. This has been taken as justification for very slow tempos (though many performances are even slower than that). But in 1908, Joplin gave three other rags in "Slow March Tempo" the brisker metronome marking of $\downarrow$ = 100 (*Fig Leaf, Sugar Cane*, and *Pine Apple*). Which indication is more appropriate for slow march time, 72 or 100?

Marches were integral to the musical life of the period, heard as much on the dance floor and in concert as on parade grounds. The tempo range for "march time," occasionally used for rags, was pretty much standardized between 110 to 128. Stepping along to that tempo shows it to be a brisk marching pace.

"Slow march time" was generally between 90 and 110. That would be a suitable tempo to accompany the two-step, which was the basic ragtime dance of these years.

"Very slow march time" (such as Joplin's *Wall Street Rag* of 1909) would be 70 to 90, a bit slow for the two-step, but perhaps appropriate for the more involved cakewalk. The slowest dance step of the period was the lingering "slow drag," and this could have music as slow 50 to 70.

Performers of ragtime should be aware of these guidelines,

and also realize they are *only* guidelines, not inflexible rules. The function of a performance is a prime consideration: if it is to be used for dance, the requirements of the dance steps are paramount. If the music is to be heard in concert, there is much more leeway. Perhaps that is why Joplin stopped including metronome indications after 1908.[29]

While Joplin was in Chicago, Louis Chauvin and Sam Patterson also went there. It was during this Chicago stay that Joplin and Chauvin collaborated on *Heliotrope Bouquet,* published the following year by Stark. Sam Patterson was present and recalled that Chauvin presented two ragtime themes to Joplin. These are the only extant samples of ragtime by this legendary pianist and composer, and they fully support the glowing reports made of Chauvin's musicianship. Joplin, on hearing Chauvin's themes, immediately sketched out two more themes for the trio.[30] Some time later, Joplin wrote the whole piece out and submitted it to Stark.

Chauvin's opening strain is quite unlike anything composed by Joplin. The harmonies seem at first to be without direction, moving circularly against a habanera rhythm and existing only for sensual effect (Ex. 9-5a). The second strain expands upon a motif from the first (Exx. 9-5b, c) and has a more conventional tonality.

Ex. 9-5. Scott Joplin and Louis Chauvin, Heliotrope Bouquet *(1907): (a) A1-8; (b) A13; (c) B1-4; (d) C6-8.*

Joplin's C strain is an immediate contrast because of its strongly directed tonality. Then, after presenting his familiar chromatically descending fourth (C to G in the bass, measures C 4-6), he reprises Chauvin's cadential pattern of A 7-8, even using the habanera rhythm to make the connection more readily apparent (Ex. 9-5d). Once again Joplin pays tribute to a younger colleague by using his music.[31]

A work that was undoubtedly done for hire was Joplin's arrangement of the coon song *Good-by Old Gal Good-bye,* with words by H. Carroll Taylor and music by Mac Darden. The song was published by the Foster-Calhoun Co. of Evansville, Indiana; the music displays a 1906 copyright notice, but it was not registered. Nothing is known of either Taylor or Darden. The melody is certainly not by Joplin—the opening augmented second in the verse is too awkward to have been his writing—but the piano harmonization is well within his style.[32]

Back in St. Louis, by the summer of 1906 Tom Turpin had closed his Rosebud Saloon.[33] Chauvin had returned the same summer and was appearing in "refined vaudeville" at the Douglass Theatorium, situated at Beaumont and Lawton. The program beginning on July 30 included two operatic-styled singers—Miss Vella Crawford, "The Renowned Prima Donna," and Richard D. Barrett, "Premier Baritone, late of Black Patti Troubadours"— which may be what prompted Chauvin to assume a billing with classical connotations: "The Black Paderewski."[34]

Sam Patterson was no longer with Chauvin, having chosen to go on the road with the newly-formed group of Spiller, Patterson and Crawford, soon to be called the Spiller Musical Trio. William N. Spiller was to become a prominent and successful figure in black vaudeville, as well as a good friend of Scott Joplin. He was a well-educated man who came from the segment of the black community that strongly opposed ragtime and the theater. In going on the stage, he became estranged from his father, a respected educator and minister, and it was not until several decades later that they were reconciled.[35]

John Stark, his wife, and his daughter were living in New York, but he made trips back to St. Louis. To keep up with increasing demand for his publications, in June he opened a larger printing facility at 1516 Locust Street.[36] During the summer he corresponded with Arthur Marshall, and several of his letters are extant.[37] These letters are a unique window revealing Stark's way of doing business and his negotiation processes.

Marshall had spoken in an interview in 1960 of Stark's fairness and willingness to pay his composers royalties. "That's one thing about old man Stark. He was pretty fair with us people that he bought rags from. He'd give 'em a royalty on it—and that's the only way it should be."[38] Marshall may have remembered it that way, but the letters he received from Stark in 1906 tell a different story.

Early in 1906, Stark had published Marshall's *Kinklets,* which sold well. In July, Marshall, still living in Chicago, sent Stark a song and another piano rag. Stark replied in a letter from New York on July 28th. Before making any decisions about the pieces, which were not named, Stark wanted to consult with Eleanor the following week. However, he expressed reservations about the song and did not mention it again in subsequent letters.[39] He also noted serious notational errors in the rag, which suggests that Marshall did not gain much from his formal musical studies at Smith College.

Stark's next letter was on August 24th, from St. Louis. He had shown the rag to Joplin, who liked it, but noted that the notation would have to be corrected. Stark concluded that the piece was too difficult to be a big seller and offered Marshall a mere ten dollars and two hundred copies of the publication. He made no mention of royalties.

Marshall wrote back two days later refusing the offer, and the music was returned to him. Two weeks later, on September 9th, he again wrote to Stark and tried to reopen negotiations. Stark, back in New York, replied on September 13th and suggested that Marshall try the Chicago publishers. Two weeks later, Marshall accepted Stark's offer.

The Stark revealed by these letters does not fit the generous description given by Marshall a half-century later. Nor does he seem to be, as some writers have portrayed him, the altruistic lover of ragtime who was willing to put aside business principles in favor of the higher calling of art. We do not doubt Stark's love of ragtime and his appreciation of the skills of Scott Joplin and other superior composers. But at the bottom line, John Stark was a businessman who would do his utmost to negotiate to his best advantage.

Joplin was back in St. Louis by August. The Stark letters show that. On November 18, he appeared again at one of William Blue's Sunday afternoon concerts, this time at the Pythian Temple, 3137 Pine Street. A brief review of the event does not mention Joplin, so we have no information on whether he performed solo or with the band, or what he played. As with the joint concert-literary lecture of November 1905, it is doubtful that ragtime would have been programmed.[40]

In October, the *Maple Leaf Rag* was again recorded, this time by the U. S. Marine Band.[41] It is my guess that Joplin was not aware of this recording; if he had been, he probably would have sent a notice of the event to the *St. Louis Palladium* or the *Indianapolis Freeman.*

Stark registered for copyright two more works by Joplin late in December (but possibly publishing them several months earlier, which was his usual pattern). These were *The Rag-Time Dance* (a shorter, piano version of the 1902 vocal publication) and the march *Antoinette.*

The piano adaptation of *The Rag-Time Dance* excludes the opening 32-measure *parlando* section of the song, but retains the other sections and leaves the original piano accompaniment unaltered. There is no way of knowing if Joplin was involved in the transformation from song to piano solo.

We have already discussed *Antoinette* in Chapter Seven, proposing that it may have been drawn from *A Guest of Honor.* However, its publication in 1906 raises another issue: Prior to 1905, Joplin had named only one piece after a woman, this being *Cleopha* (1902). After 1906, there was also only one piece so named, his opera *Treemonisha.* But in the years 1905 and 1906 he had five pieces bearing women's names: *Bethena; Sarah Dear,* dedicated to Sarah Jarret; *Leola,* dedicated to Minnie Wade; *Eugenia;* and *Antoinette,* dedicated to Marie Antoinette Williams. Even if we exclude the song *Sarah Dear,* which may have been named and dedicated by lyricist Henry Jackson, we see in 1905–06 an uncharacteristic emphasis on titles with women's names. This suggests there was something occurring in his emotional life. Was he, after the death of Freddie, becoming involved with a quick succession of women? Whatever the situation was, it has escaped our investigations.

The 1907 St. Louis directory has Joplin again at 2221 Market Street, now listed not as a composer, but as a laborer. We doubt the accuracy of this entry. Even if Joplin was working as a laborer, it is not likely that he would have listed himself that way. Tom Turpin, now at 2816 Morgan, was also listed as a laborer. Though Turpin no longer ran a saloon, as an outstanding pianist and a bartender he would have had no need to become a laborer. What probably happened in both entries is that the canvasser for the directory automatically entered "laborer" for black men.

Despite the address in the directory, there is testimony that Joplin had moved to Clayton, a suburb of St. Louis. The account tells of Joplin and a 25-year-old white friend named Harry LaMertha.

LaMertha was a self-educated and highly accomplished individual of extremely diverse interests. At the time of his friendship with Joplin, he was working in the new communications industry as a telephone installer. Within two years, he was to join the *St. Louis Post-Dispatch,* with which he would remain for the next 42 years, half of the period as an editor.

How the friendship came about is not known, but it was reportedly close. LaMertha, a self-taught musician with perfect pitch, would spend weekends at Joplin's home in Clayton so that he and Joplin could make music together, playing piano and singing. One result of this friendship was that Joplin provided a piano arrangement for a coon song that LaMertha had composed: *Snoring Sampson. A Quarrel in Ragtime.* According to LaMertha's daughter, her father also drew the cartoon cover, depicting a black woman glaring at her snoring husband. The song was published with a copyright date of May 6, 1907, by University Publishing Company, which has no other music copyrights.[42]

Another collaborative work from this period was *When Your Hair Is Like the Snow,* copyrighted May 18. The lyrics were by the publisher Owen Spendthrift, who had already published four other songs that year. Nothing is known about his relationship with Joplin.

The song is nonsyncopated and is in the nostalgia tradition into which Joplin's two songs of 1895 fall. Within that context, it is a very good song, treating with imagination the melodic and harmonic gestures of the style.

One other publication from this period was the rag *The Nonpareil.* This was issued by Stark and bears a 1907 date, but was uncopyrighted. Stark's business ledgers show that the music was available by April 1907.

We know nothing of Miss Mildred Ponder, to whom the work was dedicated, and we do not know if the title had a specific reference. "Nonpareil" was a term in frequent use at the time (and was almost as frequently misspelled as "Nonpariel"). In 1899, a new saloon on Main Street in Sedalia was named "The Nonpariel"; in 1903, Joplin's friend Billy Kersands called his company "The Nonpariel of Minstrelsy"; and in 1905 the *Freeman* had an item on a group called "The Nonpariel Jubilee Singers."[43] If Joplin had an external reference for the music, it was probably for band performance, for the B and C sections appear to have been written with a bass instrument in mind (Ex. 9-6).[44]

Ex. 9-6. Scott Joplin, The Nonpareil *(1907), B1-4.*

In June, a writer for the New York based *American Musician and Art Journal* visited the Stark office in St. Louis and met Joplin there. In an article telling of the visit, the writer introduces Joplin in the usual way, as the composer of the *Maple Leaf Rag,* and then proceeds to describe him in terms not generally associated with ragtime musicians. He portrays Joplin as a romantic artist whose creative impulses can occur at any time during the day or night. The composer, according to the report, keeps a music pad with him at all times in order to preserve the ideas of his moments of inspiration.

Joplin was his usual reticent self in the interview, but did speak of his new theater project and played the overture:

> Scott Joplin has been working a considerable time on a grand opera which will contain music similar to that sung by the negroes during slavery days, the music of today, the negro ragtime, and the music that the negro will use in the future. . . . The writer . . . heard Mr. Joplin play the overture of his new opera, and to say that it was exceptionally good would be putting it mildly.[45]

We see once again how Joplin was respected by those he met. Whereas this journal was more open to ragtime than most music periodicals, it did not generally have articles on black musicians. Yet Joplin was not only featured, with a photograph (the one used on *Cascades*); he was also shown the respect of being referred to as "Mister." This could have been taken as a positive sign of what he could expect in New York.

The two and one half years immediately following Freddie's death had been difficult for Joplin. In this time he had only three new rags published— *Leola, Eugenia,* and *The Nonpareil.* (I do not include in this category the reissue of *The Rag-Time Dance.*) Of the remaining works, only *Bethena* can be considered quality Joplin. The others include two marches (*Rose-bud* and *Antoinette*), an unsyncopated waltz (*Binks' Waltz*), and four songs. Of the twelve works issued, seven were with little-known, insignificant publishers. One additional song remained unpublished and is lost. There is anecdotal testimony that his finances were not in good order, and he apparently wrote several works for hire.

His career seems to have gotten off track. But he had big plans, as indicated in the June interview: he was at work on another opera. If conditions were not suitable in the Midwest, he would go where there was more promise. He would go to New York.

Before leaving the Midwest, he made one last visit to his family in Texarkana. Accounts of this visit come from two relatively late sources: his nephew Fred Joplin (son of Monroe), in 1971, and Texarkana resident George Mosley, in 1975.

Joplin returned as a celebrity, played the piano in a local music store and at a local dance. He also spoke of having gone to Europe, but the stories differ on the details of this trip. According to the earlier interview, Joplin had "a triumphal tour of Europe in 1907." The later interview had him going to

Germany, possibly earlier than 1907, in the company of a German associate.[46]

If Joplin had gone to Europe, it is doubtful that there was a "triumphal tour." Such an event would have prompted a dedication or commemorating title—perhaps a *Berlin Rag,* a *Kaiser Two-Step,* or a *Brandenburg Gait.* Further, African-American performers received good press coverage in Europe, so had Joplin enjoyed any success, something would have appeared in the newspapers. Nothing has been found in European papers, and an effort to find him in England has been unsuccessful.[47] Even if he had been unfairly ignored by the press, he would certainly have sent notices himself to such papers as the *St. Louis Palladium* and the *Indianapolis Freeman.* Searches for him among lists of ship passengers entering New York and in the Passport Office have also been unsuccessful. In the absence of any supporting evidence, we consider it unlikely that Joplin made a European trip.

In the summer of 1907, Joplin went to Chicago, where he visited briefly with Arthur Marshall. This was their last meeting. From there, he went on to New York, never again returning to St. Louis.

CHAPTER TEN

New York, 1907–1910

S COTT JOPLIN ARRIVED IN NEW YORK in the summer of 1907, probably in July. Planning a short stay, he moved into a small rooming house, called the Rosalline, at 128 West 29th Street. His selection of this residence may have been influenced by the presence of a piano as well as by the boarding house's location.[1] The address was between Sixth and Seventh avenues, a block from Broadway, and right in the heart of the city. It was an area seething with action and excitement: it was the Tenderloin, the theater district, Tin Pan Alley, Black Bohemia.

The Tenderloin was New York's center for entertainment, legitimate and otherwise. Here were many of the major theaters, restaurants, saloons, gambling houses, and brothels. It was an area in which police and political corruption was rampant, where payoffs, grafts, and extortion were a way of life. For regular fees, illicit businesses could function undisturbed, restaurants and saloons could remain open beyond legal closing hours, liquor could be sold without restrictions. Even the most law-abiding business people were affected, for a failure to pay tribute would result in harassment. Should a boarding house operator refuse to pay corrupt police the going rate, the operator might be arrested for running a brothel, and the tenants could face arrest. These conditions were discussed and denounced in the daily newspapers, and committees were continually being formed to stamp out corruption. Reforms were inevitably short-lived.

These conditions were responsible for the Tenderloin's name. It was 1876 when police captain Alexander C. Williams was transferred to the area. He made the most of the opportunities to enrich himself, commenting, according to legend, that previously he had been eating chuck steak, but that since moving to the 19th Precinct he dined on tenderloin.[2]

Scott Joplin's New York: The Tenderloin and Midtown. (1) John Stark & Son, 127 East 23rd St.; (2) Madison Square Park; (3) Madison Square Garden; (4) Ike Hines's Professional Club, 118 West 27th St.; (5) Tin Pan Alley; (6) Nail Brothers' Restaurant and Cafe; (7) Scott Joplin's residence from 1907 to ca. 1911–13: the Rosalline, 128 West 29th St.; (8) Cafe Wilkins, a.k.a. Little Savoy Club, 253 West 35th St.; also 269 West 35th St.; (9) Seminary Music, Crown Music, and Ted Snyder Music, 112 West 38th St.; Jos. W. Stern Company, 102–104 West 38th St.; (10) Waterson, Berlin & Snyder, 1437 Broadway; (11) Scott Joplin's residence, ca. 1911–13 to 1915: 252 West 47th St.; (12) Marshall Hotel, 127 and 129 West 53rd St.; (13) Clef Club, 134 West 53rd St.

Not everything in the Tenderloin was illicit. Here also were major restaurants and theaters, most illuminated by the new electric lights. The Broadway theater district, soon be known as "The Great White Way," was already called "The White Light District."

Within a mile of Joplin's residence were at least twenty-six theaters, five

The Five Musical Spillers. Sam Patterson is in the center; William N. Spiller is on the right.

of them within two or three blocks on Broadway between West 28th and 30th streets. Although seating was generally segregated, Broadway theaters had already learned of the public acceptance of black musicals. *In Dahomey,* by Will Marion Cook and Paul Laurence Dunbar (with musical interpolations by many others), had been a hit in 1903 at the New York Theatre (44th and Broadway). In August 1907, Cole and Johnson's black musical *The Shoo-Fly Regiment,* conducted by Joplin's friend Joe Jordan, reopened nearby at the Bijou (Broadway and 30th).[3] In October, the Five Musical Spillers, an enlargement of the Spiller Musical Trio that Sam Patterson had helped form in 1906, was playing New York vaudeville. For Joplin, the success of his friends and associates in New York theater must have been an encouraging sign.

Also important to the Tenderloin life were the many eating and drinking establishments. Among them were several black-owned saloons and clubs that featured an "open floor," i.e., performance space for anyone who cared to use it. These early cabarets were the meeting places for major black entertainers and sportsmen, with the result that the area became known as "Black Bohemia." The clubs were also frequented by a white clientele that included both curious sightseers and professional performers out to observe and gather material on black performers. The first of these clubs and the one that set the tone that others followed was Ike Hines's Professional Club at 118 West 27th Street. This club was used as a setting in a novel by James Weldon Johnson, one third of the Cole and Johnson Brothers songwriting team and one of the intellectual and cultural leaders of black America. Though Hines had moved his club to Harlem by the time Joplin reached New York, Johnson's description remains valid and vividly captures the flavor of the Tenderloin's black cabarets.

We stopped in front of a house with three stories and a basement. . . . From the outside the house bore a rather gloomy aspect, the windows being absolutely dark, but within, it was a veritable house of mirth. When we had passed through a small vestibule and reached the hallway, we heard mingled sounds of music and laughter, the clink of glasses, and the pop of bottles. We went into the main room and I was little prepared for what I saw. The brilliancy of the place, the display of diamond rings, scarf-pins, ear-rings, and breast-pins, the big rolls of money that were brought into evidence when drinks were paid for, and the air of gaiety that pervaded the place, all completely dazzled and dazed me. . . .

The floor of the parlour floor was carpeted; small tables and chairs were arranged about the room; the windows were draped with lace curtains, and the walls were literally covered with photographs or lithographs of every coloured man in America who had ever "done anything." There were pictures of Frederick Douglass and of Peter Jackson, of all the lesser lights of the prize-fighting ring, of all the famous jockeys and the stage celebrities, down to the newest song and dance team. . . .

No gambling was allowed, and the conduct of the place was surprisingly orderly. It was, in short, a centre of coloured Bohemians and sports. These notables of the ring, the turf, and the stage, drew to the place crowds of admirers, both white and colored. . . .

There was at the place almost every night one or two parties of white people, men and women, who were out sight-seeing, or slumming. . . . There was also another set of white people who came frequently; it was made up of variety performers and others who delineated "darky characters"; they came to get their imitations first-hand from the Negro entertainers they saw here.

In the back room there was a piano, and tables were placed round the wall. The floor was bare and the centre was left vacant for singers, dancers, and others who entertained the patrons. . . .

There was a young fellow singing a song, accompanied on the piano by a short, thickset, dark man. After each verse he did some dance steps, which brought forth great applause and a shower of small coins at his feet. After the singer had responded to a rousing encore, the stout man at the piano began to run his fingers up and down the keyboard. This he did in a manner which indicated that he was master of a good deal of technique. Then he began to play; and such playing! . . . It was music of a kind I had never heard before. It was music that demanded physical response, patting of the feet, drumming of the fingers, or nodding of the head in time with the beat. The barbaric harmonies, the audacious resolutions, often consisting of an abrupt jump from one key to another, the intricate rhythms in which the accents fell in the most unexpected places, but in which the beat was never lost, produced a most curious effect. And, too, the player—the dexterity of his left hand in making rapid octave runs and jumps was little short of marvellous; and with his right hand he frequently swept half the keyboard with clean-cut chromatics which he fitted in so nicely as never to fail to arouse in his listeners a sort of pleasant surprise . . .

This was rag-time music, then a novelty in New York, and just growing to be a rage, which has not yet subsided.[4]

Other prominent black clubs in the area that featured popular entertainments, including ragtime, were Sam Moran's at 26th Street and Seventh Avenue, William Banks's Cafe and Restaurant at 206 West 37th Street, Walter Herbert's cafe at 331 West 39th Street, and, the most famous of them all, Cafe Wilkins at 253 West 35th Street, operated by Barron Wilkins. Of Wilkins, lyricist and singer Noble Sissle wrote: "It was around such places as Baron [sic] Wilkins . . . that the romance and the fantastic birth of the Blues and Rag Time in New York City were made possible." Among the pianists who worked for Wilkins were Eubie Blake, "One-Leg" Willie Joseph, and John Europe, brother of James Reese Europe. Wilkins also helped to foster the development of African-American music in New York by serving as treasurer for the black-owned publishing company of Gotham-Attucks Music Co.[5]

An important meeting place in Black Bohemia that had no entertainment was the Nail Brothers' Restaurant and Cafe on Sixth Avenue near 28th Street. John "Jack" Nail was an enterprising black businessman who, in addition to the restaurant, operated the Afro-American Realty Company. With the migration of blacks to Harlem beginning at this time, Nail made his fortune. Nail also had a keen interest in black show business and gained the trust of many black performers, acting as both a counselor and an informal banker for many. Regarding most performers as inherently irresponsible or naive in financial matters, he urged them to deposit their pay with him, to be disbursed later as they required. And for those who had no money but required an advance to travel to out-of-town performances, he was always willing to make a loan.[6]

An important black establishment that was outside the Tenderloin proper, but still nearby, was the Marshall House, a small but relatively expensive hotel and restaurant at 127 and 129 West Fifty-third Street. Several major figures, such as James Weldon Johnson and his brother Rosamond, took up residence there, but its major significance was as a meeting place of the most important celebrities in New York's black entertainment community. Among those who were frequently seen there were Ernest Hogan, Joe Jordan, James Reese Europe, Bob Cole, Will Marion Cook, Abbie Mitchell, poet-lyricist Paul Laurence Dunbar, Bert and Lottie Williams, George Walker, Theodore Drury (founder of a Negro opera company), and concert singer and composer Harry T. Burleigh. The main dining room had an elegant "palm court" setting and featured subdued string music. A club was later opened, as revealed in a series of five paintings by Charles Demuth. Several of these pictures depict instrumental combos of piano, banjo, and drums accompanying a singer.[7]

A block from Joplin's residence was the original Tin Pan Alley, West 28th Street between Broadway and Sixth Avenue. The Alley got its name from the numerous popular music publishing companies on this one street. At various times from the 1890s through early 1900s, this single block contained at least twenty-one music firms between the addresses of 36 (Feist) and 57 (New-York Music), housed in three-, four-, and five-story buildings. Included were such major firms as Harry Von Tilzer (37 and 42), Vandersloot (41), Shapiro-Bernstein (45), Remick (45), Witmark (49–51), Chas. K. Harris (51), Paul Dresser (51), and the black-owned company of Gotham-Attucks (42). Walking down this street, with music offices in almost every building, one was greeted with the continuous jangling of pianos on which songs were being sampled. In the summer, with the windows open, the jangle became a cacophony, like the banging of tin pans. Hence the name.

By the time Joplin moved to the area, the larger music companies had moved to more spacious offices, but most remained within easy walking distance. Thus, the whole area became known as "Tin Pan Alley," and eventually the term became generic for the popular music publishing industry.[8]

We know from a comment Joplin made in 1913 that he attended the theater frequently during his years in New York.[9] Presumably he used his central location in the Tenderloin also to attend concerts and socialize at the meeting places of Black Bohemia. He did spend time at the offices of music publishers. He must have made an early stop at the Jos. W. Stern Company, 102–104 West 38th Street, an important firm that was developing a list of African-American composers, among them the popular team of Cole and Johnson Brothers. Stern immediately bought two of Joplin's new rags, copyrighting *Searchlight Rag* on August 12, 1907, and *Gladiolus Rag* on September 24, 1907.[10] Since it was the practice of this firm to pay royalties,[11] Joplin's agreement probably included this provision.

Edward B. Marks, a partner of Jos. W. Stern, became personally acquainted with Joplin and wrote of him in later years. He noted that Joplin would socialize with other songwriters at the publishing office and overheard him commenting upon his lack of deserved recognition.[12]

Joplin also visited with Stark, whose office was a few blocks southeast at 127 East 23d Street, between Fourth Avenue and Lexington Avenue. Stark's ledger indicates Joplin was there on September 6, 1907, and received 150 copies of *Heliotrope Bouquet,* his collaborative work with Louis Chauvin. The copies were probably part of the payment for the piece; we do not know if the agreement included a financial payment and royalties.

A small company that Joplin visited was W. W. Stuart, 48 West 28th Street, on the original Tin Pan Alley block. There, he arranged for publication of Arthur Marshall's *Lily Queen,* registered for copyright on November 7. This piece lists Joplin as co-composer, but Marshall claimed in a letter to

Blesh and Janis that Joplin's name appears on the music only to promote sales.

> Joplin told me he had a party that would publish that piece of music so I let him handle it. But for him having any part in the composing, he did not. Now he was the more popular as a composer and that is why his name was mentioned in the writing of *Lily Queen.* I got about $50 in all for it at the time. I was living in Chicago.[13]

It is obvious that Marshall was unhappy with his negotiations with Stark the previous year, and he was looking for another publisher. However, he never returned to Stuart, so he could not have been pleased with the terms of this publication either. As for Marshall's claim that the piece is his composition alone, we suspect this is not the entire story. We learned from the Stark-Marshall letters of 1906 that Marshall was unable to notate his scores correctly. Joplin must have rewritten the manuscript. In the process, we believe he smoothed out some of the rough edges and awkward spots evident in Marshall's other scores. The composition is therefore Marshall's, but with editorial assistance from Joplin.

A final new publisher that Joplin made contact with in 1907 was Joseph M. Daly of Boston, who registered for copyright his *Rose Leaf Rag* on November 15. Whether this publication means that Joplin was in Boston at the time remains an open question, but this is the only Joplin work that Daly ever issued.

Gladiolus Rag, one of the Stern publications of 1907, is clearly based on the *Maple Leaf* prototype, although it might be more accurate to say it is based on *Leola,* making it a third generation copy. The opening strain, with its wide range of a thirteenth, its many changes in direction, and its use of the blues third takes the embellished melodic concept of *Leola* one step further (Ex. 10-1a). The second strain is also in the *Maple Leaf* mold, but the first trio theme strikes out into new realms of chromatic ragtime, suggesting even the chromatic progressions of Chopin (Ex. 10-1b). The final strain (Ex. 10-1c) recalls the insistent, driving rhythmic mode first introduced in 1902 with *Elite Syncopations,* so effective as a closing section. Enhancing its pungent character are dissonant harmonies created by the moving inner voices and the melodic base line. In the final phrase of this strain, he again uses a favored attention-getting device of dropping into the flatted-sixth, here showing his recognition of the harmonic theory by spelling it not as an A-major chord, but correctly as a B-double-flat chord (Ex. 10-1d). Though clearly derived from *Maple Leaf, Gladiolus* is no pale copy. While working with the same patterns, Joplin created a distinctively new work. It was even the Joplin rag that Joe Lamb liked the best.[14]

Rose Leaf Rag is another outstanding ragtime composition. It would

Ex. 10-1. Scott Joplin, Gladiolus Rag *(1907): (a) A1-8; (b) C5-8; (c) D1-4; (d) D13-16.*

seem that after the brief hiatus of 1905 to the middle of 1907, when Joplin published few rags, he was now back in the ragtime groove, producing masterpieces of the genre.

Rose Leaf has a most unusual beginning in which the left hand suggests

Ex. 10-2. Scott Joplin, Rose Leaf Rag *(1907): (a) A1-5; (b) C1-4; (c) D1-4; (d) D12-16; (e) analytic reduction showing inner voice descent in D12-16.*

a light polyphony, moving parallel and contrary to the right. The anticipated ragtime bass is absent until measure 5, bringing with its entrance a resolution of the tension generated by the first four measures (Ex. 10-2a). The opening measures of the third strain, while adhering to the ragtime convention of bass-chord alternations, repeats a single bass note as a four-measure, tension generating pedal point against the changing harmonies of the right hand (Ex. 10-2b). The finale is once again of the type that repeats incessantly a single, driving rhythm against changing harmonies and a melodic base line (Ex. 10-2c). The final phrase features another inner-voice descending chromatic fourth, this time extended to the augmented fourth, E-flat down to A (Exx. 10-2d, c).

Joplin was not relying exclusively on publishers for income. Apparently his work for hire during the past few years was sufficiently lucrative for him to continue the practice, and in November he advertised in the *American Musician and Art Journal,* the magazine that had praised him so highly the previous June:

SCOTT–JOPLIN

If you want a first-class, up-to-date rag-time arrangement made send to the composer of that famous rag, "MAPLE LEAF,"

SCOTT–JOPLIN

128 West 29th Street, New York.[15]

The advertisement does not seem to have been effective, for only one arrangement is known to have been published subsequent to this notice, and that was in 1911. But he did bring himself to the attention of the magazine editors again, and in December another glowing article appeared:

SCOTT JOPLIN

Apostle of High Class Ragtime Is a Serious Worker.

Ragtime of the better order still has a hold on the public. Of the higher class of ragtime Scott Joplin is an apostle and authority. Joplin doesn't like the light music of the day; he is delighted with Beethoven and Bach, and his compositions, though syncopated, smack of the higher cult.

"Why do you call it ragtime?" some one asked him long ago.

"Oh!" replied Joplin, "because it has such a ragged movement. It suggests something like that."

But those who play Joplin's music carefully will find a suggestion of profound thought in it. His melodies are intensely sentimental and distinguished from most other modern creations by being new. Their harmonic treatment is masterly and as he turns out the finished products they appeal to the cultured as well as the amateur, and will bear repetition without number, growing more popular with age. Joplin is stopping in New York for the present at 128 West Twenty-ninth street.[16]

Here, again, we have confirmation of Joplin's aspirations to be associated with high art, and of the recognition in certain quarters of white music journalism that he was a composer with special qualities.

It was probably in late 1907 that Joplin and Joseph F. Lamb had their momentous meeting. Lamb, a young white man, ragtime enthusiast, and amateur pianist and composer, had published several works, including one rag—*The Lilliputian Bazaar*—with the Toronto publisher Harry H. Sparks. Lamb recognized the superiority of the rags Stark published and dreamed of being part of "The House of Classic Ragtime." Stark, however, rejected whatever Lamb submitted. Lamb's meeting with Joplin changed that.

Lamb was at the store on 23rd Street, looking through music, and mentioned to Mrs. Stark how much he admired Joplin's compositions. She asked if he would be interested in meeting Joplin, and when he answered affirmatively, she pointed to him, sitting in a corner of the store. Lamb and Joplin spoke for a while and strolled down the street to Madison Square Park, on 23rd between Fifth and Madison. Joplin, learning that Lamb was also a ragtime composer, invited the younger man to visit him.

A few evenings later, Lamb visited Joplin at the boarding house where he was staying and demonstrated several rags he had composed: *Sensation, Dynamite,* and *Old Home Rag.* The first strain of *Dynamite* originally ended with a two-octave, chromatic ascent in both hands. On hearing it, Joplin suggested it would work better with the left hand descending, in contrary motion to the right hand. Lamb followed the advice and rewrote the passage. This is the only known account we have of compositional advice that Joplin gave. Slight as it is, it demonstrates his immediate grasp of a musical context and his quick perception of alternatives.[17]

When Lamb performed *Sensation,* one of the boarders walked over to Joplin and asked if it was one of his works. Joplin said it was by Lamb, and the man put his hand on Lamb's shoulder and exclaimed, "That's a regular Negro rag." This was gratifying to Lamb, for it indicated to him that he had succeeded in composing a quality rag.[18]

Joplin then spoke to Stark about publishing *Sensation,* suggesting that he add his own name as arranger to encourage sales. (In support of Lamb's contention that he alone was the composer of *Sensation,* we note that Stark excluded Joplin's name when advertising the work in later years.) Lamb says that it was several weeks later that the rag was published, but it first appears in Stark's ledgers on May 18, 1908.[19] Lamb received an immediate payment of $25, and a month later, after the first thousand copies were sold, another $25.[20] Lamb made no mention of a royalty agreement, but the arrangement seems better than that offered in 1906 to Arthur Marshall.

Lamb offered Stark other rags, which he accepted for publication. Through 1919, Stark published a dozen of Lamb's rags, and it was by these publications that Lamb's name became known to ragtimers. None of the working ragtimers knew Lamb personally, for he did not associate with

Joseph F. Lamb.

professional musicians. Except for a few months in 1910 when he worked as an arranger for the J. Fred Helf Company on West 37th Street, Lamb did not make music his livelihood. Nevertheless, on the basis of these Stark publications, Lamb because known, along with Joplin and James Scott, as one of the "Big Three of Classic Ragtime."[21]

Lamb and Joplin developed a close personal friendship, close enough for

Lamb to bring his girlfriend to visit the older composer.[22] In later years they also collaborated in composing a rag.

The New York Sheet Music War

 As the center of the popular sheet music publishing industry, New York was also the scene of the greatest price competition. As a basis of comparison, we note that the *Maple Leaf Rag* contract prohibited sales at less than twenty-five cents. In St. Louis in 1903, Joplin's *Elite Syncopations, Weeping Willow,* and *Something Doing* were selling for twelve cents. In New York in 1904, prices had gone even lower: one major department store offered such pieces as Tom Turpin's *St. Louis Rag* (1903) for seven cents.[23] By October of 1907, soon after Joplin's arrival to the city, the price competition had heated up to a new intensity.

Macy's department store initiated this battle by sharply discounting prices on books and music, a policy that was challenged legally but upheld by the Supreme Court. Because of the large volume of business at Macy's, the store was able to induce publishers to sell it music at greatly reduced prices, as low as six cents for sheets that usually wholesaled for twenty-three cents.

In response, a consortium of major publishers, including Witmark, Feist, Mills, Harris, and Haviland, established the American Music Stores, an organization that set up its own stores and managed music departments in other large stores. Macy's continued its highly competitive policy and advertised to meet any price.

The consortium decided to use Macy's latest ploy to break the retailing giant. On October 11, 1907, the consortium offered to sell its biggest hits—music marked at fifty and sixty cents—for one cent at their 14th Street store. They also had people to go to Macy's to demand that the one-cent price be matched. Macy's finally relented. Though the most devastating part of the war ended, some smaller stores continued to discount prices. Soon Woolworth's would be selling all of its sheet music stock for ten cents.[24]

I have no specific information on how this war affected John Stark. The music most directly involved was popular song, whereas Stark's catalogue emphasized instrumental pieces. Still, the war altered the public perception of the price one paid for sheet music, and this change probably had a depressing effect on Stark's prices.

The *American Musician* article of December 13, 1907, suggests, in its last sentence, that Joplin's move to New York was intended to be brief. By

January, he had apparently changed his plans, for a columnist in the same magazine noted:

> Nobody has heard Scott Joplin say anything about returning to St. Louis. The exponent of high class rag has found out it is just as pleasant to be at the receiving end of good fortune in this old burg as it is anywhere else.[25]

Things were apparently going well. Joplin was making new and profitable contacts with publishers: he was getting good press; and, though he earned nothing from the infant recording industry, recordings brought him additional notice. The previous year, the banjo virtuoso Vess Ossman made two recordings of *Maple Leaf,* one with accompaniment by the famous Prince's Orchestra.[26]

He also looked good. "Chauf" Williams, a musician acquaintance from St. Louis, reported that it was about this time that he saw Joplin in New York. The composer, said Williams, "was dressed up like a Fancy Dan of ragtime, sporting diamonds."[27]

Early in 1908 the *New York Age,* the city's major black newspaper, started an entertainment page under the editorship of Lester A. Walton. Walton was an occasional performer in black theater and variety shows, but was more important as an organizer within the black theatrical world and as an opinion-maker. He strongly supported African-American musical dramatic endeavors, especially those that reflected well upon the race. Although disparaging stage portrayals of blacks were commonplace in the acts of black performers, as well as white, he vehemently argued against such depictions. He also urged that black entertainers, theatrical staff, and patrons demand equitable treatment from theater managers.

He probably saw in Joplin an individual whose ideals matched his own and decided to give him good press. On March 5, Walton published his first of several articles about Joplin, here revealing some of Joplin's reasons for moving to New York:

> Since syncopated music, better known as ragtime, has been in vogue, many Negro writers have gained considerable fame as composers of that style of music. From the white man's standpoint of view he at present is inclined to believe that after writing ragtime the Negro does not figure.
>
> There are many colored writers busily engaged even now in writing operas. Music circles have been stirred recently by the announcement by Scott Joplin, known as the apostle of ragtime, [that he] is composing scores for grand opera.
>
> Scott Joplin is a St. Louis product who gained prominence a few years ago by writing the "Maple Leaf Rag," which was the first ragtime instrumental piece to be generally accepted by the public. Last summer he came to New York from St. Louis and it was the opinion of all that his mission was one of

placing several of his ragtime instrumental compositions on the market. The surprise of the musicians and publishers can be imagined when Joplin announced that he was writing grand opera and expected to have his scores finished by summer.

From ragtime to grand opera is certainly a big jump—about as great a jump as from the American Theatre to the Manhattan and Metropolitan Opera Houses. Yet we believe that the time is not far off when America will produce several S. Coleridge Taylors who will prove to the public that the black man can compose other than ragtime music.

The composer is just in his thirties and is very retiring in manner. Critics who have heard a part of his new opera are very optimistic as to his future success.[28]

News of Joplin's brother Robert was also appearing in the press; he had become extremely active in several areas of black theater. In 1906 he led a company of sixteen performers in an "extravaganza" called "Cuban Belle." In December they were in Indianapolis and were heading for Boston. Late in 1907 he become stage director at the white-owned but black-oriented Lincoln Theatre in Knoxville, Tennessee. There he directed shows (a new bill every Monday and Thursday), wrote complete shows (along with the music), and appeared frequently as a singer, dancer, and comedian. In February 1908 a notice indicated, "R. B. Joplin called the ragtime dance in a way that had the audience spellbound." This was probably Scott Joplin's *Rag-Time Dance,* for at the end of the item Robert sent regards to his brother. A subsequent notice indicates Robert's prowess as a comic: "R. B. Joplin is sending them out howling." In March he moved on to another locale, becoming stage director at Clark's Theatre in Columbus, Ohio.[29]

If Scott Joplin's professional life was prospering and he was receiving good news from his brother, his personal life was dealt another blow with the distressing news of Louis Chauvin's death in Chicago in March 1908. Joplin immediately wrote of it to Arthur Marshall, who had returned to Sedalia, and to Chauvin's boyhood friend and former partner Sam Patterson, who was then on the road with the Musical Spillers.[30]

Professionally, Joplin started the year 1908 with a new work that he published himself: *School of Ragtime.*[31] This is not a composition, but an instruction manual presenting four 4-measure and two 8-measure exercises that dissect the beat to demonstrate the precise manner of playing ragtime syncopations.

The concept of a ragtime instruction manual was not new. The first was Ben Harney's *Ragtime Instructor* in 1897, but this did nothing more than present examples of well-known tunes in ragtime arrangements. Axel Christensen's *Rag-Time Instruction Book for Piano* (1904, with revisions in 1906, 1907, and subsequent years) analyzed the syncopations in a manner similar to what Joplin was to do. It may have been Christensen's success with these manuals that prompted Joplin to try his hand at it. He admitted to Joe Lamb, however, that he did not anticipate much success with it.[32]

The musical interest in *School of Ragtime* is greater than that of other ragtime manuals, for Joplin chooses rich harmonies and varied rhythms. Pedagogically, however, Joplin's examples have little to offer except on the most fundamental level of reading syncopations. It is unfortunate that he chose to make the manual so short, for a longer work might have revealed more of his thinking about the genre. It is paradoxical, also, that in such a brief work he includes elements that are not characteristic of his usual style. Only one exercise, the fifth, has the typical ragtime bass of octave-chord alternations.

Of greater interest than the music are Joplin's comments. In the introductory paragraph he defends ragtime and presents the thesis that there are qualitative distinctions in ragtime, that there is a "ragtime of the higher class":

> What is scurrilously called ragtime is an invention that is here to stay. That is now conceded by all classes of musicians. That all publications masquerading under the name of ragtime are not the genuine article will be better known when these exercises are studied. That real ragtime of the higher class is rather difficult to play is a painful truth which most pianists have discovered. Syncopations are no indication of light or trashy music, and to shy bricks at "hateful ragtime" no longer passes for musical culture. To assist amateur players in giving the "Joplin Rags" that weird and intoxicating effect intended by the composer is the object of this work.

The paragraph seems to reflect wishful thinking. While Joplin himself had received a certain amount of recognition from musicians of "the higher class," it was not the case that ragtime had been accepted as a legitimate component of musical culture. Ragtime was still considered a low-grade musical expression, and there was much more criticism of it than signs of acceptance. Despite his words, Joplin recognized this. His remarks appear to be replies to the criticisms that appeared regularly: that ragtime was just a passing fad; that it was inherently inferior, lacking in any musical merit; and that there was something cheap about music that relied upon syncopations.

The aim of the manual, though, was not to argue the state of ragtime; it was "to assist amateur players in giving the 'Joplin Rags' that weird and intoxicating effect." To achieve this effect, he points out, it is necessary to execute the music precisely as written:

> It is evident that, by giving each note its proper time and by scrupulously observing the ties, you will get the effect. So many are careless in these respects. . . .
> . . . We wish to say here, that the "Joplin ragtime" is destroyed by careless

or imperfect rendering, and very often good players lose the effect entirely, by playing too fast. They are harmonized with the supposition that each note will be played as it is written, as it takes this and also the proper time divisions to complete the sense intended.

These sentiments are consistent with his frequent admonition to play rags slowly. As a composer who took extreme care in everything he wrote, he was clearly opposed to any rendition that might compromise his conceptions.

He immediately arranged for others to market the manual, and it was first advertised in the December 22, 1907, issue of the *American Musician and Art Journal:* "Get Scott Joplin's 'School of Ragtime' for piano and learn to play ragtime correctly. For sale by Enterprise Music, New York."[33] Enterprise Music, at 46 West 28th Street, was a music jobber, that is, a distributor to music stores. The Stark ledgers show that Enterprise was an occasional purchaser of Joplin rags, but was also associated with A. H. Goetting Company, a jobber that was Stark's biggest customer in 1907–08.

The advertisement appeared four more times in the *American Musician,* the last being in the February 28th issue. Stark was also selling the manual, listing it in his ledgers on February 25, 1908, and advertising it on the backs of music as late as 1911.[34] A third company, Crown Music, placed an advertisement similar to Enterprise's in the *New York Age* on February 20 and 27, 1908.[35] The importance of the Crown Music connection was to be immeasurable.

Crown Music, 12 East 17th Street, was another Stark customer and a heavy purchaser of Joplin rags. It was owned by Henry Waterson, treasurer, and Herman Snyder, who was responsible for the daily operations.[36] On April 18, 1907, Crown formed Seminary Music, a subsidiary music publishing company owned nominally by Mary Waterson and Mary Snyder, but under the management of Herman Snyder.[37]

By the spring of 1908, both Crown Music and Seminary Music had moved to a small building at 112 West 38th Street, which became known as "The Crown Music Building." On July 10, 1908, they opened a third firm in the same office: Ted Snyder Music. Ted Snyder, presumably related to Herman and Mary Snyder, was the president, and Henry Waterson was treasurer and manager.[38] In addition to running the new company, Ted Snyder was a songwriter who was becoming increasingly popular.

In February 1908, Stark registered *Fig Leaf Rag* for copyright, which was to be his last Joplin publication for several years (although he did publish Lamb's *Sensation* three months later, with Joplin listed as "arranger"). From then on, beginning in the early spring of 1908, Seminary Music became Joplin's major publisher. The first work was *Sugar Cane,* copyrighted in April 1908. It is another rag in the *Maple Leaf* pattern, this time with all four strains in common (Ex. 10-3). Next was *Pine Apple Rag,* registered for copyright by Seminary on October 12, 1908, but published several months

Ex. 10-3. Scott Joplin, Sugar Cane. A Ragtime Two Step *(1908), A1-8.*

earlier; Stark's ledger indicates he had obtained three copies from Seminary and sold them on August 29, 1908.

The dedication of *Pine Apple*—"to the Five Musical Spillers"—is not surprising, given Joplin's close friendship with Sam Patterson and William Spiller. In addition, the dedication must have been made with the anticipation that the Musical Spillers, an increasingly popular vaudeville group, would perform the music. This is what happened, as reported by Isabele Taliaferro Spiller.

Isabele, wife of William Spiller, joined the group as a performer in 1912. Sometime between 1913 and 1916 she and her husband visited Joplin, who spoke of how his music should be performed. The following is from her notes:

> I accompanied Spiller to Scott Joplin's apartment in New York City and heard Joplin explain to Spiller exactly . . . how Pineapple Rag, and Maple Leaf Rag were to be played. It was fascinating to me because it was the first time I had ever heard a composer explain in detail what he wanted done. . . .
>
> The Pineapple Rag was played by the Musical Spillers on two xylophones and on marimba accompanied by the Theatre Orchestra. These instruments attached to each other covered the stage. There were two musicians at each instrument. The first number played was Raymond Overture followed by Pineapple Rag.
>
> At this early date Spiller was doing what they said could not be done, playing classical and "ragtime" as it was called then. When managers told him to cut the classical number out, he just did not and they were amazed to hear the applause received after the overture. "Pineapple Rag" was such a favorite that every time we took it out we had to put it back.[39]

Pine Apple Rag is a good candidate for success, for it is a rousing, rollicking piece. It also has unusual features for the attentive listener to appreciate. The second strain (Ex. 10-4b) is built upon a dominating single rhythmic idea, taking the germ motive from the first strain (Ex. 10-4a). In contrast, the third strain is explicitly bluesy and forgoes the usual ragtime bass (Ex. 10-4c).

Ex. 10-4. Scott Joplin, Pine Apple Rag *(1908): (a) A13-14; (b) B1-4; (c) C1-4.*

Echoes of Scott Joplin, Part 7: Sugar Cane

We saw in Chapter Six that Al. Morton took the last strain of *Elite Syncopations* for the finale of his *Fuzzy Wuzzy Rag* (Ex. 6-10). Morton was equally blatant in stealing the first strain of *Sugar Cane* for the same piece. Since *Sugar Cane* is itself based on *Maple Leaf*, *Fuzzy Wuzzy* is another third generation copy (Ex. 10-5).

John Stark was understandably bitter over Joplin's defections to other publishers. He expressed his feelings candidly in his ledgers, as witnessed by the following passage. It is uncertain when Stark wrote the passage, but since it is a few pages after some comments on the *School of Ragtime,* I would date

Ex. 10-5. Al. Morton, Fuzzy Wuzzy Rag, *A1-8.*

it in early 1908. This may have been a draft for advertising copy, but Stark wisely chose not to use it.

> St. John [Clarence St. John, composer of *Cole Smoak*] seems to be crowding Joplin off the perch. We are sorry to see it but Joplin is verging to the sear and yellow leaf. It is the experience of all composers of popular music that from two to ten pieces sap the fountain on inspiration and pump the unhappy muse into innocuous desuetude. Joplin is now only living in the dream of Maple Leaf. No one will perhaps ever equal that ingenious piece nor his Sunflower or Cascades. His labored effort is but a rehash of these numbers which no self-respecting publisher would print.
>
> Joplin's case is pitiful—When he hawks a manuscript around and finally sells it for a few dollars the next publisher he meets tells him, "Why, I would have given you $500 for it." This torments him and keeps him scheming to beat his last publisher.
>
> We have a few manuscripts of Joplin's which we bought before the fuel of his genius was exhausted and which we will bring out from time to time as a reminiscence of Joplin's better days and the public can depend in the future as in the past that our catalogue will contain the highest class ragtime that this country affords, as also other styles of music.[40]

On a preceding page, Stark had some additional thoughts:

> Where he attempts originality, he is strained and mechanical. And where he rehashes the old ones the public will not stand for it.
>
> Joplin's pieces are difficult. While his genius for invention lasted it would pay to study and practice them, but when the life ran out and left them dry mechanical rubbish it does not justify the labor.[41]

Stark was not entirely wrong in observing that Joplin reused earlier rags. We have seen how he composed a number of works on the model of *Maple Leaf: Cascades, Leola, Gladiolus,* and *Sugar Cane.* But to suggest that these were mere "rehashes" overlooks the striking new elements Joplin added which makes each piece a fine rag in its own right.

Stark never published these views directly, but a year later he did plant seeds of doubt while denying that the seeds originated with him. In an interview, he said:

> What are our most popular pieces? I will tell you. "Maple Leaf" rag is the greatest ever written. It is by Scott Joplin. It has been asserted that Joplin's music has been pumped dry, and his later efforts are but a rehash of his famous "Maple Leaf." But this is not so.[42]

In addition to reflecting Stark's private feelings at the time, the passages suggest what Joplin's experience might have been. Publishers wanted his music, and each one promised him more than the last. He moved from one publisher to another in his search for a "better deal," but was invariably disappointed.

Classic Ragtime

 "The Big Three of Classic Ragtime" is a group label applied to Scott Joplin, Joseph F. Lamb, and James Scott. This is a status applied after the fact with the 1950 publication of *They All Played Ragtime.* Yet, the grouping was natural. Along with their individual excellence, all three were linked by a stylistic unity and by association with John Stark.

"Classic Ragtime" became a rallying call for the growing number of ragtime fans in the 1960s. As they formed into organizations, they invariably stated the aim to "preserve Classic Ragtime." Though the term has never been precisely defined, there is general agreement that Joplin, Lamb, and Scott are "classic ragtimers." In this sense, the term signifies a ragtime in which artistic quality took precedence over the demands of the marketplace; a music that was composed with the integrity, seriousness, and skill that befits a classical art. (The polarity of artistry and commercialism may well be questioned, but it is an assumption of "classic ragtime" proponents.)

Quality alone does not confer "classic" standing. Ragtime had many serious, skilled composers who are not considered "classic." Such composers as Eubie Blake, James P. Johnson, Jelly Roll Morton, and Ford Dabney, to name just a few, were stylistically different from "The Big Three" and were not published by Stark. It is therefore easy to exclude them from the category.

But others were gradually included, particularly Midwesterners, thereby muddling the issue. Rudi Blesh, who deserves credit for generating interest and support for ragtime and who was the major promoter of the "classic" concept, stretched this concept until it encompassed too much. He included figures with such diverse styles as Tom Turpin, Artie Matthews, Charles Hunter, and Percy Wenrich. None of these composers was associated with Stark, and the only thing they had in common with each other was Midwestern roots. These were among other questionable inclusions in Blesh's anthology *Classic Piano Rags*.[43] This emphasis on a geographical criterion rather than stylistic consistency makes the fit uncomfortable for Lamb, a lifelong Easterner.

Even if the term lacks a consistent application and the nicety of a precise definition, it has historical justification. We have seen that Joplin let it be known that he preferred "classical" music and wanted to be considered a serious artist. His efforts in opera certainly indicate this, as do such titles as *The Sycamore. A Concert Rag* (1904) and *Bethena. A Concert Waltz* (1905). Even more to the point, a 1911 article in the *American Musician* recognized his connection to a "classic ragtime"; "Scott Joplin, well known as a writer of music, and especially of what a certain musician has classified as 'classic ragtime' . . ."[44]

"There is no hint as to the identity of this "certain musician." The context of the remark eliminates Joplin as a possibility. It may have been Stark, even though he was not really a "musician." Stark was known to the staff of the magazine and was quoted in discussing his rag publications as "classical." For example, in an interview in 1908, in which he was at his bombastic best, he made the point in a poem:

> Some years ago, or more, 'tis said, the subtle harpist Orpheus,
> Played Hades' plutocratic Boss into the arms of Morpheus,
> Then sneaked away his Eurydice, but as he left perdition,
> He lost the radiant maiden by not heeding the condition;
> Had Orpheus played these 'classic rags,' he'd pulled the lassie
>   through—
> Not only Eurydice, but all the flabbergasted crew.
>   . . .
> "Then your rags border on the classical?" I ventured.
> "Yes, they are both classic and popular, profound and
> simple.[45]

We do not know whether the "classic ragtime" concept originated with Joplin or with Stark, and, due to the difficulty of dating advertisements on the backs of music, we can not pinpoint when it emerged. But it was an idea that was dear to Stark, and he never missed an opportunity to compare his ragtime catalogue favorably with classical music and, at the same time, to decry commercial imitators. An advertising flyer from around 1906 presents his arguments:

> We have been somewhat emphatic in describing our rag catalog mainly to make wide the chasm between this and the flood of slush on the market.
>
> So thrilling, so ingenious, and withal so profound are our rag creations that imitations are fairly rained onto the market by the commercial composers, and the majority of people have heard only these weak solutions, dilutions and shadows, "having the form but lacking the power" and it is little wonder that many good people have taken a stand against so-called ragtime.
>
> We have the most sincere respect and admiration for the great masters. No one hears more of their really great productions, and few enjoy them more than does our "bunch." In fact one of our number has . . . struggled with the *profound* under teachers in the Royal High School in Berlin, and has slaved for two years on finishing studies with Moszkowski, in Paris . . . It is from this experience that we are able to judge our manuscripts, and we think that you can see why "our ragtime is different."

In the last paragraph, Stark was referring directly to his daughter Eleanor, who had studied in Berlin and Paris with Moszkowski. But he may also have had Joplin and James Scott in mind as ragtimers who appreciated "the great masters." Joplin had spoken of this admiration, and Scott's music demonstrates his familiarity with classical music. In particular, the A strain of Scott's *Rag Sentimental* (1918) seems to be drawn from the opening of Beethoven's Piano Sonata in D minor, Opus 31, No. 2.[46]

A decade later Stark was referring to the same issues. In an advertisement in 1915, he wrote:

> We are the storm center of high-class instrumental rags. The whole rag fabric of this country was built around our "Maple Leaf" "Sunflower" "Cascades" "Entertainer" "Frog Legs" Etc.
>
> We have advertised these as classic rags, and we mean just what we say. They are the perfection of type. . . . They have lifted ragtime from its low estate and lined it up with Beethoven and Bach.[47]

Despite Stark's efforts, the idea of "classic ragtime" did not attain widespread currency. Strangely enough, the term was appropriated by a composer who surely knew Stark's publications and, had she not been published by her own father, could have fit very well into the Stark catalogue. This was May Aufderheide, a composer of a half-dozen fine rags and a self-proclaimed writer of "Classic Ragtime."[48]

Though the term *Classic Ragtime* is inexact, it remains useful. It describes the intention of John Stark with his publications, as well as the pinnacle reached by a select few composers.[49]

We have already seen that Arthur Marshall was not entirely correct in his assessment of Stark's supposed fairness with composers. To whatever extent Stark did provide royalties, his policy changed, possibly because of the New York sheet music price wars of 1907.

According to Joe Lamb, Stark's business faltered, and in 1909 he informed Joplin that he would no longer pay royalties on new works. Relations between the two were already strained, as we noted in Stark's negative musings in his ledger. In addition, Stark had refused to consider Joplin's forthcoming opera for publication. This final announcement completed the break, and thereafter Joplin refused to publish with Stark.[50] (Lamb retold this account from memory in 1949 and could have been off by a year or so. If this policy change and break had occurred in late 1907 or 1908, it would explain Stark's ledger entry and Joplin's publications with Seminary in 1908–09.)

An incident underlining the rupture between Joplin and Stark was told by Joe Lamb. Around 1909, Lamb and Joplin collaborated on a rag, Joplin writing the first two strains and Lamb the last two. Lamb brought the rag to Stark to be published, but Stark refused, saying he would publish it only if Joplin's name was taken off. Naturally, Lamb could not agree to that condition. The work was never published and is presumably lost.[51]

Joplin had established himself with Seminary Music by this time and did not have to rely on Stark. Presumably, Seminary had agreed to a better financial arrangement than Stark. Otherwise, there would have been little reason for Joplin to publish so many works with the firm. In 1909, he published exclusively with Seminary, having six works issued. The first of these, registered for copyright February 23, was *Wall Street Rag,* a piece that may reflect on employment he had at the time. According to unconfirmed but plausible information, Joplin had been playing piano in a restaurant, possibly the historic Fraunces' Tavern, in New York's financial district.[52]

Wall Street Rag is a highly unusual work. First, it has a programmatic narrative that assigns a different mood to each of the four strains:

A—"Panic in Wall Street, Brokers feeling melancholy";
B—"Good times coming";

C—"Good times have come";

D—"Listening to the strains of genuine negro ragtime, brokers forget their cares."

Modest as this narrative is, it reveals several points about Joplin's perception of ragtime. 1) Not all ragtime is "genuine"; only the African-American creation is authentic. 2) Genuine ragtime is a happy music, endowed with the power to alter moods.

The obvious musical device to suggest melancholy is the minor mode. Joplin avoids the obvious in his A strain. Instead, while in the usually "happy" key of C major, he introduces ambiguities and dissonances. Over a C pedal point he presents tonal vagueness with diminished chords (d-sharp and d-natural), a prominent repetition of the dissonant and tonally ambiguous tritone interval of c–f-sharp, and a suggestion of the minor mode with a flatted sixth degree (a-flat), (Ex. 10-6a).

Ex. 10-6. Scott Joplin, Wall Street Rag *(1909): (a) A1-4; (b) D1-4.*

The final strain is dominated by unprecedented dissonances. These are off-beat, discordances—at times, actually tone clusters—placed in a high register (Ex. 10-6b). The programmatic intent is probably to suggest the twanging sound of ragtime banjo strumming.

The brief habañera bass in *Wall Street Rag* links the piece to Joplin's next publication, *Solace. A Mexican Serenade* (registered for copyright April 28, 1909), which is a habañera throughout (Ex. 10-7a). The similarities between ragtime and Latin-American dance rhythms had long been noted. As early as 1897, Ben Harney linked ragtime with Mexican music, and several contemporaneous writers had also commented upon it.[53]

The first strain of *Solace* presents a theme that was not original with Joplin. The same theme appeared four years earlier in Will H. Etter's *Whoa!*

Ex. 10-7. (a) Scott Joplin, Solace. A Mexican Serenade *(1909), A1-4; (b) Will Etter,* Whoa! Maud, *A1-4.*

Maud, published in Galveston, Texas (Ex. 10-7b). A comparison between the two treatments reveals, once again, how Joplin stood out among his contemporaries. Etter's piece goes clomping along, failing to bring out the potential of this lightly chromatic melody. In Joplin's hands, the theme becomes a sinuous delicacy with graceful melodic and harmonic turns.

As we have seen in previous chapters, subsequent themes in *Whoa! Maud* were drawn from earlier Joplin works: Etter's C theme resembles the A of Joplin's *Original Rags;* Etter's D resembles the D of *Maple Leaf Rag.* It is not surprising that Etter knew Joplin's music, but how did Joplin come across Etter's melody? Were the two composers personally acquainted? Did Joplin knowingly copy from Etter? Did they both draw upon a common musical source, perhaps Texas or Mexican? We have no answers to these questions, and they may be unanswerable, but researchers should keep alert for clues that might shed light on the issue.[54]

The four other works from 1909 also had unusual features. Two registered for copyright on May 11 were *Pleasant Moments* and *Country Club.* *Pleasant Moments,* like *Bethena,* is a ragtime waltz, though not as ethereally beautiful as the earlier work. Still, it is far from conventional with a number of interesting features, such as the dialogue between bottom and top lines in strain C (Ex. 10-8).

Ex. 10-8. Scott Joplin, Pleasant Moments. Ragtime Waltz *(1909), C1-4.*

Two rags registered for copyright on October 30 are *Euphonic Sounds* and *Paragon Rag*. This date is probably close to the time of publication, for a newspaper item from November 20 indicates that *Paragon* had just been completed.[55]

The cover of *Euphonic Sounds* reproduces the photograph that first appeared in the Rosenfeld article of 1903 and then on the *Cascades* cover of 1906. The photograph sufficed for the earlier usages, but was too old for a 1909 publication. Joplin, with a stiff, high collar, now looks out-of-fashion when compared with the more stylishly dressed men depicted in the cover drawing of *Euphonic Sounds*. It is my guess that it was this contrast that prompted him to have another old photograph touched up and printed.

The touch up, which puts Joplin into a low, rounded collar and a narrower tie, was clumsily executed. The drawing was done on the negative, so that the outlines show up in white on the print. We can see by the inscription—"King of Ragtime. Scott Joplin. Composer of 'The Maple Leaf Rag'"—that he used the print as a publicity photo. This out-of-focus, botched photograph is the last image we have of him.

Musically, *Euphonic Sounds* is an unconventional work. Perhaps it is significant that Joplin refers to it not as a rag, but as "A Syncopated Two Step," for he avoids the left-hand octave-chord "oom-pah" alternation pattern that is so closely associated with ragtime. Instead, the left hand is unusually busy with scalar and chordal patterns (Ex. 10-9). It would seem that he was trying to break with some of the stereotypical patterns and to expand the language of ragtime.

Ex. 10-9. Scott Joplin, Euphonic Sounds. A Syncopated Two Step *(1909), A1-4.*

Paragon Rag was dedicated to the C.V.B.A., the Colored Vaudeville Benevolent Association. The C.V.B.A was a fraternal organization founded in June 1909 and had both social and welfare purposes, providing for its members a club room and sick and burial benefits. Its charter members included major vaudeville and theatrical figures, including William N. Spiller and songwriter-performers Chris Smith and Tom Lemonier. Columnist Lester Walton and vaudevillian Bob Slater, a close associate of Joplin's, were members of the Board of Trustees.[56] The dedication on *Paragon Rag* was probably to commemorate Joplin's election to membership, and he was to become an active member of the organization.

The last known photograph of Scott Joplin.

The *St. Louis Post-Dispatch* noted in 1909, on information provided by William Stark, that Joplin was not only publishing his rags in New York but was studying music there.[57] If the report is correct, Joplin's studies must have been to enhance the writing of his opera. As noted in the *Age* on August 12, 1909, "Scott Joplin is busy finishing his opera, which will likely be given a public hearing before long."[58]

In the fall of 1909 Joplin became friends with Harry Bradford, a well-known vaudeville entertainer living in New York who also wrote for the *Indianapolis Freeman*. Beginning with the September 18th issue of the *Freeman,* Bradford, it seems, allowed no opportunity to pass without praising Joplin. In writing of a young composer, he compared his modesty to that of "my friend from the West, Scott Joplin." In the October 16 issue he praised Joplin's new "stop time instrumental," apparently referring to *Stoptime Rag,* which was soon to be published by Stern (registered for copyright January 4, 1910). Two weeks later, he again mentioned this work, lavishing additional praise on the composer:

> Scott Joplin has got a new bunch of his own original stop-time music. It is real Scott Joplin music. He has a way all his own in writing ragtime music, and you all know that there is only one Scott Joplin.

Not having substantive news did not stop Bradford, where Joplin was concerned. In the November 6th issue, he wrote, simply "Scott Joplin is still ragtime king of Greater New York." Around this time, Joplin played portions of *Treemonisha* for Bradford, and the latter commented on it on November 13:

> Mr. Scott Joplin, the composer of the Maple Leaf Rag, has got a new opera. I heard the overture; it is great as anything written by Mr. Wagner or Gounard [Gounod] or any of the other old masters. It is original Scott Joplin Negro music. Nothing like it ever written in the United States. The public will hear all about it in a few weeks. The music loving public of New York claim that Mr. Scott Joplin is the greatest composer of all American Negroes.

In the same issue, in another column, he noted that Joplin was present at a prize fight, sitting ringside with the seconds of one of the boxers. With the defeat of this fighter, it was Joplin who threw the sponge into the ring.[59]

The following week, on November 19, Bradford unexpectedly died.[60] Joplin lost a good and valuable friend. For a few weeks, Bradford's columns were continued by Harry Brown, but Brown soon discontinued them, and for the next few years the *Indianapolis Freeman* had only occasional coverage of events in New York and virtually nothing on Joplin.

The period from mid-1907 through 1909, beginning with his arrival in New York, was one in which Joplin made a new start. After the death of

Freddie, he seemed to flounder for a few years; he published little, especially in the way of ragtime, and he worked with insignificant publishers. In New York, he displayed a renewed vigor. In a span of two and a half years he had sixteen publications: twelve independent rags, one collaborative rag, one syncopated waltz, one habañera, and one instruction manual. Many of these publications were artistic successes, both developing ideas introduced earlier and pointing to new directions in ragtime.

His fresh start extended to the business side of his career, as well. He left John Stark and tried several other publishers, finally settling with Seminary Music. This publisher was part of a closely knit trio of music firms (Crown and Ted Snyder Music being the others) that were to evolve into a Tin Pan Alley legend. The crucial step in this evolution was innocuous, but momentous. Ted Snyder, firm president and a composer of growing popularity, hired a clever young lyricist with whom, in 1909, he collaborated on at least fourteen songs. This lyricist was Irving Berlin, whose connection with Seminary was to give a strange twist to the Joplin story.

CHAPTER ELEVEN

Treemonisha,
1910–1911

He was a great man, a *great* man! He wanted to be a real leader. He
wanted to free his people from poverty, ignorance, and superstition, just
like the heroine of his ragtime opera, *Treemonisha.*

Lottie Joplin, 1950.[1]

W E SEE AN IMPORTANT SIDE of Joplin's character
with his membership in the Colored Vaudeville Benevo-
lent Association. He was not just a member; he was an active participant. As
the time came for the organization to have its first winter fund-raising event,
Joplin joined the Arrangement Committee to help plan the affair. Others on
the committee included Chris Smith, Tom Fletcher, and executive board
member Bob Slater.

Held at Madison Square Garden on January 28, 1910, the event featured a vaudeville show followed by ballroom dancing. The dancing, which
took up the major part of the evening, was to music composed exclusively by
black musicians, among them Will Marion Cook, Chris Smith, J. Leubrie
Hill, Will Vodery, W. H. "Black" Carl, Tom Lemonier, and Tim Brymn. But
Scott Joplin was the most represented composer of the evening, with five
dance works performed. Of these, reports identify three: *Paragon Rag* (which
was dedicated to the C.V.B.A.), *Binks' Waltz,* and *Gladiolus Rag.*[2]

A *Scott Joplin Schottische,* credited to Chris Smith and Abe Lincoln
Jones, was also performed. The piece was never published, but probably
consisted of a selection of Joplin melodies in a schottische arrangement. This
tribute to Joplin suggests that he must have enjoyed cordial friendships in the
organization.[3]

The co-composer crediting of Abe Lincoln Jones was actually a joke, one that must have been thoroughly enjoyed by those present. Chris Smith was a popular songwriter and performer, but Jones was not a real person. He was a comic creation of Smith and Cecil Mack in their song *Abraham Lincoln Jones or (The Christening),* published the previous year. The song character's full name is based on a string of celebrated historical and contemporaneous figures: "George Washington, Christopher Columbus, Roosevelt, Douglass Lee, Jack Johnson, Joe Gans, Dixon, Booker T., Major Taylor, Chrispus Attucks, Jackson, Sherlock Holmes, Hezekiah, Obediah [*sic*], Abraham Lincoln Jones."[4]

Joplin had new friendships, but he also maintained contact with some of his former colleagues. Early in 1910 his St. Louis friend Tom Turpin joined the C.V.B.A. as an out-of-town member, suggesting that he and Joplin were corresponding. Turpin was now back in business, having opened the Eureka Club, at 2208 Chestnut. Among the pianists who worked at the club was Arthur Marshall.[5] The Eureka also served as a local Republican club, and in June 1910 Tom's brother Charles announced he was running for the office of district constable. No black man had ever been elected to office in St. Louis, and the candidacy was dismissed as a joke by many. But in November, Turpin won. There were immediate protests and challenges to prevent Turpin from taking office, but the courts finally ruled in his favor. Four years later Charles stood for re-election and was initially defeated by six votes. However, after claims of fraud and a recount, the court declared him the winner by 27 votes.[6]

Joplin's brother Robert was getting good notices. In one, he was referred to as "the noted song and play writer." He had two songs published by the John Arnold Publishing Company in Cincinnati (*Since Emancipation Day* and *If I Were with My Thoughts To-Night, Sweetheart, I'd Be with You*), had just finished six weeks at a theater in Jacksonville, Florida, and was about to start an engagement of indefinite length in Atlanta.[7]

Sam Patterson, with William Spiller since 1906, left the group in June 1910. For the next ten months, he worked alone in vaudeville, doing comic "blackface" monologues. In April 1911, he became a member of the Watermelon Trust, a comical and musical group of two men and two women.[8]

Meanwhile, the Musical Spillers regrouped and by August 1910 were an act comprised of three men and three women.[9] A photograph in the *New York Age* (August 17, 1911, p. 6) shows all six playing saxophones.

Sometime after 1907, John Stark sketched a brief history of his business in one of his ledger books. Writing in the third person, he noted that he had been selling only instruments in Sedalia until he realized how the music publishing business worked. "He bought out his competitor in 1886. It didn't take him long to figure out that a piece of music costing $1\frac{1}{2}$ cts to produce and selling at wholesale for 15 cents would not lose the publisher anything to speak of." Thereafter, he gradually increased the music publishing side of his business; with the success of his ragtime publications after 1900, he stopped selling instruments.[10]

By the end of first decade of the 20th century, the music publishing business had changed drastically. Stark's production costs must have risen, and he could no longer wholesale his music for the fifteen cents he had cited. The large department stores and such jobbers as Crown Music exerted tremendous pressure for lower prices. The *Maple Leaf* was his most expensive number, and sometimes he could wholesale it for as much as twelve and a half cents. But for major purchasers the price was usually eight or even seven cents. Other music was lower. On September 10, 1908, he sold 1500 pieces (including 200 each of *Fig Leaf, Sensation, Frog Legs,* etc.) to the Siegel-Cooper department store for $45, a unit cost of three cents. Over the course of the 18 months covered by the ledgers, his average wholesale price for *Maple Leaf* was eight and a half cents. He averaged seven cents for James Scott's *Frog Legs,* five cents for *Sun Flower,* four cents for *Fig Leaf,* less than three cents for *Heliotrope Bouquet.*

In the spring of 1910, Stark had had enough of the conditions in New York and returned to St. Louis. The following November, his wife died. Their daughter Eleanor remained in New York and married a man named Jim Stanley, who had been a singer. I could find no sign of her continuing her concert career. The family does not know if "Aunt Nell" had children, and it is reported that she died young.[11]

The most important event in 1910 for New York's black musicians was the formation of the Clef Club. This was an organization headed by James Reese Europe, formally trained as a violinist and pianist. Jim Europe had arrived in the city in 1903 and quickly gained respect as a skilled bandleader and music director for leading black shows, including the Cole and Johnson musicals *Shoo-Fly Regiment* (1906–07) and *Red Moon* (1908–09).

The Clef Club, with offices at 134 West 53d Street, across the street from the Marshall Hotel, functioned as both a booking agency and a union for black musicians. Europe maintained high standards: members had to be skilled, reliable, and musically literate. The result was that Clef Club musicians came to dominate the New York dance and society music scene; they were hired for major restaurants, hotels, cafes and the most prestigious social affairs. They earned the highest salaries of any musicians in New York, even higher than their white counterparts.[12]

Jim Europe also formed the Clef Club Orchestra, a concert orchestra that gave a platform to black American composers. (For the first few years, as in the concert described below, works by white composers were included; the policy was later changed to perform only black composers.) The orchestra gave several concerts each year, most often at the Manhattan Casino, 155th Street and Eighth Avenue. The event on October 20, 1910, shows the format. This "Second Grand Musical Melange and Dancefest" had four parts:

1. 45 Minutes with the Clef Club Entertainers in New and Novel Songs.
   This part was essentially vaudeville, and included Joe Jordan's *Lovie Joe,* a
   hit song from 1909 made popular by Fannie Brice in her first year with the
   Ziegfeld Follies.
2. 60 Minutes with the Popular Colored Composers, who will render their
   own selections, accompanied by The Clef Club Orchestra.
   This part included works and performances by Will Tyers (who, as Assistant Director, conducted the orchestra), Ford Dabney, Will Dixon, and
   James Reese Europe.
3. 60 Minutes with the CLEF CLUB SYMPHONY ORCHESTRA . . .
   Selections included Europe's *The Clef Club March* and *Queen of the Nile,*
   Henry Creamer's *Clef Club Chant,* Edward MacDowell's *To a Wild Rose,*
   and Paul Lincke's *Unrequited Love.*
4. Ballroom dance, to music of the orchestras of W. F. Craig and Miss Hallie
   Anderson.[13]

The most unusual feature of the Clef Club Orchestra was its instrumentation. The October 20th concert had an orchestra of more than one hundred:
8 violins, 9 cellos, 3 basses, 3 percussions (traps and tympani), 27 first
mandolins and bandoris, 10 second mandolins, 8 banjos, 23 harp guitars, and
11 pianos. In later years, with this and other orchestras, Europe added
woodwinds, brass, and one or more organs. This is what he had to say on his
instrumentation in 1914:

> Although we have first violins, the place of the second violins with us is taken
> by mandolins and banjos. This gives that peculiar steady strumming accom-
> paniment to our music which all people comment on, and which is something
> like that of the Russian Balalaika Orchestra, I believe. Then, for background,
> we employ ten pianos. That, in itself, is sufficient to amuse the average white
> musician who attends one of our concerts for the first time. The result, however,
> is a background of chords which are essentially typical of negro harmony.
>
> Other peculiarities are our use of two clarinets instead of an oboe, because,
> as I have said, we have not been able to develop a good oboe player. As a
> substitute for the French horn we use two baritone horns, and in place of the
> bassoon we employ the trombone. We have no less than eight trombones and
> seven cornets. The result, of course, is that we have developed a kind of sym-
> phony music that, no matter what else you may think, is different and distinc-
> tive, and that lends itself to the playing of the peculiar compositions of our
> race.[14]

The Clef Club made history on May 2, 1912: it performed a concert of
black music at Carnegie Hall. America's greatest concert hall, a bastion of
high art, opened its doors to African-American artists. For this performance,
even the rules of segregated seating were suspended.

It was a concert for the benefit of the Music School Settlement for Colored People, and other celebrated figures also appeared, among them Will Marion Cook, leading a chorus, and J. Rosamond Johnson, playing piano and singing. The program included music of various degrees of "refinement," the high point represented by works of the black British composer Samuel Coleridge-Taylor. Only slightly lower were art songs by Will Marion Cook, Harry T. Burleigh, and Rosamond Johnson. At the other end of the scale was ragtime, not ignored, but minimized.

James Reese Europe was ambivalent about ragtime. He considered ragtime an artificial category, invented to pigeonhole and restrict black musicians: "In my opinion there never was any such music as 'ragtime.' 'Ragtime' is merely a nick-name, or rather a fun name given to Negro rhythm by our Caucasian brother musicians many years ago."[15] At the same time, he valued ragtime as the true and natural expression of African Americans. "The negro plays ragtime as if it was a second nature to him—as it is."[16] He frequently programmed Joe Jordan's song *That Teasing Rag* and Wilbur Sweatman's instrumental *Down Home Rag,* but ragtime was never prominent in his concerts. Ford Dabney, who succeeded Europe as head of the Clef Club in 1914, followed the same policy.

Given Joplin's renown, his goal of idealizing ragtime into a "classic" or art music, and his composition of operas that celebrated the race, one would think he would find natural allies in the leadership of the Clef Club. This was not the case. Joplin is conspicuously absent in Clef Club programs. I have been unable to discover the reason, but James Reese Europe, Ford Dabney, Will Marion Cook, J. Rosamond Johnson and others in the black musical elite seem to have ignored Scott Joplin.

In 1910 Joplin still lived at the Rosalline boarding house on West 29th Street. In the U.S. Census taken on April 22, 1910, he listed himself as a

Thirteenth Census of the United States: 1910 Population . . . New York City . . . Enumerated on the 22 day of April 1910.

The building number, "128," is on line 71. Joplin's entry is on line 77. In the columns after his name are "Lodger," "M" [male], "Mul" [mulatto], "40" [age], "W" [widowed].

widower, age 40; actually he was 42. His listing as "mulatto" (third column after his name) is an obvious error. All persons in the building were so listed, probably because of a confusing Bureau of Census directive that only black persons "of full blood" be considered Negro, whereas those with "perceptible traces of Negro blood" be categorized as mulatto—and thereby of "mixed parentage." This instruction resulted in an undercount of Negroes and was protested by the black community.[17]

The U.S. Marine Band had made its second recording of the *Maple Leaf* in 1909, and in 1910 Haenschen's Banjo Band recorded it. Also in 1910, the Zonophone Orchestra recorded *Wall Street Rag,* this being the only other Joplin rag recorded in his lifetime.[18] In May, Seminary advertised *Euphonic Sounds* for fifty cents in the *American Musician,* and *Euphonic, Wall Street,* and *Pine Apple,* in the *New York Age.*[19] Seminary also published a song version of *Pine Apple,* with lyrics by Joe Snyder, obviously another in the family that included Ted, Herman, and Mary. There are significant changes in this version, and since they are competently executed they could have been made by Joplin, but we cannot say that this is definitely so. The music is cast into a verse-chorus song form, using rag strains A and B for the verse and C for the chorus; D is dropped. The piece is entirely in E-flat, the piano part is transformed considerably to function as an accompaniment, and some of the original pianistic melodic figurations that could not comfortably be translated to the voice are altered.

This is Joplin's last publication with Seminary. Ted Snyder had developed a reputation of dealing fairly and courteously with black composers,[20] but Joplin was not satisfied. He returned to Stern to have *Stoptime Rag* published.

Stoptime, completed by Joplin in the fall of 1909 (Harry Bradford mentioned it in the *Freeman* on October 16, 1909), was Joplin's only other publication in 1910 (registered for copyright January 4). It is a curious work, consisting of seven strains, most having eight rather than sixteen measures. All sections except the third use the stoptime rhythm, in which the music pauses just before a cadence:

The only other piece in which Joplin had used this rhythmic device was the *Ragtime Dance.*

It must have been in the summer or fall of 1910 that Joplin finished *Treemonisha* and brought it around to publishers. Stark had turned it down. Seminary held it for several weeks before rejecting it. Stern also refused, as did unnamed others. Joplin rags had a ready market, but a 230-page opera score, especially one by a black man, was too much of a risk. Finding no takers, Joplin finally published it himself.

With his work on the opera essentially completed, Joplin seems to have relaxed. We come across several notices of his attendance at small social gatherings, many including C.V.B.A. members.

On January 27, 1911, Bob Slater hosted a gumbo dinner for several notable vaudevillians at the C.V.B.A. headquarters. Joplin was present and after dinner he played several selections at the piano. Others present included singer John Rucker, who also performed, songwriter Alex Rogers, and columnist Lester Walton.[21]

Two months later, on March 29th, Joplin was the one to host a party. This was a luncheon for William Spiller and others at the home of a Miss Christie Hawkins, 145 West 95th Street.[22] We know nothing of Miss Hawkins, but we can surmise she must have been a good friend; since Joplin's boarding house was unsuitable for the affair, she extended the use of her home.

On Friday, May 19, 1911, Joplin published the piano-vocal score of his opera *Treemonisha* and offered it for sale for $2.50. The following Monday he went to the Library of Congress in Washington to file the copyright application personally (both his handwritten date and the stamped Copyright Office date are the same), paying the fee with a one-dollar bill. He arranged for a British copyright through B. F. Wood Music Co., a British publisher with a New York office.[23]

Ever conscious of the value of publicity, Joplin notified Lester Walton of the *New York Age* of the opera's completion. On Thursday, the *Age* contained the following notice:

> "Treemonisha," an opera in three acts, is the latest contribution to the music world by a colored composer. Scott Joplin, who wrote "The Maple Leaf Rag" and other syncopated compositions, is responsible for "Treemonisha," words and music, which is being published by him also.
>
> The story deals with a colored waif who was found under the trees by Ned and Monisha, and because of her inclination to play under the trees she was named "Treemonisha." The scene of the story is laid on a plantation in Arkansas. The book provides for eleven people in the cast and a large chorus. Composer Joplin characterizes the music as "strictly Negro." There are twenty-seven musical numbers and there are 230 pages to the score of the opera.[24]

To celebrate the opera's publication, Joplin's friends held a party for him on June 14th, again at the home of Miss Christie Hawkins. Among the 28 men and woman present were William Spiller, Mildred Creed (of the Musical Spillers), and Bob Slater. The presence of one Hans Yorgensen suggests that there were white guests as well as black.[25] We do not come across Miss Hawkins again, but must wonder, in view of the two parties at her home, whether she and Joplin were romantically involved.

An extraordinarily favorable review in the *American Musician and Art Journal* highlighted this post-publication period of celebration. Considering

Copyright application for Treemonisha.

Joplin mistakenly gave the building number as "12" instead of "128" on the face of the application, but listed it correctly on the reverse side.

the good press this magazine had previously given Joplin, it is not surprising that he brought a copy of the score to the editor. He had to have been pleased with the response. The writer begins with a discussion of the abilities and "the progress of the colored race," and then presents Joplin's work as an example of this progress. He praises its musical worth and recognizes Joplin's aims of presenting African-American life and his mission in pointing to a better way through education. The writer's conclusion is startling and bold: he suggests that Joplin, more than any other American composer, has succeeded in creating a purely *American* opera.

Scott Joplin, well known as a writer of music, and especially of what a certain musician classified as "classic rag-time," has just published an opera in three acts, entitled "Treemonisha," upon which he has been working for the past fifteen years. This achievement is noteworthy for two reasons: First, it is composed by a negro, and second, the subject deals with an important phase of negro life. The characters, eleven in number, and the chorus are also negroes. . . .

A remarkable point about this work is its evident desire to serve the negro race by exposing two of the great evils which have held his people in its grasp, as well as to point them to higher and nobler ideals. Scott Joplin has proved himself a teacher as well as a scholar and an optimist with a mission which has been splendidly performed. Moreover, he has created an entirely new phase of musical art and has produced a thoroughly American opera, dealing with a typical American subject, yet free from all extraneous influence. He has discovered something new because he had confidence in himself and in his mission and, being an optimist, was destined to succeed. . . .

The principal theme of the opera is one of entrancing beauty, symbolic of the happiness of the people when they feel free from the spells of superstition animated by the wiles of the conjurors:

Scott Joplin has not been influenced by his musical studies or by foreign schools. He has created an original type of music in which he employs syncopation in a most artistic and original manner. It is in no sense rag-time, but of that peculiar quality of rhythm which Dvořák used so successfully in the "New World" symphony. The composer has constantly kept in mind his characters and their purpose, and has written music in keeping with his libretto. "Treemonisha" is not grand opera, nor is not light opera; it is what we might call character opera or racial opera. . . .

There has been much written and printed of late concerning American opera, and the American composers have seized the opportunity of acquainting the world with the fact that they have been able to produce works in this line. Several operas by American composers have been produced recently, and there is promise of several others being heard next year, among which will be Professor Parker's "Mona," which won the $10,000 Metropolitan Opera prize. Now the question is, Is this an American opera? And a correlative question is, Are the American composers endeavoring to write American operas? In other words, are they striving to create a school of American opera, or are they simply employing their talents to fashion something suitable for the operatic stage and satisfactory to the operatic management? If so, American opera will always remain a thing in embryo. To date there is no record of even the slightest tendency toward the fashioning of the real American opera, and although this work just completed by one of the Ethiopian race will hardly be accepted as a typical American opera for obvious reasons, nevertheless none can deny that it serves as an opening wedge, since it is in every respect indigenous. It has sprung from our soil practically of its own accord. Its composer has focused his mind upon a single object, and with a nature wholly in sympathy with it has hewn an entirely new form of operatic art. Its production would prove an interesting and potent achievement, and it is to be hoped that sooner or later it will be thus honored.[26]

The closing paragraph is breathtaking. The writer asserts that Scott Joplin succeeded where Horatio Parker and other distinguished American composers had failed. Parker's opera, for which he was awarded $10,000, lacks American character. Joplin's opera, in contrast, "is in every respect indigenous" and is "an entirely new form of operatic art."

We can imagine what Joplin thought of this review. We wonder how other readers reacted.

The editor of the *American Musician and Art Journal* was right. *Treemonisha* is a fine opera, certainly more interesting than most operas then being written in the United States. It is evident that Joplin had studied the form. He knew operatic conventions and understood how to introduce his own special vision and qualities to fashion a uniquely African-American work.

But Joplin was wrong when he told the *New York Age* that the opera was "strictly Negro." It achieves that quality only in the opera's most glorious moments.

Treemonisha is not a ragtime opera. Joplin wanted to demonstrate his ability to transcend the limitations of ragtime and black music. Except for the texts, many of the numbers he composed could fit into more conventional operas. He used ragtime and other black music sparingly, but effectively; such music would occur when it was important to convey racial character.

Treemonisha takes place in September 1884 in an Arkansas community of former slaves. Joplin must have had an actual place in mind, for he gives a fairly specific location: "Northeast of the Town of Texarkana and three or four miles from the Red River." Though the community is near Joplin's boyhood town of Texarkana, it is isolated within a dense forest, symbolizing both the isolation of African Americans from the rest of American society and a place that has not been reached by the light of reason and education.

The community is led by Ned and Monisha, who had remained childless until 1866 when they found an infant girl under a tree outside their home. They take the child in as their own and name her Monisha; because of the child's fondness for the tree under which she was found, they prefix her name with "Tree." They are determined to have her educated and arrange, in exchange for household and yard chores, for a nearby white woman to teach her to read.

At age 18, Treemonisha is ready to lead her community out of ignorance. Her major attack is against superstition, represented by the conjurors who victimize the gullible villagers. The conjurors, in an effort to preserve their way of life and means of income, abduct her. They are about to throw her into a wasps' nest when she is rescued by her friend Remus. Her neighbors recognize the liabilities of their ignorance and acknowledge her as their teacher and leader. Her first lesson, one that meets resistance, is that the remorseful abductors be forgiven.

The opera depicts events in a small, isolated community, but the story is an allegory with wider ramifications. The subject is really the African-American community which, as seen by Joplin fewer than fifty years after emancipation, was still living in ignorance, superstition, and misery. The way out of this condition, he tells his intended audience, is with the education that can be provided by white society.

Joplin wrote his own libretto, with mixed results. We learned from Alfred Ernst's remarks in 1901 that Joplin greatly admired the operas of Richard Wagner. We see evidence of this admiration in Joplin's adherence to the Wagnerian operatic ideal of a single artist being responsible for both the music and the libretto. But Joplin was not a competent dramatist, and his book does not measure up to the quality of his music. Although he wrote some very good individual sections, such as "Superstition," in which the conjurors recite their credo, he was unable to unite the sections into a comprehensive book that engages the listener's attention. We never develop concern for the characters. Sam Patterson, who had been on the stage since his early teenage years, recognized this shortcoming. He commented to Blesh and Janis, "There were a few nice songs. No action. All he had was an album of music."[27]

Had Joplin taken on an experienced collaborator, *Treemonisha* could probably have made stronger theater. But in gaining a greater stage work, we would have lost the biographical clues in the story. The libretto of *Treemonisha* is not just a fictional vehicle for Joplin's music; it is a medium through which he commemorated people in his life and expressed some of his most personal feelings and innermost convictions. So concerned was he that certain details not be omitted, he included them in an explanatory Preface that must be read before the opera begins. That the Preface is needed underlines the opera's dramatic shortcomings. Theater's loss becomes biography's gain.

Treemonisha uses beliefs in superstition and mysticism, themes that are common enough in opera. The tree under which Treemonisha was found is referred to as "The Sacred Tree," and Monisha sings an extended, lullaby-like aria about it. But the sacredness of the tree plays no function in the story; we suspect Joplin simply wanted to include a familiar operatic device, perhaps having in mind the tree from which Siegmund draws his enchanted sword in Wagner's *Die Walküre*. Wagner is invoked also in another parallel in Monisha's aria: her retelling of Treemonisha's origins recalls the corresponding circumstance in *Siegfried* of Mime's explanation to his orphaned ward. (Of course, the parallels do not extend to the characterizations; the deceitful, greedy Mime is not comparable to the loving Monisha.)

More integral are superstitions that reflect upon African-American folklore, such as the conjurors' talismans and the litany of beliefs they recite concerning luck and omens. Joplin must have been familiar with these superstitions since early childhood and could have been reminded of them by the many articles on the subject in popular magazines of the time.

African-American folk tales appeared in numerous magazines and were popular among both blacks and whites, especially as transmitted in the Uncle Remus stories of Joel Chandler Harris. In this connection, the threat of Treemonisha being thrown into the wasps' nest recalls the story of Brer Rabbit and the briar patch. The ballet "The Frolic of the Bears" in Act 2 may not refer to a specific story but corresponds to the emphasis on animal tales found in African-American folklore.

Joplin celebrates his heritage also with musical forms he knew from childhood, giving the opera additional value as a record of black, rural musical practices of the 1870s–90s. The sections with folk types of music are, in affect, re-creations by a skilled and sensitive participant. The exhilarating ring dance "We're Goin' Around" is easily recognized as a country dance. The quartet "We Will Rest Awhile" (Ex. 11-1) underlines a significant historical point about the turn-of-the-century "barbershop quartet." Though

Ex. 11-1. Scott Joplin, "We Will Rest Awhile," measures 3–6, No. 16 from Treemonisha *(1911).*

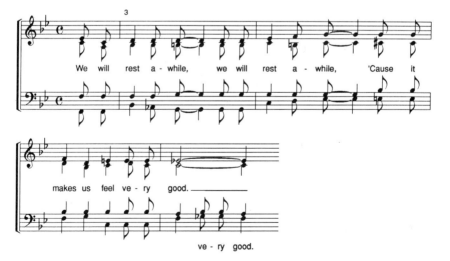

this number is typical of the style, it is sung by farm workers in the fields rather than by barbers and others in town. In Joplin's day it had not yet been forgotten that close-harmony quartet singing was characteristic of black culture, both as a religious expression and as a secular entertainment.[28] Joplin's quartet experience stemmed from his earliest days, the first contact likely to have been in church. As an adolescent performer he was part of a vocal quartet and he continued working with such groups at least until 1904. Here Joplin confirms that the quartet was integral to black rural practice of the 1880s.

Education is central to the story, as it was to Joplin's life. "Ignorance is criminal," he tells his audience (p. 209).[29] Education was widely regarded as the means by which black Americans would earn the respect and acceptance of the rest of American society. He could reasonably have viewed the degree of respect he himself enjoyed as the working of this concept.

Education, as he saw it, had to be obtained from white America. The people in the opera lived "in dense ignorance, with no-one to guide them, as the white folks had moved away" (p. 5). Treemonisha was educated by a white woman, just as Joplin received his education from a white music teacher. When he attended Smith College, there were some black instructors, but the school was administered by whites.

Treemonisha's educated state is indicated by her manner of speech, which is "white," that is, without African-American dialect. Notice the difference between Treemonisha's speech and that of the conjuror Zodzetrick:

Treemonisha: You have lived without working for many years,
              All by your tricks of conjury.
              You have caus'd superstition and many sad tears.
              You should stop, you are doing great injury.
Zodzetrick: You 'cuse me wrong for injury I'se not done,
              An' it won't be long 'fore I'll make you from me run.
              I have dis bag o' luck, 'tis true,
              So take care, I'll send bad luck to you. (pp. 30–31)

Similarly, others in the community speak with dialect. Note, for example, in the call-and-response between the parson and the congregation:

Parson: Aint yer glad yer have been redeemed?
Chorus: O yes, om glad ah have been redeemed.
Parson: Does yer always aim ter speak de truth?
Chorus: O yes, ah always aim to speak de truth. (pp. 81–82)

The only other person to speak without dialect is Remus, who, through Treemonisha's efforts, has also learned the "white" skill of reading (p. 33).

> Remus: Shut up, old man
>   enough you've said;
>   You can't fool Treemonisha
>   She has a level head.
>   She is the only educated person of our race,
>   . . .
>   To read and write she has taught me,
>   and I am very thankful,
>   I have more sense now, you see,
>   and to her I'm very grateful. (pp. 32–33)

Non-dialect speech as a sign of education was a trait Joplin himself cultivated. As Monroe Rosenfeld remarked in 1903, "He is attractive socially
because of the refinement of his speech."[30]

One facet of respectable society that Joplin seems to have rejected is
organized religion, represented in the opera by Parson Alltalk. Joplin does
not succeed in making the Parson as comical as the name might suggest, but
the Parson lives up to his name: all he does is talk, exhort the people to be
good. He is totally ineffectual, unable to see his congregation's real needs and,
being uneducated (as his dialect tells us), unable to provide leadership.

But Joplin is not an atheist. He has Treemonisha invoke the Deity on
several occasions, as when she is saved from the wasps' nest and exclaims,
"But thanks to the great Creator, You came in time to save me" (pp. 137–38).
It is probably significant that she uses the impersonal "Creator" rather than
the more familiar "Lord God" or "Jesus" of Spirituals and Gospel hymns.
The absence of any suggestion of religious song, in the context of an opera
that celebrates music of the rural black community, underscores Joplin's
ambivalence about the church.

Still, while he may have rejected the organized black church, Joplin did
not reject Christian ideals of moral behavior. When the community is about
to punish the conjurors, Treemonisha and her protégé Remus urge that they
be forgiven: "Wrong is never right" and "Never do wrong for revenge" (No.
22). She wins her way with a personal appeal: "Will all of you forgive these
men for my sake?" (No. 25).

If Joplin was Christian in his views, why should he have rejected the
Christian churches? Because the churches rejected what was important to
him—ragtime, dance, and the theater. We have seen how the black ministers
of Sedalia had attacked ragtime and dance. They were not alone in their
negative view of the music. As Willie "The Lion" Smith wrote:

> Back in those early days churchgoing Negro people would not stand for rag-
> time playing; they considered it to be sinful. . . . Yeah, in the front parlor,
> where the neighbors could hear your playing, you had to sing the proper
> religious words and keep that lilting tempo down![31]

The Methodist-run Smith College that Joplin attended proscribed dance and theater for its students. The governing body of the A. M. E. (African Methodist Episcopal) Church issued similar prohibitions for its members, a discipline that seemed unreasonably strict even to some of its clergy.[32] Ernest Hogan, one of black America's most beloved theater personality, commented on this issue:

> Ernest Hogan, the well-known colored comedian, sets out some good thoughts in a recent article in his brief discussion of the church and the stage. He can't understand just why the preacher has a warm reception for every other class of people excepting stage people. Mr. Hogan says:
>
> "The church goes after the drunkard, the gambler and so on, but for some reason it has a tendency to shun the prominent Negro of the stage, who has opened a field and is doing much to lessen race prejudice and who gives to hundreds employment. There are many places in the United States today where the preacher will get up and admonish the members for attending a theatrical performance which has in every respect been clean and instructive."[33]

The church that Joplin knew attacked the things that were important to him, making what he knew to be false accusations of inherent evil. This attitude by the religious community could explain why he might reject the churches.

By setting the opera near his childhood home of Texarkana, Joplin alerts us to watch for autobiographical references. Lottie, as quoted in the epigraph opening this chapter, drew the connection between Joplin's personal aspirations and the traits he gave his operatic heroine. Does Treemonisha, then, represent Joplin?

On the intellectual level, she personifies his ideals. A few biographical parallels also exist. Her education by a white woman in exchange for her parents' chores recalls Joplin's situation and tells us that Monisha commemorates his mother. Since Treemonisha begins her education at age seven, we assume Joplin had a reason for being so specific, that he, too, began his education at that age.

That is where the parallels end. If Treemonisha was to be his alter ego, why give her a birth date of September 1866, making her a year and a half older than himself? Why make the main character a woman? His history with women would seem to rule out a feminine identification. In most respects, it is difficult to see how Treemonisha could represent Joplin.

Does Joplin's choice of a woman leader demonstrate sympathies with feminist views? Not necessarily. His decision to have a heroine rather than a hero has many operatic precedents. African-American history also has female leaders, such as Harriet Tubman.

Joplin recognized the problem of a woman leader, for Treemonisha asks whether the men, as well as the women, would be willing to follow her (p. 213). The men say they will. A woman is capable of leading men.

At the same time, he expressed the thought that women inherently had less intellectual vigor than men: "All of the Negroes, but Ned and his wife Monisha, were superstitious, and believed in conjuring. Monisha, being a woman, was at times impressed by what the more expert conjurers would say" (p. 5).

The Preface's conclusion is particularly perplexing: "The opera begins in September 1884. Treemonisha, being eighteen years old, now starts upon her career as a teacher and a leader" (p. 7).

There is nothing in the story that requires that it take place in a specific month and year. Joplin must have been commemorating something. One attractive hypothesis suggests that Julius Weiss, Joplin's early music teacher, left Texarkana between April and August of 1884. With his teacher no longer available to him, the 16-year-old Scott saw no reason to remain in town. In September 1884, he then set off upon his career as a musician, perhaps with hopes of eventually becoming a teacher and leader of his people, the course he ascribes to his heroine Treemonisha.[34]

But there is another, more compelling observation: Freddie was probably born in September 1884.

We do not have Freddie's exact date of birth, but we know she was nineteen at the time of their marriage on June 14, 1904, and Joplin said she was twenty at the time of her death the following September. It does not necessarily follow that her birthday was in September; it could have been anytime between mid-June and September. But her precise birth date is unimportant. We have seen that Joplin was extremely casual about dates: every time he was asked for his own birth date or age, he gave a different answer. It would be sufficient that he *thought* she was born in September. The ending of the Preface points directly to Freddie and links her with Treemonisha, suggesting that the opera was intended as a memorial.

In an effort to explore this hypothesis, we search for parallels between Freddie and the operatic character. Was Freddie an orphan? Did she, or her family, originate from the Red River area northeast of Texarkana? Did she aspire to be a teacher?

We have no answers. But we can draw a linkage between *Chrysanthemum,* the piece Joplin dedicated to Freddie, and *Treemonisha. Chrysanthemum* was subtitled "An Afro-American Intermezzo," and *Treemonisha* celebrates African-American culture. These are the only works in which Joplin specifically addresses his racial heritage. From this observation we now add to our list of conjectures that Joplin's thinking on racial pride was influenced by Freddie.

Much of the music in the opera is attractive, but within standard practices. Joplin purposely followed conventions. But occasionally he strayed beyond what was expected. In one place, he created an innovative, nonpitch notation to portray the wailing and confusion of the moment (Ex. 11-2).

Ex. 11-2. Scott Joplin, "Confusion," measure 35, No. 10 from Treemonisha.

The musical highlights tend to be not with the arias, pleasing as they may be, but with the ensembles and choruses. Probably the most rousing moment in the opera is the Act Two finale "Aunt Dinah Has Blowed de Horn," a rag to which a chorus of cotton pickers dances home at the end of day.

In the opera's penultimate number, the townspeople confirm in chorus: "We want you as our leader." This sentiment sets up the ensemble finale, in which Treemonisha gives her second lesson. She leads the townspeople on two levels: on the literal level, she calls the steps for "A Real Slow Drag"; metaphorically, she guides the people to a better life—"Marching onward." But they march not to the military strains of a John Philip Sousa. Rather, they march to a characteristically African-American music—a rag, a *slow* rag.

This finale is a fitting and glorious conclusion, summing up Joplin's philosophy that African Americans choose education as their guide to a brighter future. But in a perverse twist of fate, Joplin witnessed the misappropriation of his call to "march onward." He saw it taken by a white man and transformed into a non-syncopated ragtime song directed to white America, a song that became the greatest financial success of the time and a virtual symbol of the era. Lottie outlined what happened:

> After Scott had finished writing it, and while he was showing it around, hoping to get it published, someone stole the theme, and made it into a popular song. The number was quite a hit, too, but that didn't do Scott any good. To get his opera copyrighted, he had to re-write it.[35]

Sam Patterson told a similar story, reporting that it was none other than Irving Berlin, then working in the Crown-Seminary-Snyder offices, who had stolen Joplin's "Mayflower Rag" and "Slow Drag." Members of the Stark family confirmed the story to Blesh and Janis in 1949, and to Trebor Tichenor some years later. They further claimed that the result of the theft was *Alexander's Ragtime Band,* and that, on first hearing it, Joplin was brought to tears.[36] As reported by Tichenor:

> This is a story that circulated in the Stark family for years; that's what the grandson told me. Joplin took some music to Irving Berlin, and Berlin kept it for some time. Joplin went back and Berlin said he couldn't use it. When *Alexander's Ragtime Band* came out, Joplin said, "That's my tune."[37]

The verse of Berlin's song does resemble the "Marching Onward" section of Joplin's "A Real Slow Drag" (Ex. 11-3a, b). The resemblance is not close enough to charge Irving Berlin with plagiarism. Yet, while the recognizable melodic linkage does not prove that Berlin had taken Joplin's melody, it does explain Joplin's *perception* of a theft. And the resemblance between the two pieces may originally have been closer, for Lottie had said Joplin rewrote his theme after hearing the melody used by someone else. The "Real Slow Drag" that we know may not be as Joplin had originally composed it. In addition, in 1913 Joplin went to the trouble and expense of revising and republishing this number, distancing himself even more from the Berlin classic (Ex. 11-3c).

Ex. 11-3. (a) Irving Berlin, Alexander's Ragtime Band *(New York: Snyder, 1911), Verse 1-4 (transposed from key of C); (b) Scott Joplin, "A Real Slow Drag," measures 21–24, No. 27 from* Treemonisha; *(c) Scott Joplin, Revision of "A Real Slow Drag" (1913), measures 21–24.*

The stories present no glaring inconsistencies with established facts. *Alexander's Ragtime Band* was registered for copyright on March 18, 1911, two months before the copyright of *Treemonisha.* Despite Lottie's statement that Joplin had to rewrite the opera, only a few measures of the final number would have been affected. The revision could easily have been done in the intervening time.

While Irving Berlin certainly had the opportunity to hear Joplin's theme before it was published, there is no conclusive evidence that he consciously stole it. "A Real Slow Drag" is one of the more memorable moments in *Treemonisha,* and the theme could simply have lodged itself in Berlin's memory, to be drawn out inadvertently a few months later. The resemblances between the two themes could also have been coincidental.

But neither can the evidence of a misappropriation be dismissed. Certainly, something had happened, and Joplin was angry. A columnist writing "Timely Tattle" in the *American Musician* in November 1911 reported: "Scott Joplin is anxious to meet Irving Berlin. Scott is hot about something."[38] Then there were rumors heard throughout Tin Pan Alley, scandalous rumors that *Alexander's Ragtime Band* had actually been written by a black man. These continued for several years, and Berlin could no longer ignore the accusations. Finally, in the April 1916 issue of *Green Book Magazine,* he gave his reply:

Two years or so ago, when "Alexander's Ragtime Band" was a big hit, some one started the report among the publishers that I had paid a negro ten dollars for it and then published it under my own name.

When they told me about it, I asked them to tell me from whom I had bought my other successes—twenty-five or thirty of them. And I wanted to know, if a negro could write "Alexander," why couldn't I? Then I told them if they could produce the negro and he had another hit like "Alexander" in his system, I would choke it out of him and give him twenty thousand dollars in the bargain.

If the other fellow deserves the credit, why doesn't he go get it?[39]

Our knowledge of the incident is too incomplete to draw any firm conclusions. We do not know for sure that the rumors Berlin addressed even concerned Joplin. Lukie Johnson, for example, is another black composer who has been named as the originator of Berlin songs, although in later years Johnson denied that there was any truth to the stories.[40]

In the course of his phenomenal career, Irving Berlin was often accused of stealing from other writers.[41] To my knowledge, no charges have ever been substantiated. It seems unlikely that the greatest of all Tin Pan Alley empires could have been built upon a succession of misappropriations. Irving Berlin's success can be explained only by his own creative genius.

But could Berlin have pirated a theme once, early in his career? Did he take from Scott Joplin, the first King of Ragtime, a jewel for his own coronation?

CHAPTER TWELVE

The Elusive Production, 1911–1917

A N INTERESTING DESCRIPTION OF JOPLIN dur-
ing this period comes from William Sullivan. Sullivan, a
young white Canadian living in New York, studied piano with Joplin from
about 1910 to 1912. He was one of several students that Joplin had. At first
he paid fifty cents for lessons; later, seventy-five cents. He studied *Maple Leaf*
and *Cascades* but could not remember the other titles. He described Joplin's
playing as very slow and methodical. As a teacher, Joplin regularly admon-
ished Sullivan to place a strong accent on the first beat of each measure.

Sullivan and Joplin formed a friendly relationship and after lessons
would go to a nearby saloon for sandwiches. Another black man, whose name
is now forgotten, frequently joined them, and Sullivan and this man would
talk. Joplin was usually silent and withdrawn, although he would occasionally become animated and join the conversation in a high-spirited fashion.
Sullivan perceived Joplin as being depressed—almost morose—because of
money problems, concerns with his opera, and poor health.[1]

Joplin was still active with the C.V.B.A. and was again on the Arrangements Committee for a carnival and minstrel show held on August 10.[2] He
also spent time at the office of the *American Musician* and spoke with W. A.
Corey, one of the columnists. Corey was very much taken with Joplin as a
composer and with his slow, deliberate style of playing ragtime:

> It takes Scott Joplin to play ragtime on the piano. There are ragtime players,
> but when it comes to playing in a musicianly way, Joplin is there with the goods
> every time.

When it comes to playing ragtime in a musicianly manner, Scott Joplin has got the other fellows beaten by a mile. Joplin is a wonder in his way.

Scott Joplin, the ragtime composer, has no trouble in placing his efforts. A rag with his name attached to it means it will sell.[3]

He had two more music publications in the summer, but neither were of any particular importance. Stark retrieved *Felicity Rag* from his catalogue and registered it for copyright in July. This was another Joplin–Hayden collaboration from earlier days. It has only three strains rather than the usual four, and Joplin was probably responsible only for a single trio theme. Stark omitted Hayden's name from the cover and from all subsequent advertising for the piece, fueling speculation that the publisher did not get along with the younger composer.[4]

Also in July, Al. R. Turner, a black man living on West 37th Street, published *Lovin' Babe,* a coon song that he had written with Joplin's assistance as arranger. Turner is otherwise unknown and has no other copyrights.[5]

Joplin's major concern was with getting his opera staged. In late September he found a willing producer: Thomas Johnson, a well-considered black man who had formerly headed the Crescent Theatre, a small vaudeville and movie house at 36–38 West 135th Street in Harlem. Johnson arranged for a showing in Atlantic City in November, and Joplin sent an announcement to the *Age:*

Arrangements are being made to present Scott Joplin's grand opera, "Treemonisha," with a colored cast. Some of the leading colored singers in this country will be members of the company, which will consist of eleven principals and a chorus of forty people. Thomas Johnson, formerly president of the Crescent Theatre Company, will be the producer. Mr. Johnson has already booked the company for an engagement in Atlantic City early in November. When the organization will play in New York has not been decided. Rehearsals will commence at once.[6]

But Joplin had only his published piano-vocal score and the manuscript for a two-piano version; he had not yet completed the orchestral score. With the prospect of the coming performance, he worked feverishly on an orchestration, with Sam Patterson assisting by copying the parts. He also assembled and rehearsed a cast.

In the midst of their efforts, the plans for the Atlantic City showing fell through. In an attempt to salvage something of their work, Joplin rented a hall in Harlem and mounted an informal performance, to which he invited friends and potential backers. William Sullivan was present and reported that there were about seventeen people in the audience. The performance, without scenery or costumes and with Joplin accompanying the singers from the piano, failed to impress the right people.

Blesh and Janis describe this informal performance as having occurred in 1915, but they were probably mistaken. They may have compressed events for a more lively narrative, or Sam Patterson, who had told them of the incident, could have been off by a few years. But the orchestral score that Joplin had worked on in anticipation of a full staging would have been completed long before 1915. Also by that year, Joplin's health had declined so seriously that he would have been unable to render the opera score at the piano. Eubie Blake reported that in 1915 he had witnessed a pitiful event in which Joplin attempted to play his *Maple Leaf Rag,* but was so infirm he played worse than a child of five.

The year 1911 is much more likely. He had a cast ready for a performance, and at this earlier date it is more plausible that the orchestral score might not have been fully prepared. It is also the year Sullivan had said he had witnessed the event, and the year Lottie gave for a performance Joplin had put on in Harlem at his own expense.[7]

The year ended without firm plans for a staging of the opera, but Joplin was still hopeful. He placed a Christmas greeting in the *Age,* identifying himself as composer of the *Maple Leaf Rag, Euphonic Sounds,* and *Treemonishā.*

In 1911 and 1912, a new era of popular dance had begun. No longer was dance restricted to private ballroom affairs, or even to the public, but notorious, prostitute haven at the Haymarket on West 30th Street and Sixth Avenue. The most select and prestigious restaurants in New York, among them, Delmonico's, Bustanoby's, Rector's, Martin's, and Reisenweber's, opened dance floors. It was an innovation that outraged some and delighted others. Almost invariably, the black musicians of the Clef Club provided the music.

New dance steps, steps that outraged some while delighting others, were the inspiration for the changed attitudes at the restaurants. The Texas Tommy, a dance described in later years as being similar to the Lindy Hop of the 1930s and 1940s, was especially popular. But most of the new steps were

A Merry Christmas and a Happy New Year to All.

SCOTT JOPLIN

COMPOSER OF

"Maple Leaf Rag," "Euphonic Sounds" and "Treemonisha" (an opera in 3 acts)

Christmas Greeting from Scott Joplin. New York Age, *Dec. 21, 1911, Magazine Section, 7.*

identified as "animal dances": the turkey trot, grizzly bear, bunny hug, and others. The phenomenon, for that is what it was, soon generated its own subgenre of sheet music. In 1912 a young Jerome Kern published *The Ragtime Restaurant,* which tells of waiters serving in rag time. The team of Nat D. Ayer and Seymour Brown wrote *Gee! But I Like Music with My Meals* (1911), in which knives and forks are used to keep time on the dishes, waiters have musical feet, and the cook dances while stirring the stew. In *Ragtime Eating Place* (Manley and Trombley, 1913), the chicken does a turkey trot, and "Tango fiends cleared the table away, and danced till break of day." In *That Ragtime Dinner-Time Band* (Chandler and Haines, 1913), the waiters do the turkey trot, the chickens do the grizzly bear, and "You eat your chop to the Kangaroo Hop."

Dance instructors, specialists in the waltz, schottische, and (reluctantly) two-step, were furious as they viewed their profession slipping away from their control. At their meetings they proposed bans on the new dances and demanded that music publishers cease production of ragtime, replacing it with "properly syncopated music."[8]

Even more far-reaching than the dance revolution was the new era that had begun in Harlem, the area of Manhattan north of 110th Street. It was turning black.

There had long been isolated pockets of African Americans living in this otherwise white and affluent suburb, but the migration there had increased in recent years. Due to the efforts of John Nail and other enterprising business-men, unoccupied housing in the West 130s was opened for black residency.

Accompanying black residents were black institutions and businesses. The C.V.B.A. moved its offices to 109 West 133rd Street in September 1912.[9] Theaters, such as the Crescent and the Lincoln, a 200-seat house at 58 West 135th Street, began to program to the tastes of black patrons. Joplin's friends Charles Williams and Gus Stevens, pictured on the cover of the Joplin-Jackson song *Sarah Dear,* were booked at the Crescent for eleven weeks.[10]

The black cabarets and ragtime clubs began leaving the Tenderloin for Harlem. Ike Hines had long since moved his Professional Club to Harlem, to 135th Street in 1903, and then to 23 West 134th Street in 1906. In 1910, Leroy Wilkins, the older brother of Barron Wilkins, opened his own caba-ret, Leroy's, at 513 Lenox Avenue (between 135th and 136th streets). By 1914, Leroy had moved to 2220 Fifth Avenue, near 135th Street. Unlike his brother, who had both black and white patrons (and in later years catered primarily to whites and to blacks who were celebrities), Leroy adhered strictly to a "no whites" policy.[11]

Barron Wilkins closed his Tenderloin club around 1910 and worked briefly for his brother. In 1911, he opened his own club, Cafe Wilkins No. 2 (later called the Astoria Club, and then the Exclusive Club) at 134th Street and Seventh Avenue.[12] Ragtime and jazz great Jelly Roll Morton made an

Scott Joplin's New York: Harlem (1) Lottie Joplin, furnished rooms, 163 West 131st St.,; (2) Lafayette Theatre, 2227 Seventh Ave.; (3) Scott Joplin, music teacher, 160 West 133rd St. (1916); (4) C.V.B.A., 109 West 133rd St.; (5) Barron Wilkins, Astoria Cafe (a.k.a. Cafe Wilkins No. 2; later The Exclusive Club), 2275 Seventh Ave. (1914); (6) Ike Hines's Professional Club, 23 West 134th St. (1906); (7) Lincoln Theater, 58 West 135th St.; (8) Crescent Theatre, 36–38 W. 135th St.; (9) Leroy's, 2220 Fifth Ave. (1914–), 135th St. & 5th Ave.; (10) Connor's Royal Cafe, 71 West 135th St.; (11) Leroy's, 513 Lenox Ave. (ca. 1910); (12) Martin-Smith Music School, 139 West 136th St.; (13) Lottie Stokes, boarding house, 259 West 137th St.; (14) Scott Joplin, 212 West 138th St. (1917); (15) Scott Joplin, 133 West 138th St. (1915–16).

appearance there in 1911.[13] Wilkins's club was immensely popular, and Barron was considered one of Harlem's leading and most beneficent citizens. He met an untimely death in 1924 when he was shot in front of his club for refusing to assist a murderer escape from the police.[14]

Joplin's friends and close associates continued to appear in the news. Joe Jordan had been enjoying success as a performer in Germany when he returned to the United States in February to be with his dying mother.[15] Sam Patterson received good notices in his appearances as a musician and comic with the Watermelon Trust. When the group disbanded in July 1912, he went to Chicago and worked for a while as an entertainer at the Pompeii Restaurant and other spots.[16] In April 1911, Tom and Charles Turpin opened a theater in St. Louis, the Booker T. Washington Airdrome, which became one of the main showplaces for black vaudevillians.[17] Robert Joplin became manager of the Ruby Theater in Louisville in April 1912, but was dismissed after two months. He then returned to the traveling life of vaudeville singer and dancer. In one town in Pennsylvania, he took first prize in a cakewalk contest, winning $100.[18] Fred Van Eps, the new king of the ragtime banjo, recorded *Maple Leaf,* this being the eighth and last recording of Joplin's masterpiece during his lifetime.[19]

A report that Joplin had been "advised by musicians of ability to rewrite the opera, which he enthusiastically set out to do," probably referred to this period, after the failure to produce *Treemonisha* in Atlantic City the previous November.[20] In 1912, Joplin met with one such "musician of ability," this being Harry Lawrence Freeman, a black composer and conductor with operatic experience.[21] Freeman had composed and produced his first opera in 1893 and in succeeding decades was active in black musical theater. In 1910 he was music director of the touring company of Cole and Johnson's *Red Moon.*[22] With these credentials, it is not surprising that Joplin consulted with him, but no details survive of their discussion. If Joplin then set about to revise the opera, it would explain why there is no mention of efforts to have it performed during the rest of the year. It would also mean that the published *Treemonisha,* the opera that we know, was not Joplin's final version.

In addition to his work on the opera, Joplin continued his activities with the C.V.B.A. He was on the Entertainment Committee for its Masquerade Benefit in February, 1912, and its Vaudeville Show in July; at the latter, he, Spiller, and other committee members were seen leading "a strenuous existence—counting the people of the audience." In June he was elected to the Executive Committee, which included also Bob Slater and William Spiller.[23]

Joplin had only one publication during 1912, registered for copyright by Stern on May 1 as *Scott Joplin's New Rag.* The simplicity of the title— almost a non-title—belies the piece's content: it is a rag bursting with life and ideas.

The introduction begins inauspiciously with a syncopated C-major scale played in sixths (Ex. 12-1a). But this beginning foreshadows the importance

Ex. 12-1. Scott Joplin, Scott Joplin's New Rag *(1912): (a) Introduction,
A1 8; (b) A15 16; (c) C1 4.*

(a) [Introduction]

(b)

(c)

that scales play in the rag. The B strain is based on the A-minor scale and
chromatic scales are used in both the D strain and in the coda.

While the introduction begins with the strongest possible statement of a
tonality, it ends ambiguously. Does the e-g interval (measure 4) mean C-major or E-minor? The ambiguity is intentional, for Joplin adapts the ensuing
lead-in to the A strain to both tonal implications: at measure 8 it follows
E-minor; at measure 16 it follows C-major (Ex. 12-1b). In a sense, C-major
and E-minor are the rag's tonal poles.

The first strain, in C-major, uses the tonal ambiguity of the diminished chord for unexpected shifts. In the first phrase, the chordal surprise (at measure 3) is mild; in the second, the surprise (at mm. 7-8) is much heightened with a cadence not to the usual dominant (G-major), *but to E-minor.*

The trio continues the surprises. The C strain is in the tonic rather than the expected subdominant, and is richly harmonized with diminished and applied dominant chords (Ex. 12-1c). Against the quick harmonic rhythm and the syncopations, the bass line presents distinct scale fragments.

The D strain is in E-minor, and entirely forgoes the usual ragtime bass in favor of a banjo-like strumming, as Joplin had done also in the trios of *Euphonic Sounds* and *Paragon Rag.* The syncopated chromatic scale of the coda is more than a tag ending. With its inherent tonal ambiguity, the scale acts as a counterbalance to the tonal certainty of the syncopated major scale in the introduction. Together, the introduction and coda, enclosing the piece's two ends, summarize the rag's fluctuations between tonal certainty and ambiguity.

In considering this and the preceding three rags—*Euphonic Sounds, Paragon,* and *Stoptime*—we note several trends in Joplin's writing. He was moving away from the mechanical octave-chord alternation in the bass, and he was trying tonal patterns other than the stereotyped tonic-subdominant design. Joplin was not discarding ragtime; he was stretching its boundaries, looking for new ways to broaden and enrich the language.

Chicago in 1913 was a great city for ragtime. This was the assertion of Perry "Mule" Bradford, a popular vaudevillian who in later years was to become an important songwriter (his most successful song being *Crazy Blues,* 1920) and arranger. Perry mentioned Arthur Marshall ("Sparrow Marshall, Scott Joplin's old buddy") as one of the leading players. Other Joplin associates in the city at that time were Sam Patterson, playing violin and leading the Patterson Trio; Joe Jordan, playing violin, piano, leading a large orchestra, singing, and dancing; and Otis Saunders, Joplin's friend from the Chicago World's Fair and Sedalia in the 1890s, playing piano in a four-piece combo.[24] As reported earlier, Joplin and Saunders had parted in a dispute over authorship of the *Maple Leaf Rag.*

In St. Louis, Charles Turpin's success with the Booker T. Washington Airdrome encouraged him to build a larger Booker T. Washington Theater on Market at Twenty-third Street. At its opening on July 14, 1913, its five-piece house orchestra performed several works specifically written for the occasion: the *New Booker Washington Theater March,* by Tom Turpin; *Song Without Words—Summertime,* by Artie Matthews and Charles Hunter; *Battle Axe Company March,* by Dink Coppridge, the orchestra's violinist and leader; and *Cream of Society,* by Ruth Heath, the house pianist.[25]

In New York, the mayor decreed that supper clubs must adhere to a 1:00 A.M. curfew. This policy particularly hurt the clubs in the theater district and was a major cause for the Marshall Hotel, the favorite meeting place for black entertainers and intellectuals, to close on September 4, 1913.[26]

The elegant restaurants that had opened dance floors and featured dance exhibitions added another innovation: afternoon "tango teas," at which patrons (sipping beverages stronger than tea) were taught the steps of two South American dance sensations: the Argentinian tango and the Brazilian maxixe. The dances and the afternoon "teas" were roundly condemned by moralists, who tried to have them banned. The noted African-American minister Adam Clayton Powell complained, "The race is dancing itself to death."[27] In 1914 and 1915, the black cabarets in Harlem brought the afternoon tango teas uptown. John Connor's Royal Cafe, at 71 West 135th Street, was the first. Barron Wilkins was next, and soon "teas," tango picnics, and tango contests were held all over Harlem.[28]

In February, Stark retrieved from his files another Joplin–Hayden collaboration, *Kismet Rag*. As with *Felicity*, Stark omitted Hayden's name from the cover. Joplin's contribution to this joint effort seems to be the C theme (as with the other collaborations, he uses the right-hand sevenths and builds upon a motif presented by his younger colleague) and possibly the D theme. However, the link between the C and D strains does not work harmonically and could not have been composed by Joplin (Ex. 12-2). This awkwardness suggests that if the D strain is by Joplin, it probably was not intended for this work.

Ex. 12-2. Scott Joplin and Scott Hayden, Kismet Rag *(1913), C15-16, D1.*

In April, Lester Walton reported on a talk he and Joplin had on ragtime and on the music of the popular theater. Joplin observed that many African Americans were ashamed that ragtime was a creation of their race. This shame is unfounded, he noted, for only ragtime lyrics deserve censure, not the music itself.

> There is no harm in musical sounds. It matters not whether it is fast ragtime or a slow melody like "The Rosary."
>
> Scott Joplin, who is recognized as one of the world's greatest writers of ragtime, in discussing the question, said:
>
> "I have often sat in theatres and listened to beautiful ragtime melodies set

to almost vulgar words as a song, and I have wondered why some composers will continue to make the public hate ragtime melodies because the melodies are set to such bad words.

"I have often heard people say after they had heard a ragtime song, 'I like the music, but I don't like the words.' And most people who say they do not like ragtime have reference to the words and not the music.

"If some one were to put vulgar words to a strain of one of Beethoven's beautiful Symphonies, people would begin saying: 'I don't like Beethoven's Symphonies.' So it is the unwholesome words and not the ragtime melodies that many people hate.

"Ragtime rhythm is a syncopation original with the colored people, though many of them are ashamed of it. But the other races throughout the world are learning to write and make use of ragtime melodies. It is the rage in England to-day. When composers put decent words to ragtime melodies there will be very little kicking from the public about ragtime.

"There has been ragtime music in America ever since the Negro race has been here, but white people took no notice of it until about twenty years ago."[29]

Joplin himself had written "coon" lyrics, as in *The Rag-Time Dance,* and had set such lyrics by others to music. We see that he now had misgivings about the practice.

By the late spring of 1913, after a year of rewriting, Joplin was again ready to mount a production of *Treemonisha.* In June and July he announced that *Treemonisha,* now referred to as a "comic opera," was in rehearsal and would have its first showing on July 14th at the Washington Park Theatre in Bayou, New Jersey.[30]

There is no Bayou, New Jersey. Joplin had probably written "Bayon," misspelling *Bayonne,* a city across the river from New York. The newspaper editor mistook the "n" for a "u," compounding the error. Checking the Bayonne newspapers, we find a notice for Scott Joplin's Colored Comedians and an advertisement for Scott Joplin's Merry Makers, performances to be on July 14th, 15th, and 16th. The company would be sharing the program with "five reels of the latest photoplays."[31]

The planned performances were apparently not of the full opera. But they must have consisted of more than just "A Real Slow Drag," for that number is too brief to have been listed as a complete show. There are no reviews or further notices in the Bayonne paper, and Joplin makes no mention of performances in the *Freeman* or the *Age.* The performances were probably cancelled, or, if they did occur, must have been so poor that Joplin preferred to have them go unnoticed.

In anticipation of the performances, Joplin issued a separate publication of "A Real Slow Drag." The copyright application shows a new address: 252 West 47th Street.[32] Since 128 West 29th Street was on the *Treemonisha*

WASHINGTON PARK
THEATRE Avenue C and First Street ... ...

MONDAY, TUESDAY and WEDNESDAY
SCOTT JOPLINS MERRY MAKERS
"In a Real Slow Drag"
20 Colored Comedians Including 10 Pretty Girls Singers and Dancers.

Doors Open 7:30.   ADMISSION, 5, 10 and 20c.

Advertisement, The Evening Review (Bayonne, N.J.), *July 14, 1913, 2.*

copyright application (May 29, 1911), Joplin moved between June 1911 and July 1913. The new residence was a three-room apartment in a narrow, six-story building, and the reason for the move, we can safely assume, was a marital relationship established with Lottie Stokes.[33]

The first sign we have of Lottie is on a business agreement she signed on October 20, 1913, establishing her as co-owner of the newly formed Scott Joplin Music Publishing Company. Both she and Joplin list the same address, but since Lottie used the name "Stokes," they were not married. (A woman during that period would not have retained her maiden name.) Not surprisingly, in view of other inaccurate statements Lottie had given interviewers, this document conflicts with her testimony regarding their relationship.

In an interview in 1950, Lottie said that Joplin had arrived in New York in 1904 and that, from 1907 on, they had "lived together as man and wife."[34] Her wording suggests they may have had a common-law relationship rather than a contractual marriage. There is no record of a marriage in New York.

We have found no sign of Joplin in New York in 1904, but we cannot rule out the possibility that he was there briefly. We lose track of him after Freddie's death in September 1904, so he could have been in New York for a few months before going to St. Louis early in 1905. We strongly doubt, however, that Joplin and Lottie established a marital relationship in 1907. She was evasive in what she told Blesh and Janis, for they wrote that a marriage took place "about 1909."[35] Arguing against a marriage in either 1907 or 1909 is the census of April 1910, in which Joplin listed himself as a widower. Nor is it likely that they were cohabiting in 1910, for the census shows only male lodgers at the 29th Street boarding house.[36]

Copyright application for A Real Slow Drag.

Lottie gave Brun Campbell a specific marriage date: June 18, 1910.[37] They may possibly have established their relationship by that time, but again, known facts suggest otherwise. If they were "together as man and wife" by June 1910, we wonder why Lottie was not present at the two parties in 1911 at Christie Hawkins's home, especially the second one, which was held in Joplin's honor.

The business contract signed in October 1913, making Lottie co-owner of the Scott Joplin Music Publishing Company, shows that Joplin had total trust in her. There was certainly a personal relationship by that time, regardless of whether it was contractual or common-law. The importance of the

Borough of Manhattan, City, County and State of New York, under
the Name of

SCOTT JOPLIN MUSIC PUBLISHING COMPANY,

and that the true or real full names of the persons conducting
and transacting such business, with the post office addresses
of the said persons, are as follows:

NAMES | POST OFFICE ADDRESSES.

SCOTT JOPLIN | No. 252 West 47th Street, Borough of Manhattan, City of New York.

LOTTIE STOKES | No. 253 West 47th Street, Borough of Manhattan, City of New York.

Dated, New York City, October 20, 1913.

IN PRESENCE OF:

CITY AND COUNTY OF NEW YORK, ss.

On this Twentieth day of October, One Thousand Nine

Certificate of Scott Joplin and Lottie Stokes Conducting Business Under the Name of Scott Joplin Music Publishing Company, Oct. 20, 1913.

partnership is that it relieved Joplin of business responsibilities, something he was very poor at. (Patterson commented, "He did not know the first thing about saving a dollar.")[38] In addition, Lottie developed a separate source of income; as Belle had done years earlier, Lottie operated boarding houses.[39] Real-estate records show that neither Lottie nor Joplin owned property, but they could have rented rooms within their own apartment and rented other apartments for sublease. This was an income source that Lottie was to use for the rest of her life.

Through July and August of 1913, both the *Freeman* and the *Age* carried notices that Joplin wanted to add cast members to his company, and that he had signed an agreement with Benjamin Nibur for a production at the Lafayette Theater in late September or the first of October.[40] The wording in one of the notices ("Scott Joplin, the ragtime king, has signed several contracts, iron-clad, for the appearance of his big comic, ragtime opera in some of the larger theaters in and around New York") suggests there were promised productions that never materialized.[41] On August 14th he again expressed defensiveness about his status as a ragtime composer, saying *Treemonisha* was grand opera:

> "Treemonisha" is a grand opera—Scott Joplin, the well known composer, says: "I am a composer of ragtime music but I want it thoroughly understood that my opera "Treemonisha" is not ragtime. In most of the strains I have used syncopations (rhythm) peculiar to my race, but the music is not ragtime and the score complete is grand opera." Mr. Joplin has made arrangements with Benjamin Nibur for the production of this opera at the Lafayette Theatre early in the fall with a company of forty singers, supported by an orchestra of twenty-five musicians.[42]

On the same page he advertised for singers:

Singers Wanted at Once
CALL OR WRITE
SCOTT JOPLIN
252 West 47th St., New York City
State Voice You Sing

The first of October passed without a performance. Instead, the *Freeman* contained a notice that "Scott Joplin has again started rehearsing his comic opera. He now had twenty-two people."[43] Still, there was no production.

In December 1913, Joplin registered for copyright a separate publication of the "Prelude to Act III," using the imprint "Scott Joplin Music Publishing Company" for the first time. He made some minor changes in expressive and performance markings, but these revisions are too insignificant to justify a new edition. More likely, he wanted to create interest in the opera by making an inexpensive selection available. He then turned it over to Henry Waterson's Crown Music Company for distribution.[44]

Theater Segregation and Harlem's Lafayette Theater

 In trying to open *Treemonisha* at the Lafayette Theater, Joplin was aiming for the top black-oriented house in New York. Most New York theaters had segregated seating, policies that naturally rankled the black theatergoing public. When the Dunbar and Cook show *In Dahomey* opened in 1903 at the New York Theatre, at Broadway between 44th and 45th streets, the first time a black musical was to be shown in a

Copyright application for Prelude to Act III.

quality house, it was a major event for the African-American community. Surely, segregated seating would be suspended for an all-black show. Blacks prepared for a festive event showed up at the theater in livery cabs, dressed in formal evening wear, prepared to pay top dollar for box seats . . . and were ushered to the side entrance into the gallery. "Orchestra and boxes of this theatre are for our white patrons and no others," declared the manager.[45]

A week later, after vehement protests, presumably supported by the show's cast, the theater management "compromised." There would still be racial separation, but blacks would be permitted into selected priority seats—sections of the dress circle and top tier of boxes—as well as the balcony and gallery. The lower boxes and orchestra were still the exclusive preserve of white patrons. "No colored person will be admitted to the sections reserved for the whites, and no white person will be admitted to the parts of the house reserved for colored people. In this way all question of race discrimination is avoided."[46] That blacks accepted this arrangement illustrates the state of race relations at that time.

After the black community took hold in Harlem, the Johnson Amusement Company was formed in 1911 to build a non-segregated theater at 138th Street between Fifth Avenue and Lenox Avenue. The corporation officers included Thomas Johnson (who had tried to arrange an Atlantic City performance of *Treemonisha* in 1911), Lester Walton, and Barron Wilkins. The theater was to be called the Walker-Hogan-Cole Theatre, in honor of the three black stage stars who had passed away since 1909. There was great excitement over the project, and a drawing of the theater was printed in the *Age*. Plans then foundered.[47]

In the meantime, Benjamin Nibur and Henry Martinson, two successful white businessmen, saw their opportunity. They operated Harlem's most successful liquor dealership, had a large black clientele, and were known in the community as "friends of the Negro," always ready to support charities and assist black businessmen. In the summer of 1911 they decided to try their hands at theater. They purchased the Crescent Theatre at 36-38 West 135th Street, a small movie and vaudeville house that had many black patrons. By the fall of that year, they realized that the potential market of black theatergoers exceeded what the Crescent could handle and decided to build a larger Harlem theater. They organized a consortium of investors to build the 1500-seat Lafayette Theater at Seventh Avenue and 132nd Street, on the boundary of the black and white neighborhoods. They hoped to attract patrons from both communities.[48]

The Lafayette opened on November 4, 1912, and immediately had problems. It was opened as a "white theater." "Highly respectable colored people" were permitted to sit in the orchestra, but the "riff-raff" were restricted to the balcony. This policy outraged the black community, and attendance at the theater was consequently low.

Nibur and Martinson recognized their error and tried to institute open seating, but were voted down by the board of directors. They then resigned from the board and organized a new group of investors, who purchased the Lafayette for close to $100,000. In February 1913 they reopened the theater, advertising a policy of equal seating and courteous treatment for all. (The theater was now entirely within the boundaries of the black community, which had expanded another block south, to 131st Street.) Their example was praised in the black press, which pointed out that blacks and whites could, indeed, occupy adjoining seats without disruption.[49]

The theater prospered initially, but then frequently ran into problems of poor attendance. Blame was placed on the types of vaudeville acts that were booked, and different managers were employed. Lester Walton, as entertainment editor of the *Age*, had long urged blacks to become more active as businessmen and entrepreneurs, especially in Harlem. It was the way, he maintained, that the African-American community could take control of its own destiny. He had been part of the Walker-Hogan-Cole Theatre project that failed, but in 1914 he saw another opportunity to demonstrate how a black theater should be run. In May, he leased the Lafayette Theater (with backing from J. Morganstern) for a yearly sum of $25,000.[50] This effectively put the Lafayette into the control of a black man who was intimately familiar with theater and entertainers, but profits were still irregular. Nevertheless, until surpassed by the Apollo Theater on 125th street two decades later, the Lafayette was to remain Harlem's—and New York's—foremost black theater.

James Reese Europe resigned from the Clef Club on December 30, 1913, and formed the Tempo Club, a smaller, more select organization. At the same time, he became music director for Vernon and Irene Castle, America's favorite dance team, providing music with his Society Orchestra. His association with the Castles gave him national prominence, increasing the popularity of black dance music and dance musicians.[51] The Clef Club continued functioning without Europe, and at a concert on June 4, 1914, Sam Patterson, a member since at least 1912, was one of the thirty pianists who performed.[52]

At Harlem's Lafayette Theater, Wilbur Sweatman took over leadership of the orchestra in late April for about a month. It was Sweatman who had made the first reported recording of *Maple Leaf Rag,* around 1903. Since that time, he had become known as an outstanding musician and bandleader. He played trombone, piano, and organ, but established an enviable reputation among musicians for his virtuoso playing of the clarinet. In vaudeville, he developed a novelty routine of playing two clarinets at once. When the public tired of this, he managed a simultaneous playing of three clarinets. As a composer, his biggest hit was *Down Home Rag* (1911), which was frequently included in Clef Club concerts.[53]

Sweatman and Joplin were friends, although we have no detailed information on the relationship.[54] After Joplin's death, Sweatman and Lottie remained friends for the rest of their lives, and Sweatman was to play an important role in the Joplin legacy.

Joplin was also associated with Jelly Roll Morton, but, again, the extent of this relationship is not known. Lottie said Joplin and Morton corresponded, but she did not know if they had ever met. Further, she claimed that Joplin had helped Morton with his popular *King Porter Stomp,* a claim we cannot confirm or evaluate.[55] This piece was reportedly composed in 1906, but the earliest documented sign of it is in a recording by Morton in 1923.[56]

Otis Saunders was now in New York, accompanying singers at the Dunbar Hotel, which had replaced the Marshall at 129 West 53rd Street.[57] Whether he and Joplin renewed their friendship at this time is not known.

In July, Joplin self-published *Magnetic Rag.* This was the last rag of his that he was to see in print; since he published it himself, we can assume it reflects his wishes in every respect.

Its appearance is different. Instead of the usual ragtime meter of 2/4, he puts it in "common time," or 4/4. Since the note values are all doubled, the changed metric marking has no practical effect on the way the music is performed.

Revealing Joplin's aims and attitude, he put the tempo indications in Italian, as is customary in classical music. (He used Italian also in *Tree-monisha* and in his previous rag publication, *Scott Joplin's New Rag.*) At the beginning, instead of the customary "Slow March Time," he indicated "Allegretto ma non troppo" (a little fast, but not too much); before the fourth strain, he redundantly repeated the meter and indicated "Tempo l'istesso" (tempo remains the same). The latter was probably a warning that one should not slow down for the minor-mode D strain. That Joplin should use Italian in a publication over which he had complete control suggests that he was making a statement regarding the seriousness of his work: it should be considered a classical piano piece.

Perhaps, too, the black-and-white cover design, typical of classical publications, was intended to impart this message. However, one cannot rule out the cost factor. The cover used may have been a stock cover that the printer had on hand. Adding the lettering would have been less expensive than creating an original, colorful, popular music cover. Regardless of the factors that went into the decision, the "classical" cover is consistent with the Italian tempo indications.

The music reveals that, in this last rag, Joplin was still striking out in new directions. As with *Scott Joplin's New Rag,* he ignores the conventional tonal patterns for rags and puts the four themes in tonic, relative minor, and parallel minor keys. This use of two minor keys is unique in his output, and both of these strains have a distinct Hebraic cast. Martin Niederhofer, a Joplin student, had commented to his wife that it was strange for a black man to write such "Jewish sounding" melodies.[58]

Copyright application for Magnetic Rag.

I disagree. I do not think it was at all strange. By this time, Joplin had been in New York for seven years and was absorbing the sounds around him. The sounds included the music of a large, bustling immigrant Jewish population and a vibrant Yiddish theater. Joplin had shown in his opera that he was open to musical influences other than African-American. It was perfectly natural for him to extract from his urban environment what was useful to him.

He also extracted from the most current African-American music, shaping it to his needs. This is apparent in the C section, which follows a blues harmonic pattern, but extended beyond the conventional 12-measure boundary. The blues had been known to the black community for more than a decade, but had just recently become popular, beginning with W. C. Handy's *Memphis Blues* (1912). Joplin had hinted at blues with "flatted-3 to 3" appogiaturas as far back as *Original Rags,* but this was the first time he used the full blues harmonic pattern (Ex. 12-3.).

Ex. 12-3. Scott Joplin, Magnetic Rag *(1914): (a) C1-14; (b) harmonic analysis of C1-14, revealing blues pattern.*

Harmonic Analysis of C 1–14

I	I	I	$\flat 7$		
IV	IV	IV	IV	I	I
V	V	I	V^7		

Also notable in this section is Joplin's asymmetrical phraseology, virtually unknown in ragtime literature (although present in the B strain of *Paragon Rag*). In addition to the usual 4-measure phrases, he has extended phrases of 6 measures. The entire strain is of 24 measures, but is not evenly divided into 12 and 12, as would be expected for a blues, but 14 and 10.

This is not the last Joplin work to be published; pieces composed earlier appeared in print later. But *Magnetic Rag* is Joplin's last composition to survive. It is our last musical view of him.

And what a view it is. Though the 46-year-old Joplin was almost at the end of his career, in seriously declining health, he was not declining as a musician. He was still master of his craft, at the height of his creative powers. He was continuing to respond to his musical environment, thinking in new ways, creating a new type of ragtime. This was a growing, dynamic composer betrayed by a failing body.

At the time of publication, Joplin offered to sell copies of *Magnetic Rag* to professional performers for twenty-five cents in stamps. Did he actually expect performers, accustomed to receiving free professional copies of music, to pay for *Magnetic Rag,* and to pay a premium price? Was he so unrealistic? Or was his announcement simply designed to appeal to the nonprofessional pianist who might aspire to stage performance? He issued an orchestration in October, originally turning the marketing and distribution over to Henry Waterson's Crown Music and to Enterprise Music. In late December and throughout January he offered to sell the orchestration himself for twelve cents.[59]

Silver Swan Rag, uncopyrighted and known only from piano roll publication, was issued by the QRS Piano Roll Company in July 1914. It was issued also by the National Music Roll Company, date unknown.[60] It did not reach printed form until 1971, when it was included in the *Collected Works of Scott Joplin* as a two-hand transcription that eliminates some of the octave doublings on the roll.[61]

The printed edition stops short of ascribing the piece definitely to Joplin, listing it as "attributed to," but there is no reason to doubt its authenticity. It is included in the QRS catalogues of the period and has the technical polish of a Joplin rag. Furthermore, it has features of Joplin's late rags, such as a B strain in the relative minor, like the B strains of *Magnetic Rag* and *Scott Joplin's New Rag.*

Piano Rolls and the Mystery of "A Princeton Tiger"

 The discovery of *The Silver Swan* piano roll encouraged collectors to believe that other unpublished Joplin pieces might also exist on piano roll. There was special interest in a curious listing in the catalogue of one of the major roll companies.

At the end of 1904, the Q.R.S. Company listed in its piano roll catalogue seven Joplin selections: *Original Rags, Maple Leaf Rag, Swipesy, The Easy Winners, The Entertainer, Palm Leaf Rag, . . .* and *A Princeton Tiger, March and Two Step,* catalogue number QRS X3075. Roll collector Michael Montgomery noted this listing in the 1960s and called it to the attention of other collectors. The search began for an unknown Joplin work.

Through the next twenty years, various facts surfaced: Q.R.S. catalogues continued ascribing *A Princeton Tiger* to Joplin at least until 1911; a piece by the same title was registered for copyright by one Gerald Burke on December 20, 1902; the Burke piece was issued by six different piano roll companies and did not appear to be in Joplin's style. It would seem either that the Q.R.S. catalogue, usually carefully edited, had an error that went uncorrected for seven years, or that Joplin had composed a piece by the same title. A conclusive answer could be reached only by locating the roll.

In June 1993, thirty years after the search began, the mystery was solved. Paul Johnson, another collector, found QRS X3075. Both the roll and the box it came in ascribed *A Princeton Tiger* to Joplin, but it was the Burke piece.

Why had Q.R.S. allowed its catalogue to perpetuate an error for such a long time? Probably because copyright laws were slow in catching up with technology. Until 1909, piano rolls were not covered by composers' or publishers' copyrights. Since neither Burke nor his publisher was due a royalty for the march, they probably did not bother to complain—or if they did complain, it was to no avail. As long as Q.R.S. had rolls in stock erroneously ascribed to Joplin, a change in the catalogue would have been confusing.

This brief story would have been much more dramatic had the roll been a previously unknown Joplin piece. But even with the disappointing conclusion, the tale illustrates our indebtedness to collectors. Without this thorough and tenacious breed, many of the artifacts of ragtime would have long since disappeared.[62]

At the end of 1914, Joplin placed a notice that he had moved to Harlem and would be teaching: "Scott Joplin, the composer, has moved from 252 West 47th street to 133 West 138th street. He will devote a part of his time to the instruction of pupils on the violin and piano."[63]

The building in Harlem that Joplin and Lottie moved to was a solid row house in an established black neighborhood, one block from a particularly attractive street (the 200 block), with houses designed by some of the nation's most distinguished architects. This latter block became the home of some of the most successful African Americans, many in the entertainment field, and in a few years acquired the name "Strivers' Row."

That Joplin and Lottie were able to move to the more desirable area of Harlem was probably due to Lottie's earnings in operating the boarding house, for Joplin was showing signs of financial desperation. In the last months of 1914, he was trying to sell his music through the mail. In February 1915, he advertised he would sell not only his own music, but any music published in New York.

> Six Copies of Music for One Dollar.
> Send one dollar in money order to Scott Joplin and receive in return six piano
> copies of any of his compositions you may select or six assorted of any popular
> pieces published in New York. 133 W. 138th St., New York.[64]

Poor though he may have been, he did not stint on the tools of his trade.
Quality sound was important to him, and sometime after 1908 he purchased
a large Steinway grand piano. Unlike most Steinways, its sound was mellow
rather than brilliant, somewhat like that of a Mason-Hamlin, making it especially suitable for vocal accompaniment.[65]

Joplin had not given up on *Treemonisha* and succeeded in arranging for
an orchestral performance of the Act 2 ballet, "Frolic of the Bears." Renamed
"Dance of the Bears" for the performance, it was included in the annual
concert of students of the Martin-Smith Music School on May 5, 1915.

The Martin-Smith Music School, at 139 West 136th Street, two blocks
from Joplin's home, enrolled some 300 black children, and the recital was
designed to show "the progress . . . colored people of New York are making
along serious musical lines. Wonderful exhibition of colored children of ex-
ceptional talent."[66] The concert was given at the Palm Garden, 150 East
58th Street, and the orchestra was augmented by one hundred professional
musicians. The major work was *Kubla Khan,* for chorus and orchestra, by the
black British composer Samuel Coleridge-Taylor. Of "The Dance of the
Bears," the reviewer stated that the conductor was unable to bring out "the
full value of Joplin's tone poem." The only orchestrally performed selection
from his opera that Joplin was ever to hear was apparently short of success.[67]

To capitalize on this performance, he issued a separate, slightly revised
piano version of the "Frolic of the Bears," placing notices in newspapers. In
July, Crown, Enterprise, and A. H. Goetting of Springfield, Massachusetts,
a major distributor of Joplin rags, also advertised the piece.[68]

In September, Joplin announced two new works. The first was *Morning
Glories:* "Scott Joplin, the composer, will soon put another catchy number on
the market entitled 'Morning Glories.'"[69] The piece remained unpublished.
Thirty-five years later, Rudi Blesh saw *Morning Glories* in manuscript; he
described it as an incomplete song.[70]

The other piece was a complete vaudeville act, reportedly in rehearsal,
with staging by his friend Bob Slater:

> Scott Joplin's vaudeville act called "The Syncopated Jamboree," with music by
> Mr. Joplin and staged by Bob Slater, will open in New York. Rehearsals have
> been going on for quite sometime [*sic*]. Many good voices, including Mme.
> Marguerite Scott are in the cast.[71]

Syncopated Jamboree was never performed, and no sign of it has been found.
The Marguerite Scott mentioned in the item was probably Margaret Scott,

who was a singer of operatic quality. She had studied with Dvořák at the National Conservatory in New York, along with such black musicians as Will Marion Cook, Harry Burleigh, and Theodore Drury.

It may have been in 1915 that Eubie Blake had a meeting with Joplin. In different interviews, Blake gave years varying from 1907 to 1915, but from his description of Joplin's health, I would guess that the latest date is most likely.[72]

They met at a reception in Washington, D.C., at which many ragtime pianists were present. Some of the best ones played, and then the crowd called upon Joplin to perform. He refused at first, mentioning to Blake that he was too sick. Finally, in response to the demands, Joplin sat at the piano and played *Maple Leaf Rag:* "So pitiful. He was so far gone with the dog [syphilis] and he sounded like a little child tryin' to pick out a tune. . . . I hated to see him tryin' so hard. He was so weak."[73] "He was dead but he was breathing. I went to see him after but he could hardly speak he was so ill."[74]

Blake saw the end coming. According to Joplin's death certificate, he experienced the onset of his final illness in October 1915.

In 1916, and possibly also in 1915, Martin Niederhofer, a young bachelor still living with his parents on West 106th Street, studied ragtime piano with Joplin. Information about him and his studies comes from his widow, Bertha Niederhofer, whom I interviewed by telephone in 1989. Though Mrs. Niederhofer was then ninety, she was articulate and fairly knowledgeable about Joplin's music. In discussing rags she and her husband used to play at the piano, she was able to sing the first strain of *Magnetic Rag,* music that does not easily lend itself to vocalization.[75]

Martin Niederhofer was a bookkeeper for Waterson, Berlin & Snyder Music Company. Since Joplin was still doing business with Crown Music, which was at the same address, Niederhofer and Joplin may have met there.

Niederhofer loved the way Joplin played and wanted to perform exactly like him. He could read music, but he preferred to imitate Joplin and would ask the composer to demonstrate how a rag should be played. Among the pieces he studied were *Maple Leaf, The Entertainer, Magnetic Rag,* and, the last piece he learned from Joplin, *Pretty Pansy Rag.*

Pretty Pansy Rag is a lost work that Blesh reported seeing in manuscript.[76] Niederhofer had a copy that Joplin had written out for him, but lost it. In the 1970s, after Joplin's music had again become popular, the Niederhofers looked for their copy of *Pretty Pansy Rag* but were unable to find it. Mrs. Niederhofer could not recall any portion of this piece.

Lessons were a dollar, and were usually held at Niederhofer's home. Mrs. Niederhofer laughed at my suggestion that Joplin may have had a studio. "He was too poor for that," she said.[77]

Mrs. Niederhofer met Joplin several times and remembered him as a very pleasant person, "a gentleman," about five foot seven in height. She last saw him after she and Martin had married and moved to the Bronx. Martin had offered Joplin an old coat and fedora, and Joplin took the trip of about 45 minutes to fetch them. She related this incident to demonstrate how poor Joplin was.

On one occasion, Niederhofer went to Joplin's residence for a lesson, which was frequently disrupted by noise and other distractions. He reported to his fiancée that Joplin had all of his brothers and sisters living with him, that it was a big, happy family. However, he could not help but notice one very unusual feature. He was there on a hot, summer night, and since Joplin's brothers and sisters were too poor to own appropriate clothing, they went about with only towels wrapped around them.

Mrs. Niederhofer offered this story as evidence of the close family life that Joplin enjoyed and of the poverty he lived in. What the description suggests is something quite different. Lottie, while operating boarding houses, also operated "houses of assignation," that is, establishments in which rooms were rented for several hours for illicit sex. Lottie had admitted this to Blesh and Janis.[78] In support of Blesh's report, old songwriters from the Tin Pan Alley days have told me that whenever they wanted "girls" for a party, they would call up Lottie Joplin. It would seem that Scott Joplin, who avoided working in the "district," who, "if he had any vices, he kept them to himself,"[79] and who yearned for respectability, spent his final year or so living in a house of ill repute.

In April, May, and June of 1916, Joplin made seven hand-played piano rolls, six for the Connorized Company and one—*Maple Leaf Rag*—for Uni-Record. Since hand-played rolls are cut from actual performances, one might think that these rolls are direct evidence of Joplin's playing.

If only that were so. The rolls were made by having the piano hammers mark a roll as the performer played. A technician then used the marks as a guide for punching holes in a master roll, a procedure that allowed ample opportunity for correcting, modifying, and amplifying the performance. And these rolls were obviously edited. Moreover, while rolls were routinely printed with tempo ranges, the ranges are not evidence of the actual tempo that Joplin used.

Coming so late in his life, after the onset of his final illness, the rolls probably do not reflect Joplin's abilities at an earlier and healthier time. The Uni-Record performance of *Maple Leaf Rag* is painfully bad, revealing extremely poor rhythmic coordination; it is probably the truest record of his playing at that time. The Connorized rolls are markedly better and were probably heavily edited. The extant rolls are *Maple Leaf Rag, Weeping Willow Rag, Something Doing, Magnetic Rag,* and W. C. Handy's *Ole Miss.* A sixth roll, *Pleasant Moments,* was advertised in the Connorized catalogue, but no copy has been found.[80]

In September, Joplin announced he had completed a musical comedy entitled *If* and was busy at work on his Symphony No. 1.[81] The following month, he again mentioned the symphony in a notice that stated he would be going to Chicago to recuperate from a serious illness:

> Scott Joplin, whose eminence as a composer of ragtime music has led him to the title of "Ragtime King," is convalescing from a serious illness. In order to gain a complete rest, he is now making preparations to spend some time in Chicago at his sister's home. He is expected to be back in the Metropolis, though, in time to get off several new numbers for various big productions, especially the great ragtime number which he has already commenced, "Symphony."[82]

He never got to his sister's home. He remained in Harlem, and in times of depression he destroyed many of his unpublished pieces, fearing that they would be stolen after his death.[83] In mid-January he again announced he would be going to Chicago for a rest, but within two weeks entered Bellevue Hospital in New York. On February 3rd he was transferred to Manhattan State Hospital, where he was confined to a mental ward. He died there on April 1, 1917.[84]

We do not have access to Joplin's medical records, but from available information we can surmise what his final months were like. The death certificate lists his diagnosis as "Dementia Paralytica—cerebral form," with syphilis as a contributory cause. The onset of the symptoms had begun eighteen months earlier.

Dementia paralytica, a manifestation of tertiary syphilis, is a general mental and physical deterioration, ending in paralysis and death. It usually occurs between 20 and 30 years after contraction of the disease, although onset is not invariably within that range.

The earliest symptoms may be a loss of recent memory, minor erratic behavior, and irritability; routine behavior may remain unchanged. This can be followed by a manic-depressive stage, with exalted emotional periods and delusions of grandeur alternating with periods of apathy or melancholia. There may also be severe loss of judgment and outbursts of destruction. Physically, there are speech defects, at first slurring and tremors of the tongue and lips, later an inability to complete sentences and unintelligibility; tremors and discoordination of the fingers; pupillary abnormalities. At the end, there are convulsions.[85]

We have every reason to believe that this description fits Joplin's final illness. Eubie Blake described Joplin's feeble attempts to play the piano and, on a second meeting, his inability to speak. Lottie said that, toward the end, he destroyed manuscripts. He seems to have had classic symptoms.

Scott Joplin had been determined to rise from his extremely humble beginnings to become an artist of greatness and a person of refinement and

Scott Joplin's death certificate.

sensitivity. He came a long way and demonstrated what could be achieved with intelligence, hard work, and a sense of purpose. But these admirable traits could not forestall the ravages of disease.

While he knew he had syphilis, he did not know for sure that he would suffer the insidious effects of its tertiary stage. Syphilis frequently remains dormant after the initial symptoms have passed. But as the tremors began and he lost coordination of his fingers, as he lost his ability to speak intelligibly, he must have known what was occurring. It had happened to others, many in the entertainment field. Finally, still unfulfilled, his dreams of greatness collapsed as he slipped into madness. Scott Joplin, the King of Ragtime, died a horrible death.

Joplin's death attracted little attention. Only two newspapers had notices. The *New York Age* had a brief item on page 1, and Lester Walton discussed him on the entertainment page, linking Joplin's madness and death to his inability to have his opera produced:

> Scott Joplin, known throughout the United States as the composer of syncopated music, died Sunday at the Manhattan State Hospital, where he had been confined for a number of months for mental trouble. His death was not a surprise to friends, who had been informed that his malady was incurable. Funeral service will be conducted from the undertaking establishment of G. O. Paris, 116 West 131st Street, Thursday at 1 o'clock.
>
> Scott Joplin first came into prominence as the writer of "The Maple Leaf Rag," which was published in St. Louis about eighteen years ago. He was born about 150 miles from St. Louis some forty odd years ago, and resided in New York about ten years. The deceased is survived by a widow, Mrs. Lottie Joplin.[86]

> Scott Joplin's burning desire to have produced a ragtime opera he wrote many years ago was responsible for the composer's death, is the opinion of his friends. About twelve years ago in St. Louis Joplin started to write the book and music to an opera which he had finished when he came East ten years ago. One of his missions to New York was to interest someone in producing his ragtime opera.
>
> He was advised by musicians of ability to rewrite the opera, which he enthusiastically set out to do; but even after making numerous changes in the book and score found it a herculean task to interest people with money in the opera's production.
>
> His failure to have his opera produced weighed heavily on his mind, and a few months ago he was taken to Ward's Island, where he died Sunday.[87]

John Stark went to his files one last time for an unpublished Joplin work and pulled out *Reflection Rag,* publishing it with the same cover he had used for James Scott's *Sunburst Rag* (1909) and Joseph Lamb's *Cleopatra Rag* (1915). I would guess that *Reflection* stems from around 1907, just before Joplin and Stark parted company. The B strain bears some resemblance to the B of *Magnetic Rag,* suggesting that Joplin has decided to reuse the idea from the still-unpublished rag. The C theme is the most complex, with a nice interplay of outer and inner voices.

Whatever differences Stark and Joplin may have had, they were put aside as Stark wrote a touching eulogy, summarizing Joplin's accomplishments in a single, brief sentence:

Scott Joplin Is Dead
A homeless itinerant, he left his mark on American music.[88]

CHAPTER THIRTEEN

The Legacy
of Scott Joplin

Joplin's was a curious story. His compositions became more and more intricate, until they were almost jazz Bach. "Boy," he used to tell the other colored song writers, "when I'm dead twenty-five years, people are going to begin to recognize me."

Music publisher Edward B. Marks, 1934.[1]

THOUGH HE WAS ONLY FORTY-NINE when he died, Scott Joplin was already of a generation whose time had passed. He may have been the king of ragtime, but ragtime was no longer king. Jazz was the new order of things. The Original Dixieland Jazz Band opened at Reisenweber's in January 1917 and electrified New York's popular music world. When the United States entered into the First World War, two weeks after Joplin's death, ragtime was tottering at his gravesite. At the end of the war eighteen months later, ragtime and Joplin were all but forgotten. Both would sink into near obscurity.

In 1936, Maude Cuny-Hare published her pioneering book *Negro Musicians and Their Music*. She was clearly most comfortable in discussing concert music, but did devote two chapters to the black vernacular styles during the ragtime and early jazz years. Her point of view is evident, for she restricted commentary to the most successful musicians, such as Will Marion Cook, J. Rosamond Johnson, James Reese Europe, Ernest Hogan, Joe Jordan, Williams and Walker, and the like. She made no mention of piano ragtime or Scott Joplin.[2]

Also in 1936, the distinguished black scholar Alain Locke published

The Negro and His Music, in which he included ragtime in chapters on black musical theater and jazz. He discussed piano ragtime, but did not consider it an important segment of the style:

> What passed for ragtime was not the full rhythmic and harmonic idiom of the genuine article as used, for example, by Will Marion Cook and the Negro musical-comedy arrangers who had chorus and orchestra at their disposal, but the thin, and rather superficial eccentric rhythm as it could be imitated on the piano or in the necessarily simplified "accompaniments" of popular sheet music of the day. Still, a few artists like the famous Scott Joplin wrote real rag in compositions like his *"Maple Leaf Rag"* (1898) and *"Palm Leaf Rag"* (1903). Also, Kerry Mills, with his *"Georgia Camp Meeting,"* (1897); *"Rastus on Parade,"* *"Whistlin' Rufus"* set the pace that was to catch the whole country and culminate in that instrumental classic of matured ragtime—Irving Berlin's *Alexander's Ragtime Band.*[3]

To Locke, piano ragtime was nothing but a weak imitation of the "authentic" ragtime of the black lyric stage. He granted that Scott Joplin had something original to say with his piano rags, but immediately weakened his argument by pairing Joplin with Kerry Mills, a composer of popular cakewalks that are on a decidedly unimaginative musical level. To Locke, Joplin did not represent mature ragtime; that status was reserved for Irving Berlin.

Locke's opinions were clearly based on popularity rather than musical values. Not only was he unfamiliar with Joplin's music, he did not know who Joplin was. He thought Joplin was a white musician who, along with songwriter Lewis Muir ("Waiting for the Robert E. Lee") and such jazz musicians as Bix Beiderbecke, deserved "bracketed credit with the Negro pioneers"![4]

What a blow! To think that Joplin, a child of a former slave, an artist who had struggled to overcome the barriers of discrimination, a composer who was intent on speaking to his racial peers in his opera *Treemonisha,* that this man should be so forgotten by those he addressed.

The oversight speaks not just of Cuney-Hare and Locke; they were representative of African-American intellectuals of their generation. They could accept ragtime in its relatively refined presentations by such musicians as Will Marion Cook, J. Rosamond Johnson, and James Reese Europe, but still rejected its original and more earthy guises.

If Joplin was forgotten by the general public and by the black intelligentsia, he was not totally forgotten by jazz musicians and lovers of syncopated piano music. Joe "King" Oliver, Louis Armstrong's mentor, reportedly honored Joplin by having all of his sheet music bound into a single volume.[5] Willie Eckstein, the amazing Canadian virtuoso, made the first piano recording of *Maple Leaf* in 1923.[6] It is a brilliant performance, but one that takes liberties with the score. In introducing habañera rhythms, Lisztian flourishes and other embellishments, it loses many of Joplin's carefully wrought bass

lines and inner voices. This casual attitude toward Joplin's original conceptions would be the pattern with the many *Maple Leaf* recordings that
followed.

And there were many recordings, far more than during the ragtime
years. Though Joplin's name was nearly forgotten, it was kept alive by the
Maple Leaf Rag, which never left the repertory. *Maple Leaf* was recorded at
least nine times in the 1920s, and almost double that number in the 1930s.
The only other Joplin works to be recorded in those two decades, with only a
single recording each, were *Easy Winners, Gladiolus,* and *Original Rags.*

Joplin may have died in poverty, but money was still earned on his
music. John Stark continued to thrive on the *Maple Leaf,* cranking out 5,000
copies per month well into the 1920s.[7] In 1917, blues pioneer W. C. Handy
recorded *Maple Leaf* with his orchestra, but it was not a tribute to Joplin. It
was the plagiarized version going under the title *Fuzzy Wuzzy Rag* by Al.
Morton, published by Handy two years earlier.[8]

Lottie, fearful that Joplin would soon be totally forgotten, was intent on
earning whatever small amounts she could from what Joplin had left. She put
the remaining copies of *Treemonisha* into the hands of the C.V.B.A. for quick
sale, offering it at a "popular" price, reduced from the original $2.50.[9] She
also sold rights to *Magnetic Rag* to the Winn Music Schools—schools that
specialized in teaching popular styles—which issued an orchestration the
same year. Winn transferred *Magnetic Rag* to Jack Mills in 1922, who subtitled it "Syncopations Classiques" and reissued it the following year as one of
six selections in the *Jack Mills Modern Folio of Novelty Piano Solos, Number
Four.*

Lottie continued running her rooming house at 163 West 131st Street,
specializing in boarders from the entertainment world. Wilbur Sweatman
stayed with Lottie, as did Jelly Roll Morton, and the two had furious arguments there.[10] Stride pianist Willie "The Lion" Smith began boarding with
Lottie around 1919 and continued to stop in for many years. This is how he
described it:

> I was living at the time at Lottie Joplin's boardinghouse. That was where all
> the big-time theatrical people stayed and everything was free and easy. Mrs.
> Joplin . . . only wanted musicians and theater people for tenants. The place was
> a regular boardinghouse but sometimes operated like an after-hours joint. She
> had the entire house at 163 West 131st Street and it was a common occurrence
> to step in at six in the morning and see guys like Eubie Blake, Jimmy Johnson,
> . . . sitting around talking or playing the piano in the parlor. We used to play
> Scott's *Maple Leaf Rag* in A-flat for Mrs. Joplin. Before she died she took me
> down in the cellar and showed me Scott's cellar full of manuscripts—modern
> things and even some classical pieces he had written.[11]

Though Joplin had destroyed many manuscripts, some obviously survived.

Lottie learned that there was value in Joplin's copyrights. In 1927 she renewed the copyright on the *Maple Leaf* and through the passing years renewed others as they came due. In 1938 and 1940 she renewed copyrights to *Treemonisha* and its separately published sections.[12] Still, little was earned on them, except for *Maple Leaf*.

It was in the 1940s that ragtime began its comeback. In 1941, swing reigned. This was the music of Benny Goodman, Tommy Dorsey, Count Basie, Duke Ellington, Glenn Miller. It was played by combos and by big bands, conceived both as jazz and as popular music, and many of its top musicians moved easily between both worlds. This coexistence with the commercial, noncritical and popular music world disturbed some jazz musicians and fans, making the time ripe for a reaction. In one camp the reaction came as bebop, an elitist, esoteric, anti-populist art. Another point of view was expressed with "traditional jazz," a return to earlier styles, which brought with it a new look at ragtime.

The latter view was expressed most forcibly and convincingly by Lu Watters's Yerba Buena Jass Band ("jass" being one of the early spellings of "jazz"). The band electrified its San Francisco audiences with the freshness of the rejuvenated "old sound" and, beginning in late 1941, went on to develop a national following with its recordings. The first Yerba Buena recordings included *Maple Leaf* and *At a Georgia Campmeeting,* but it was Wally Rose's solo piano performance of George Botsford's *Black and White Rag* (1908) that made the most indelible impression. Unexpectedly, it became a popular hit. Ensemble ragtime was back, and so was piano ragtime.

The enormous success of these and following recordings by the Yerba Buena and by Wally Rose put ragtime, and ragtime piano, back into the spotlight. Pee Wee Hunt's dixieland rendition of Euday Bowman's *12th Street Rag* (1914) became one of the biggest selling recordings of the late 1940s.

As jazz fans became curious about the old music, a few began to research the subject and write articles. The first piece on Joplin was "Scott Joplin: Overlooked Genius," which appeared in a small record magazine in 1944.[13] Five years later, Rudi Blesh and Harriet Janis, owners of Circle Records, started investigations that resulted in their writing "the bible of ragtime": *They All Played Ragtime* (1950), a remarkable biography of Scott Joplin and a social history of the ragtime world. The same year saw the publication of another book on a ragtime great: Alan Lomax's *Mister Jelly Roll: The Fortunes of Jelly Roll Morton, New Orleans Creole and "Inventor of Jazz".*[14]

Piano ragtime became more familiar in the 1950s. It was presented as a happy, good-time music, played on rinky-dink, honky-tonk pianos, frequently doctored to make them sound tinny. The most popular pianists were Joe "Fingers" Carr (Lou Busch) and Johnny Maddox, sometimes called "Crazy Otto" (after his most successful recording). Other notable performers were John "Knocky" Parker, who in 1965 made the first complete recording of Joplin's rags,[15] and Dick Hyman, the latter recording under the pseudonyms of "Knuckles" O'Toole, Willie "the Rock" Knox, and other names. Bob Darch was less well known but had a significant impact as he traveled the country and brought live ragtime to new audiences. He included Sedalia in his tours in the late 1950s and made Sedalians aware of two of their town's surviving musicians of the ragtime years, Arthur Marshall and Tom Ireland. Also, as a result of his concerts, Sedalians became interested in Joplin and formed a Scott Joplin Memorial Foundation Committee.[16]

At the close of the decade, Max Morath appeared on the scene. Morath was a scholar as well as a superb entertainer. He had carefully researched the history of the ragtime years, so he understood the songs and piano music in a social context. His performances were not concerts; they were lively and humorous illustrations of turn-of-the-century American culture as reflected in song, piano music, anecdotes, and antique photo slides. Without quite realizing it, his delighted audiences left with a deeper appreciation and understanding of the ragtime years. Beginning with *The Ragtime Era* and *Turn of the Century* for public television in 1959–61, Morath developed a succession of shows that spread the word in television, night-club and theater acts, recordings, and ragtime festivals.

A more discriminating audience developed for ragtime. It was now recognized as a distinct genre rather than merely a subset of traditional jazz. Ragtime aficionados formed their own fan clubs and newsletters: the *Ragtime Review* (1962–67), edited by Trebor Tichenor and Russ Cassidy; the Ragtime Society (in Toronto), with its newsletter *Ragtimer* (1962–86); the Maple Leaf Club (of Los Angeles), with *Rag Times* (1967–). The latter two organizations also held musical meetings and promoted ragtime festivals. Along with modern performers, they brought to audiences such figures of the ragtime years as Joe Lamb, Joe Jordan, Eubie Blake, and Shelton Brooks. Regardless of fluctuations of interest by the larger American public, this core of devoted fans kept ragtime flourishing—on a small scale, to be sure—and dynamic.

Scott Joplin was a focus to much of the ragtime revival. In 1959, Sam Charters and his ragtime-pianist wife Ann recorded the comments and playing of Joplin's friend and colleague Joe Lamb, releasing it as *Joseph Lamb: A Study in Classic Ragtime*.[17] A few years later, they spoke of *Treemonisha* to Ted Puffer, director of the opera workshop at Utah State University. Puffer was easily convinced. After seeing the score, he programmed excerpts for a performance by the University Concert Chorale on February 28, 1965. Joplin was put in good company that evening as his excerpts shared the program with Bach's *Magnificat*. But it was "A Real Slow Drag" that had the audience dancing in the aisles.[18]

Lottie lived to witness the beginning of the Joplin and ragtime renaissance. Brun Campbell, who as a 15-year-old runaway had gone to Sedalia to study with Joplin in 1899, befriended Lottie and engaged her assistance, as well as that of clarinetist Tom Ireland, one of the founding members of

Sedalia's Queen City Cornet Band. Campbell collected information and a few photographs and wrote several articles about Joplin.[19] While I now regard much of what Campbell wrote to be inaccurate, his intentions were honorable. He placed his letters and other materials in a Scott Joplin Collection established at Fisk University, a traditionally black college dating back well into the 19th century.

Others also sought out Lottie, including writers Kay Thompson, Rudi Blesh, and Harriet Janis. Some asked if Joplin had left any unpublished manuscripts. In response, she told an interviewer, "to get rid of them, I sometimes let them have a few scraps, but sooner or later, I always get after them, and make them bring them back."[20] One has to wonder whom she gave those "scraps" to, and whether she really got them all back. Her wording suggests an alarmingly casual attitude.

Blesh and Janis were among those who saw the cellar of music, and they wrote down titles: the songs *Morning Glories* and *For the Sake of All;* a song version of *Magnetic Rag;* incomplete orchestrations of *Stoptime Rag* and *Searchlight Rag;* and piano pieces *Pretty Pansy Rag* and *Recitative Rag;* other items they left unnamed.[21] Blesh and Janis were permitted to borrow the manuscripts for *Pretty Pansy Rag* and *Recitative Rag;* in later years Blesh regretted that they had been too ethical to photocopy the manuscripts before returning them.[22]

Before Lottie died in 1953, she formed the Lottie Joplin Thomas Trust as a means of protecting her husband's copyrights and other materials.[23] She made her long-time friend Wilbur Sweatman a trustee. After her death, Sweatman applied to the Surrogates Court to become executor of her estate, which, excluding royalty rights, had a net value of less than $4,000. As executor, Sweatman presumably distributed royalties of the Joplin works to Lottie's three siblings and her niece Mary Wormley. He also assigned *Treemonisha* to his own publishing company and retained physical possession of much Joplin sheet music and the manuscripts.

Pete Clute, successor to Wally Rose as pianist with the Turk Murphy band, visited Sweatman's office in the late 1950s in search of Joplin sheet music and found Sweatman extremely generous.[24] Leonard Kunstadt, editor of *Record Research* magazine and co-author, with Sam Charters, of *Jazz: The New York Scene,* befriended Sweatman and reported seeing in his office a duffle bag full of Joplin manuscripts, including a piano concerto. Kunstadt told me he did not bother examining the manuscripts closely because he never expected Joplin to become "so big." He also saw about 90 pages of an autobiography Sweatman was writing. Glancing through it, he noted discussions of Sweatman's friendship with Joplin, Sweatman's arguments with Jelly Roll Morton at Lottie's boarding house, revelations on Tony Jackson and Mike Bernard, and the like.[25]

In March 1961, Sweatman died, leaving no will. Barbara Sweatman, his daughter, visited him in the hospital and received his keys. After his death, she took possession of his apartment and everything in it—which included three clarinets, tape recordings, photographs, press clippings, and an immense amount of sheet music and music manuscripts, both Sweatman's and Joplin's. She immediately emptied the apartment of the music and other items she thought might be of value, and stored them with a friend, paying him a storage fee. Some things from the apartment had been stolen and she gave away or sold other items. In June she moved into the apartment herself and retrieved what she had stored with the friend.

Meanwhile in April 1961, one month after Sweatman's death, Harry G. Bragg, a black lawyer who had represented such figures in the black theatrical and musical worlds as the singer and composer Harry T. Burleigh,[26] appeared in court to challenge Barbara Sweatman's possession of the estate. He represented Wilbur's sister Eva Sweatman of Kansas City, who claimed that Wilbur Sweatman had no children.

A hearing was held on this issue and the court accepted Barbara as Sweatman's daughter, but as she was born out-of-wedlock she had no rights to the estate. Barbara was ordered to turn the property over to Bragg. She failed to do this for several months, until threatened with arrest. Finally, on September 22, 1961, she appeared in court and gave the following testimony:

Q. Will you tell us what that inventory [of the Sweatman estate] was?
A. The inventory consists of almost a room 9 × 9 of Scott Joplin's compositions and music and Wilbur Sweatman's compositions and music.
Q. How much was there of music of Joplin's?
A. I wouldn't know. I would say maybe . . . half a room as far as one side of the room is concerned.
Q. From floor to ceiling?
A. No, I would say from the floor to half the wall.
Q. And how much of Wilbur C. Sweatman's?
A. Oh, about five packages—well, say it is about like from here to there (*indicating*). I didn't measure it. . . .
Q. Say 3 × 4?
A. Yes. I would say that. . . .
Q. Scott Joplin's compositions you have?
A. Yes, I have those.
Q. Sweatman's compositions you have?
A. I have those.[27]

Barbara Sweatman turned over to Bragg the property that was in the apartment, but there is no mention of what happened to the contents of Sweatman's office. She knew her father had been writing an autobiography, but she had never seen it. Regarding the property turned over to Bragg, the lawyer appeared in court on November 6 for a hearing to determine estate taxes. He submitted an inventory of the property, with appraised values. The listing included "11 boxes unpublished and uncopyrighted music . . . no value."

Since this was a hearing to determine taxes due on the estate, the lawyer naturally attempted to undervalue everything. Still, we have to wonder whether the value of the manuscripts was recognized. In all of the hearings, concern was shown solely for royalties and published sheet music. The manuscripts did not seem to attract much attention. Pete Clute reported to me he had been told that Barbara Sweatman, not recognizing the value of the manuscripts, disposed of them as trash.

But if she had thrown the manuscripts out, how could they have appeared on the inventory of items in Bragg's possession in November? Were these only Sweatman's manuscripts, or did they still include Joplin's?

The following year, Eva Sweatman discharged Bragg as executor of the estate, choosing to handle the royalties herself. She eventually turned over the executorship to Robert L. Sweeney, of Kansas City. However, I could find no record of either Eva Sweatman or Sweeney receiving physical items; their only apparent interest was in the royalties.[28]

While the hearings over the Sweatman estate were being held in 1961, there was a parallel matter concerning the Lottie Joplin Thomas Trust. Many of the same lawyers were involved.

In August 1961, five months after Sweatman died, Mary Wormley, niece of Lottie, applied to be executrix of the Lottie Joplin Thomas Trust. Her application was granted, but she, as the others, was apparently satisfied with rights to the royalties on compositions, then coming to between $1,000 and $1,500.[29] There is no indication that she took possession of the physical items, and the following month Barbara Sweatman said she still possessed them. Furthermore, the papers relating to Mary Wormley specifically state that she receive only the royalties and the renewal rights to Joplin's compositions. Her attorney Robert Rosborne, who had been Wilbur Sweatman's lawyer, was appointed successor trustee.

We are left with several open questions. Do Scott Joplin's manuscripts still exist, and, if so, who has them? We have tried following all the trails stemming from both Sweatman and Lottie Joplin, without success. We are convinced that no one *knowingly* possessed them during the 1970s, when widespread interest in Joplin reached its apex, for that would have been the time to announce their existence and perhaps offer them for sale. But someone may have *unknowingly* possessed them, holding them, perhaps, in a warehouse.

A second question is: What happened to the materials in Sweatman's office? It is possible that there were additional manuscripts stored there, since they had been seen in his office. Sweatman's autobiography, too, would be of great interest. As a major performer during the ragtime years, his autobiography could contain much revealing information about Joplin and others.

The large-scale Scott Joplin revival was launched not by the people who had for so many years fostered, preserved, and performed ragtime, but by academic and concert musicians. Scott Joplin would have appreciated the strange course interest in his music had taken.

In the mid-1960s, composer William Bolcom taught at Queens College in New York and shared an office with Rudi Blesh. The only Joplin rag he knew was *Maple Leaf,* but he had heard that Joplin had composed an opera and asked Blesh about it. Blesh made for him a photocopy of his own piano-vocal score of *Treemonisha* as well as many piano rags. Reading through the music, Bolcom was captivated. Responding with a composer's instinct, he tried his own hand at writing rags, in the process becoming a major (if occasional) composer of contemporary ragtime. Bolcom's rags are decidedly of his own period, but preserve the spirit and gestures of the music of a half-century earlier.[30]

Bolcom spoke to colleagues about his fascination with Joplin. Among them was William Albright, who reacted much as Bolcom had and in 1967 started on his own route to becoming a leading (and also occasional) composer of contemporary ragtime.[31] Bolcom suggested to T. J. Anderson, an African-American composer and college teacher, that together they edit *Treemonisha* with a thought toward having it produced. Anderson agreed and wrote the orchestration.

When Bolcom approached Vera Brodsky Lawrence, she was editing a five-volume collection of the music of 19th-century American composer Louis Moreau Gottschalk. On hearing some Joplin rags that Bolcom played for her, she decided that Scott Joplin would be her next project.[32]

Musicologist Joshua Rifkin had long been interested in jazz and knew a few rags, including *Maple Leaf.* He heard Bolcom play several Joplin rags at a party and asked for copies. As he went through the music, he, too, grew enthusiastic about Joplin. In late 1970 Rifkin used his position at Nonesuch Records, a classical label, to record a selection of Joplin's rags himself. The results were astonishing: Rifkin's *Piano Rags by Scott Joplin* became a hit, a best-seller for several years. But a best-seller in a special category. In another peculiar twist of circumstances that Joplin would have appreciated, Rifkin's recording being on Nonesuch prompted the record industry to label Joplin as "classical."[33]

Despite the acclaim, many in the ragtime community were less than enthusiastic about Rifkin's performance style: it was too stodgy and did not swing. But his approach, radically different from that of others who had recorded ragtime, was defensible. Instead of using Joplin's rags as a vehicle for his own improvisations, Rifkin followed the principles of his classical music background and respected the score. Playing on a Baldwin concert grand piano, he performed the music almost exactly as Joplin wrote it, giving prominence to inner voices, melodic bass lines, and other subtleties that were often lost in freer interpretations. Rather than make his performance the focus of attention, he allowed Joplin's music to speak for itself. The classical status of his recording was not just a record industry label. Rifkin performed Scott Joplin's classic ragtime as Joplin would have wanted: as *classical music*. This was the context in which H. Wiley Hitchcock wrote of it in a *Stereo Review* article:

> [Joplin's] rags can only be described as elegant, varied, often subtle, and as sharply incised as a cameo. They are the precise American equivalent, in terms of a native style of dance music, of minuets by Mozart, mazurkas by Chopin, or waltzes by Brahms.[34]

At the same time that Rifkin's recording was stirring interest, Vera Brodsky Lawrence, with substantial assistance from sheet music collectors of the ragtime community, completed her project of collecting Joplin's music into two volumes. It was published by the New York Public Library in late 1971 as *The Collected Works of Scott Joplin*. ("Collected" rather than "Complete" because permission to reprint three rags was withheld by one copyright owner.) Now in addition to having a classical status based on Rifkin's recording, Joplin's music had the recognition and sponsorship of one of the great institutions of scholarship.

Rifkin followed his successful recording with additional discs of Joplin's music; Bolcom recorded other "classic" rags with Nonesuch;[35] and soon numerous recordings of Joplin were issued. Gunther Schuller's recording with Angel Records (another classical label) was notable. Schuller was at that time president of the New England Conservatory of Music, one of the nation's leading music institutions. He had previously played French horn with several major orchestras and also had impressive jazz credentials: some twenty years earlier he had recorded with such monumental jazz figures as Gerry Mulligan and Miles Davis.[36] Vera Lawrence had spoken to Schuller of her Joplin project and gave him a copy of *The Red Back Book,* a collection of ragtime orchestrations published by Stark around 1909. Impressed by the orchestrations, Schuller formed the New England Conservatory Ragtime Ensemble with which to perform them. When an informal tape of the performance reached executives of Angel Records, they approached Schuller with a contract. Lightning struck a second time as Schuller repeated Rifkin's experience. His recording of old ragtime orchestrations became another hit, a bestseller for thirty-two weeks, and won a Grammy for the best chamber recording.[37]

Movie director George Roy Hill heard the Schuller recording, reportedly from his teenage son's room, and decided to use the exuberant music in his next film, *The Sting*.[38] That the music expressed an era earlier than that of the film's 1930 setting did not disturb Hill. Nor did it disturb the millions of fans who loved the movie and the music. The film, starring Paul Newman and Robert Redford, won the Academy Award for the Best Picture, and Scott Joplin's music, adapted by Marvin Hamlisch from Joplin's originals and the Schuller arrangements, won the award for best film score. The recording of the score was one of the top ten pop albums of 1974, and the film's main musical theme, *The Entertainer,* joined rock 'n' roll hits on the charts of popular best-sellers, although in many listings the piece was misidentified as "*The Sting,* by Marvin Hamlisch." In addition, Hamlisch's piano rendition of *The Entertainer* won him a Grammy for Best Pop Instrumental Performance. In his acceptance of the award, Hamlisch named Joplin as "the real new artist of the year."[39]

The market was flooded with Joplin recordings in 1974, especially recordings of *The Entertainer.* It was heard everywhere; one could not avoid it. It was on the radio, in elevators, in supermarkets, at park concerts, in piano bars, in the homes of almost every piano student in the land. It was played, as well, on organ, on harpsichord, on electronic synthesizer, by bands of every imaginable and unimaginable combination.

As Joplin's music achieved unbelievable renown in the popular music world, its position in the classical world kept pace. Classical performers and audiences also loved Joplin. His music was heard with increasing frequency at recitals and was recorded by leading artists. As *Record World* reported in July 1974:

> During the past year, *Record World* has shown albums of Joplin's works holding the distinction of "Classic of the Week" in our Classical Retail Report 24 percent of the time. The same albums have held spots as "Best Sellers" in the same report 74 percent of the time. For several weeks over the past year, the entire list of "Best Sellers" has been comprised of nothing but the works of Scott Joplin, making him the classical phenomenon of the decade.[40]

This was acclaim without precedent. The popular and classical music worlds came together in celebrating Scott Joplin, a composer who could speak to everyone.

Before the Joplin revival had gotten off the ground, before the cascading success of Joplin's rags made such a project a certain success, a decision was made to attempt a full production of *Treemonisha.* Funded by the Rockefeller Foundation, the opera had its premiere performance at the Atlanta Memorial Arts Center on January 28, 1972, culminating a week-long Afro-American Music Workshop at Morehouse College. The production used T. J. Anderson's orchestration, Katherine Dunham's stage direction and choreography, and a cast of black operatic singers, many of significant artistic stature and renown.

The performance's impact could not have been greater. The jubilant audience could hardly keep from dancing as it left the auditorium. Harold Schonberg wrote in the *New York Times:*

> "A Real Slow Drag" . . . is amazing. Harmonically enchanting, full of the tensions of an entire race, rhythmically catching, it refuses to leave the mind. . . .
>
> The audience tonight went out of its mind after hearing "A Real Slow Drag." There were yells and great smiles of happiness, and curtain call after curtain call. If the rest of the opera were as breath-taking as this, and also the second act "Aunt Dinah Has Blowed de Horn," the opera would run forever on Broadway.
>
> . . . Joplin . . . made it ["Aunt Dinah Has Blowed de Horn"] last only a few minutes. But this could be one of the sock curtains of American Theater, and Miss Dunham should have given it a reprise. And another. And yet another.[41]

Joplin's dreams of greatness, if not fulfilled during his lifetime, were more than justified.

For a follow-up performance at Wolf Trap, outside of Washington, D.C., on August 10–14, 1972, the Lottie Joplin Thomas Trust commissioned a new version of the score, edited by Lawrence and Bolcom and orchestrated by Bolcom. Two-and-a-half years later, the Trust commissioned yet another orchestration, this time from Gunther Schuller, for a production by the Houston Grand Opera on May 23, 1975. This was the production to reach Broadway.

Early on, Vera Lawrence had expressed her opposition to a Broadway showing, feeling that the crassness of commercial theater would ruin the opera.[42] In 1975, though, she changed her mind, and a production was prepared.

It almost closed before it opened. Its scheduled Broadway premiere on September 25, 1975, was canceled by a musicians' strike, and there was no certainty that the company, with a cast of sixty-five and elaborate scenery, would survive the job action. Disaster was averted with a settlement two weeks later, but the run was shortened from six weeks to three. *Treemonisha* finally held previews at the Uris Theater on October 15 and opened officially on October 20. Audiences were delighted, but the critics gave mixed reviews. Nevertheless, public response enabled the company to move to another theater, the Palace, for an additional six weeks.[43]

The *Treemonisha* production on Broadway may have lacked sparkle, but Joplin was due one more notable victory. Because of the opera, the unquenchable demand for his rags, and the decades of neglect, the Pulitzer committee early in 1976 awarded Joplin a special Bicentennial Pulitzer Prize for his contribution to American music.[44]

There was general delight that this neglected African-American composer had finally received his deserved recognition. The revival promoted innumerable instances of genuine good will and selflessness. But the revival also had an acrimonious side. Artistic differences and monetary considerations can disrupt the best of intentions.

Scott Joplin had become an industry. Considerable money was being made on his music. For music that was still under copyright—about half of it, including the opera—the Lottie Joplin Thomas Trust was intent on protecting its rights.

The trust was nominally controlled by Mary Wormley, Lottie Joplin's niece, who had successfully become executrix in 1961. Gerald Kearney, the trust's attorney in 1986, told me about her. He traveled from Florida every two weeks to visit Mrs. Wormley in her home in the South Bronx (an economically depressed area) and bring $500 in cash. He described her as an unsophisticated woman who had spent most of her working life doing domestic house cleaning. She claimed to have met Joplin once when she was six, but knew virtually nothing about him. She was only vaguely aware of the activities surrounding the Joplin revival. She had little interest in his music and did not own a phonograph. For the portion of the trust's earnings that she shared, she was content to follow the advice of consultants and lawyers.[45]

It became clear early on that the interests of the trust, or the wishes of those who advised Mrs. Wormley, frequently conflicted with the plans of others. Disagreements arose regardless of whether intents were commercial or altruistic.

One incident concerned T. J. Anderson. He wanted to build upon the success of *Treemonisha* and create a lasting and meaningful monument for Joplin. He proposed that a foundation be established to funnel proceeds from *Treemonisha* productions into a scholarship fund for young musicians. He enrolled the support of Bolcom, Blesh, Morath, Robert Shaw, Eileen Southern, Olly Wilson, David Baker, and others. The trust rejected the proposal.[46]

Following the successful production of *Treemonisha* in Atlanta, calls came from all over the country for additional performances. Anderson wanted to use his orchestrations to bring *Treemonisha* to college audiences. The threat of lawsuits prevented him from doing so.[47]

Dick Zimmerman, who as head of the Los Angeles-based Maple Leaf Club had assisted materially in alerting the ragtime community to the sheet music needs for the New York Public Library edition, and who had provided the transcription of the *Silver Swan Rag* piano roll, was connected with another unpleasant incident. He had recorded a five-record set entitled *Scott Joplin—His Complete Works*.[48] This set had all of the piano works and piano renditions of the songs, including those that had been omitted from the NYPL edition. On one of the ten sides, he played excerpts from *Treemonisha*.

The *Treemonisha* excerpts were a tragic error. Since these were from a stage work, the recording had to receive prior license, or "grand rights."

Joseph Abend, president of Olympic Records, tried to get these, but went to the Harry Fox Agency, which said it did not license the work. Assuming that no one controlled the rights, Abend issued the album in the fall of 1974.

In January 1975, counsel representing the trust and Mary Wormley sent notification that the album infringed their copyrights. On receiving the letter, Abend contacted ASCAP and was told that the Wilbur Sweatman Publishing Company, then owned by Robert Sweeney, held the copyrights. A search of the Copyright Office records confirmed Sweatman's self-assignment of the opera on August 14, 1959. For a payment of one dollar, Sweeney granted rights for the recording. Abend continued distribution and sales.

Counsel for the trust brought the matter to court on April 23, 1975, and the court reached a determination on May 26, 1977. The court ruled that though Sweatman was administrator of the Lottie Joplin Thomas Trust, he was not administrator of the Scott Joplin Estate; that he acted improperly in assigning the rights of *Treemonisha* to his own publishing company; that the trust owned the copyrights and was entitled to a share of the profits. Abend argued that since the excerpts occupied only one tenth of the album, the trust should not receive more than that percentage of the profits. The trust argued that the album sold well because it was the only "complete" recording of Joplin's works, and that it was the inclusion of the excerpts that made it complete. The court agreed with the trust. From the total profits of $175,303.90 earned by Abend, Olympic Records, and the parent company Crown Publishers, the court awarded the trust one-half of the profits plus "in lieu" damages: a total of $177,980.63 plus interest and costs. On September 21, 1978, the Court of Appeals affirmed the lower court's decision.[49]

This was only one of several suits surrounding the Joplin properties. Lawyers were kept busy.

I was once asked by a grand-niece of Scott Joplin (a grand-daughter of his brother Monroe) why no one in the Joplin family shared in the proceeds during the Joplin revival. I lack the legal qualifications to offer a meaningful opinion, but wondered out loud what would have happened if they had challenged Mary Wormley's rights. In this regard, I note the following:

Lottie Joplin had specified three different years—1907, 1909, and 1910—as the time of her marriage to Scott Joplin, but in October 1913 used her maiden name on a legal document that required her correct name. Apparently they were not married at that time, and I have been unable to locate record of a subsequent marriage. Has anyone established that they were married? If there was no contractual marriage, was their marital relationship long enough to constitute a legally recognized common-law marriage? Since Lottie in 1913 became a partner in the Scott Joplin Music Publishing Company, which published *Magnetic Rag* and the three republished excerpts from *Treemonisha,* she had a publisher's rights to these works. But did she have a widow's rights to all of his copyrights? Was Mary Wormley, Lottie Joplin's niece, entitled to the Joplin copyrights?

At this date, 1993, almost all the Joplin publications are in public domain, so there is little monetary incentive for pursuing the issues. But in view of all the litigation that occurred, one must wonder whether raising the basic issue of ownership could have turned events in a different direction.

The frenzy of the 1970s' ragtime revival ended, but its fading at the end of the decade did not return Scott Joplin to obscurity. Seeds had been planted that have kept ragtime alive and vital, with Joplin still the main focus.

The most important seed was the music. Before the revival, ragtime sheet music was difficult to find. With few pieces still in print, performers and scholars had to search out rare original editions. In response, Max Morath in 1963 put out a collection of one hundred reprints,[50] but because of copyright challenges was obliged to discontinue sales. More successfully, the Ragtime Society provided its members with photocopies of out-of-print rags.

In the 1970s, publishers saw an easy opportunity and began reprinting music already in public domain, music on which royalty payments were not due. At the time, these consisted of rags from before 1907 and those that lacked original copyrights or renewals. Reprint collections flooded the market, but most contained the same pieces. However, as more music passed into public domain, as the demand increased, and as publishers reissued what they owned, the variety increased significantly. This greater availability of the music created an impetus for more performances of ragtime and for its increased study.

Of the vast numbers of pianists and other instrumentalists who began playing ragtime during the 1970s, of the even greater number of fans who were captivated by the music, many stayed with it, forming the core of a new, enlarged ragtime community. This new ragtime community, dispersed geographically, formed numerous local ragtime clubs, joining the already existing Ragtime Society (of Toronto, Canada) and the Maple Leaf Club (of Los Angeles).[51] The clubs hold meetings, play and discuss the music, and organize ragtime festivals. The festivals feature pianists, certainly, but also singers, dancers, and ragtime bands—of which there are more than at any time since the 1910s. With all of this activity, we are in a ragtime renaissance.

The Scott Joplin Ragtime Festival in Sedalia is a special case because of the town's historic connection with Joplin, but its history illustrates the trend in festivals during the past twenty years. Bob Darch's concerts in Sedalia in the late 1950s and early 1960s had inspired Sedalians to form a Scott Joplin Memorial Foundation Committee that sponsored annual recitals of Joplin's music. Nevertheless, interest remained low until the 1970s. Then, bolstered by the revival, Larry Melton, a local history teacher, stirred up new enthusiasm. With the cooperation and assistance of other interested Sedalians and the chamber of commerce, Melton organized a much greater event for the town: a four-day Scott Joplin festival, held in 1974.

The setting was ideal, as much of central Sedalia remained as it had been when Joplin lived there. The local committee enlisted the aid of Dick Zimmerman as music director and engaged the participation of many of the revival's key figures: Rudi Blesh, Wally Rose, Max Morath, Bob Darch, Bill Bolcom, the New England Conservatory Ragtime Ensemble, and Eubie Blake.[52] With such headliners, the festival could hardly be anything but a resounding success. It was repeated the following year but then died due to the cost and flagging community interest.

If the festival died, the idea of it survived. The festival re-emerged in 1983, timed to coincide with the first day issue of a Scott Joplin postage stamp. It has been an annual event ever since, continually growing, a major musical and scholarly affair that is today regarded as the model for up to two dozen ragtime festivals held in the United States, Canada, and, most recently, in Hungary.

The festivals are primarily events for the performance of ragtime. Attendees revel in the music whether as listeners or performers, dance to it, and talk about it. This last should not be understated. There is such a thirst for information about the music, its composers, its performers, and the times that produced it, that festival organizers have been encouraged to provide platforms for professional and amateur scholars, as well as major collectors of sheet music, piano rolls, and period recordings. Scholarship has become an integral part of many festivals.

Early in 1971, Harold C. Schonberg, the influential music critic of the *New York Times,* saw the possibilities for scholarship. Awakened to Scott Joplin by Joshua Rifkin's first recording, and not finding enough information on him, he urged: "Scholars, Get Busy on Scott Joplin."[54]

They have been busy. Through the 1970s and 1980s there were seventeen books and doctoral dissertations on ragtime, including three specifically on Scott Joplin.[55] The number of articles is beyond counting. It is true that the vast majority simply repeat what Blesh and Janis had set out, but a

Scott Joplin postage stamp, issued June 1983.

respectable number reveal new discoveries and interpretations. Ragtime scholarship has become a flourishing, ongoing activity.

The general public is not aware of the clubs, the festivals, or the scholarship. It has again forgotten ragtime. At most, it recognizes "The Sting." But while Joplin is not the constant presence he was at the height of the 1970s revival, he is still a feature of everyday life. Despite its lack of awareness of ragtime, the public still likes it, a fact recognized by the producers of radio and TV commercials. Repeatedly, they use the upbeat strains of Scott Joplin's music, or of some thinly disguised copy of it, to sell products. Unknowingly, the public still listens to ragtime. Joplin has become almost subliminal.

More openly, he remains a subject for stage productions. His infectious syncopations have a natural appeal to dancers, and choreographers are forever developing and staging new steps to those rhythms. In 1974, seventy-five years after he mounted *The Ragtime Dance* at Wood's Opera House, the magazine *Dance and Dancers,* referred to Joplin as "the year's most popular ballet composer."[56] Similarly, Joplin inspires drama. Playwrights seem never to tire of his story. Hardly a year goes by that his life is not staged to probe vital questions of art, race relations, and life. Joplin is a constant in today's world.

And so we come to the end of our examination of Scott Joplin, an inquiry that traverses 125 years. The story is quintessentially American, beginning in the humblest circumstances and culminating in achievement, recognition, and honor. In between, we witness the struggles for personal and artistic achievement, the conflicts between classes, between concert music and popular, and between races. Joplin was dedicated to his art and steadfast in his efforts. He surmounted some of the barriers against him, forced a degree of recognition and respect from a society not ready to concede African-American ability and competence, but stumbled as he approached his final goal. His was a human failing.

We may lament that Joplin did not live to enjoy the glory that would ultimately flow to him, that society and the artistic world of his time were not ready for him, but he understood what he had accomplished, and he knew that the world would, some day, appreciate his vision. As a perceptive writer in the *American Musician* noted in 1911, Joplin "had confidence in himself and in his mission and . . . was destined to succeed."[57]

Scott Joplin's music is no longer a fad; it is no longer a craze. But on a less fevered and more even pitch, his influence is with us. We honor him mostly for his music, but his life also has meaning for us. Scott Joplin has become ingrained in the American musical and cultural landscape.

Appendix A: The Music

Publication and Copyright Information

Copyright notices appear on all of Scott Joplin's published music, but not every piece was registered for copyright. For some unregistered pieces, copyright applications were made but the required copies of the music were not received by the Copyright Office. Even when a piece was correctly registered, the registration date did not necessarily coincide with the publication date. In several instances, specified in the listings, publication occurred many months prior to registration. The reverse was also possible, with registration preceding publication.

Despite these shortcomings, copyright dates are frequently the only guide we have to determine when a piece was published. We list all of the dates and, unless otherwise indicated, consider them as approximations of first printing. There are up to three dates connected with a copyright: the copyright entry date, which we indicate with an "e"; the date of receipt by the Copyright Office, indicated with an "r"; and, least common, the publication date, indicated with a "p."

A title listed on the first page of the music may differ from that on the cover. In these cases, we list the music page title first, followed by a slash and the cover title. The title on the music page, not being subject to the cover artist's modifications, is the more accurate reflection of the composer's wishes.

For the sake of a uniform and accessible reference to Scott Joplin's music, we consider the "standard copy" to be the reprints contained in *The Complete Works of Scott Joplin,* 2 vols., 2d ed., ed. Vera Brodsky Lawrence (New York: New York Public Library, 1981). These volumes are indicated as CWSJ. They were originally published in 1971 as *The Collected Works of Scott Joplin,* "Collected" rather than "Complete" because permission to reprint three rags was refused. The second edition includes these three rags, but is otherwise unchanged. In instances in which these reprints differ from the copyright copies or from other significant early printings, we discuss the distinctions.

The first listing below presents the music alphabetically. The second listing is chronological, according to the following criteria: copyright registration date; copyright notice on music; known or assumed date of composition. The third listing is of spurious works and false or mistaken attributions.

Alphabetic Listing of the Works of Scott Joplin

TITLE: *Antoinette. March and Two-Step*
PUBLISHER: Stark Music Co., New York and St. Louis.
COPYRIGHT: Dec. 12, 1906, e, r.

TITLE: *The Augustan Club. Waltzes / Augustan Club Waltz*
PUBLISHER: John Stark & Son, St. Louis.
COPYRIGHT: March 25, 1901, e; March 26, 1901, r.
COMMENT: Composed January 1900.
  The title is misspelled; the club name was "Augustain."

TITLE: *Bethena. A Concert Waltz*
PUBLISHER: T. Bahnsen Piano Mfg. Co., St. Louis.
COPYRIGHT: March 6, 1905, e, r.

TITLE: *Binks' Waltz*
PUBLISHER: Bahnsen Music Co., St. Louis.
COPYRIGHT: August 11, 1905, e, r.
COMMENT: Announced July 22, 1905, in the *Indianapolis Freeman*.

TITLE: *A Blizzard*
COMMENT: Unpublished and lost. Mentioned in the *Indianapolis Freeman* on Nov.
  16, 1901, 5.

TITLE: *A Breeze from Alabama. March and Two Step / A Breeze from Alabama. A
  Ragtime Two Step*
PUBLISHER: John Stark & Son, St. Louis.
COPYRIGHT: Dec. 29, 1902, e; Jan. 8, 1903, r.

TITLE: *Carnation*
COMMENT: A rag by Clyde D. Douglass, published by Stark in 1902. Mistakenly
  attributed to Joplin on the back of Edward Hudson's *Sandella Rag* (Stark,
  1921).

TITLE: *The Cascades. A Rag*
PUBLISHER: John Stark & Son, St. Louis.
COPYRIGHT: Aug. 22, 1904, e, r.

TITLE: *The Chrysanthemum. An Afro-American Intermezzo*
PUBLISHER: John Stark & Son, St. Louis.
COPYRIGHT: August 22, 1904, e, r.
COMMENTS: Published before April 4, 1904.
  First printing has dedication to Freddie Alexander on the cover. Covers of subse-
    quent printings (including reprint in CWSJ) lack dedication and "American" in
    subtitle.

TITLE: *Cleopha. March and Two Step / Cleopha. March and Two-Step*
PUBLISHER: S. Simon, St. Louis.

COPYRIGHT: May 19, 1902, e, r.
COMMENT: Bottoms of pages have erroneous title: 'The "Conductor"'.

TITLE: *Combination March*
PUBLISHER: Robt. Smith, Temple, Texas.
COPYRIGHT: Nov. 16, 1896, e, r.

TITLE: *Country Club. Ragtime Two Step / Country Club. Rag Time Two-Step*
PUBLISHER: Seminary Music Co., New York.
COPYRIGHT: May 11, 1909, e; Oct. 30, 1909, r.

TITLE: *The Crush Collision March / Great Crush Collision March*
PUBLISHER: John R. Fuller; Agent Rob't Smith, Temple, Texas.
COPYRIGHT: Oct. 15, 1896, e, r.

TITLE: *Dudes' Parade*
COMMENTS: Lost. From *A Guest of Honor*. Mentioned by Joplin in the *Indianapolis Freeman* on Sept. 12, 1903, 5.

TITLE: *The Easy Winners. A Rag Time Two Step / The Easy Winners. A Ragtime Two Step*
PUBLISHER: Scott Joplin, St. Louis.
  Republished by Shattinger Music Co., St. Louis, ca. 1903, in different key and with other changes.
  Republished by Stark before 1908, using Joplin's original plates.
COPYRIGHT: Oct. 10, 1901, e, r.
COMMENT: The tie in the middle of B15 in CWSJ is not found in the copyright copy and many early printings, but is present in the Shattinger edition.

TITLE: *Elite Syncopations*
PUBLISHER: John Stark & Son, St. Louis.
COPYRIGHT: Dec. 29, 1902, e; Jan. 8, 1903, r.

TITLE: *The Entertainer. A Rag Time Two Step*
PUBLISHER: John Stark & Son, St. Louis.
COPYRIGHT: Dec. 29, 1902, e; Jan. 8, 1903, r.

TITLE: *Eugenia*
PUBLISHER: Will Rossiter, Chicago.
COPYRIGHT: Feb. 26, 1906, e, r.

TITLE: *Euphonic Sounds. A Syncopated Two Step / Euphonic Sounds. A Syncopated Novelty*
PUBLISHER: Seminary Music Co., New York.
COPYRIGHT: Oct. 28, 1909, r; Oct. 30, 1909, e.

TITLE: *The Favorite. A Ragtime Two-Step / The Favorite. Ragtime Two Step*
PUBLISHER: A. W. Perry & Sons' Music Co., Sedalia.
COPYRIGHT: June 23, 1904, e, r.

TITLE: *Felicity Rag. A Ragtime Two Step*
CO-COMPOSER: Scott Hayden
PUBLISHER: Stark Music Publ. Co., St. Louis, New York.
COPYRIGHT: July 27, 1911, p; July 31, 1911, e, r.
COMMENTS: Probably composed by 1903.

TITLE: *"Fig Leaf". A High Class Rag / Fig Leaf Rag*
PUBLISHER: Stark Music Printing & Pub. Co., New York & St. Louis.
COPYRIGHT: Feb. 24, 1908, e, r.
COMMENT: There are two entirely different engravings of the music. They contain the
    same musical content, but have cosmetic differences: phrase lines of different
    lengths and different letter type faces. Advertisements indicate they are both
    early, and we cannot determine which came first. The copyright version has a
    metronome indication, and before strain C has a broken double bar. The other
    version, reprinted in CWSJ, lacks the tie connecting B-flats in the middle of B13,
    and lacks the metronome indication.

TITLE: *For the Sake of All*
COMMENT: Unpublished and lost.

TITLE: *Frolic of the Bears.* Revised excerpt from *Treemonisha*
PUBLISHER: Scott Joplin Music Pub. Co., New York.
COPYRIGHT: June 22, 1915, e, r.

TITLE: *Gladiolus Rag*
PUBLISHER: Jos. W. Stern & Co., New York.
COPYRIGHT: Sept. 4, 1907, e, r (a preliminary copy, without cover, was sent; second
    copy, with cover, r. March 30, 1908).
COMMENT: On the final eighth of C6 there is an F-natural, which is probably an error.
    The F-flat in later printings (and in CWSJ) is more logical.

TITLE: *Good-bye Old Gal Good-bye*
MUSIC: Mac Darden; arranged by Scott Joplin.
WORDS: H. Carroll Taylor.
PUBLISHER: Foster-Calhoun Co., Evansville, Ind.
COPYRIGHT: 1906 notice on music; not registered.
COMMENT: Not included in CWSJ.

TITLE: *Great Crush Collision March.* See *The Crush Collision March.*

TITLE: *A Guest of Honor*
COPYRIGHT: Feb. 18, 1903, e; music not received.
COMMENT: Lost.

TITLE: *Harmony Club Waltz*
PUBLISHER: Robt. Smith, Temple, Texas.
COPYRIGHT: Nov. 16, 1896, e, r.

TITLE: *Heliotrope Bouquet. A Slow Drag Two Step*
CO-COMPOSER: Louis Chauvin.

PUBLISHER: Stark Music Co., St. Louis and New York.
COPYRIGHT: Dec. 23, 1907, e, r.
COMMENT: Published by Sept. 6, 1907.

TITLE: *I Am Thinking of My Pickanniny Days / I Am Thinking of My Pickaninny Days*
WORDS: Henry Jackson.
PUBLISHER: Thiebes-Stierlin Music Co., St. Louis.
COPYRIGHT: April 9, 1902, e, r. Copyright notice on music is 1901.

TITLE: *If*
COMMENT: Lost. A musical comedy mentioned in the *New York Age* on Sept. 7, 1916.

TITLE: *Kismet Rag*
CO-COMPOSER: Scott Hayden.
PUBLISHER: Stark Music Co., St. Louis, Mo.
COPYRIGHT: Feb. 21, 1913, p.; Feb. 24, 1913, e; Feb. 26, 1913, r.
COMMENTS: Joplin probably composed C, and possibly D, but he did not write the link connecting the two strains. Probably composed by 1903.

TITLE: *Leola. Two-Step / Leola. Two Step*
PUBLISHER: American Music Syndicate, St. Louis.
COPYRIGHT: Dec. 18, 1905, e; no copies received.
The music indicates a copyright in England, but registration can not be found.

TITLE: *Lily Queen. A Ragtime Two-Step*
CO-COMPOSER: Arthur Marshall.
PUBLISHER: W. W. Stuart, New York.
COPYRIGHT: (by Willis Woodward) Nov. 7, 1907, r; Nov. 8, 1907, e.
COMMENT: Marshall claimed to be sole composer; Joplin probably edited Marshall's original.

TITLE: *Little Black Baby*
WORDS: Louise Armstrong Bristol.
PUBLISHER: Success Music Company, Chicago.
COPYRIGHT: Oct. 7, 1903, e, r.

TITLE: *Lovin' Babe*
WORDS & MUSIC: Al. R. Turner; arranged by Scott Joplin.
PUBLISHER: Robin Press, New York.
COPYRIGHT: (Al. R. Turner), Aug. 4, 1911, p; application Aug. 16, 1911; Sept. 8, 1911, e, r.
COMMENT: Not in CWSJ.

TITLE: *Magnetic Rag*
PUBLISHER: Scott Joplin Music Pub. Co., New York.
Transf. 1917 to Winn School of Popular Music.
COPYRIGHT: July 21, 1914, e, r.

TITLE: *Maple Leaf Rag*
PUBLISHER: John Stark & Son, Sedalia, Mo.
COPYRIGHT: Sept. 18, 1899, e, Sept. 20, 1899, r.

TITLE: *Maple Leaf Rag. Song*
WORDS: Sydney Brown.
PUBLISHER: John Stark & Son, St. Louis.
COPYRIGHT: Aug. 22, 1904, e, r; 1903 notice on music.

TITLE: *March Majestic. March and Two-Step / March Majestic*
PUBLISHER: John Stark & Son, St. Louis.
COPYRIGHT: Not registered; 1902 notice on music.

TITLE: *Mint Leaf Rag*
COMMENTS: Probably an error. Listed in an undated advertisement for the Winn
    School of Popular Music. Reprinted in *Rag Times* 7/5 (Jan. 1974), 8.

TITLE: *Morning Glories*
COMMENTS: Unpublished and lost. Mentioned in the *Indianapolis Freeman* on Sept.
    4, 1915, 6.

TITLE: *The Nonpareil. A Rag & Two Step / Nonpareil, (None to Equal)*
PUBLISHER: Stark Music Co., New York and St. Louis.
COPYRIGHT: Not registered; 1907 notice on music.

TITLE: *Original Rags*
ARRANGER: Chas. N. Daniels.
PUBLISHER: Carl Hoffman, Kansas City, Mo.
    Transferred to Whitney-Warner in 1903; republ. by Remick. Claimant's Card in
    Copyright Office credits Chas. N. Daniels as composer.
COPYRIGHT: March 15, 1899, application; no copy received.
COMMENT: Chas. N. Daniels was probably credited as "arranger" as a condition of
    publication; his active participation is doubtful.

TITLE: *Palm Leaf Rag / Palm Leaf Rag. A Slow Drag*
PUBLISHER: Victor Kremer Co., Chicago, New York.
COPYRIGHT: Nov. 14, 1903, e, r; 2d copy received Jan. 2, 1904.

TITLE: *Paragon Rag*
PUBLISHER: Seminary Music Co., New York.
COPYRIGHT: Oct. 29, 1909, r; Oct. 30, 1909, e.

TITLE: *Peacherine Rag*
PUBLISHER: John Stark & Son, St. Louis.
COPYRIGHT: March 18, 1901, e; Sept. 3, 1901, r.

TITLE: *Patriotic Patrol*
COMMENTS: Lost. From *A Guest of Honor.* Mentioned by Joplin in the *Indianapolis
    Freeman* on Sept. 12, 1903, 5. Listed on an advertising flyer distributed by
    Stark.

TITLE: *Pepper Rag*
COMMENT: Probably an error. Mentioned in Frederic Ramsey, Jr., and Charles Edward Smith, eds. *Jazzmen* (New York: Harcourt, Brace, 1939), 22.

TITLE: *Piano Concerto*
COMMENT: Unpublished and lost.

TITLE: *A Picture of Her Face*
WORDS: Scott Joplin
PUBLISHER: Leiter Bros., Syracuse, NY.
COPYRIGHT: July 3, 1895, e, r.

TITLE: *Pine Apple Rag*
PUBLISHER: Seminary Music Co., New York.
COPYRIGHT: Oct. 12, 1908, e, r.

TITLE: *Pine Apple Rag. Song* / *Pine Apple Rag*
WORDS: Joe Snyder.
PUBLISHER: Seminary Music Co., New York.
COPYRIGHT: No separate copyright for the song; 1910 notice on music.

TITLE: *Pleasant Moments. Ragtime Waltz* / *Pleasant Moments. Rag-Time Waltz*
PUBLISHER: Seminary Music Co., New York.
COPYRIGHT: May 11, 1909, e, r.

TITLE: *Please Say You Will*
WORDS: Scott Joplin.
PUBLISHER: M. L. Mantell, Syracuse, NY.
COPYRIGHT: Feb. 20, 1895, e, r; 2d copy received Dec. 28, 1898.

TITLE: *Prelude to Act 3*. Revised excerpt from *Treemonisha*
PUBLISHER: Scott Joplin Music Pub. Co., New York.
COPYRIGHT: Dec. 15, 1913, e, r.

TITLE: *Pretty Pansy Rag*
COMMENTS: Unpublished and lost.

TITLE: *A Princeton Tiger*
COMMENTS: Mistaken attribution, listed in QRS piano roll catalogues beginning in 1904. The piece is by Gerald Burke.

TITLE: *The "Rag Time Dance"* / *The Ragtime Dance*
WORDS: Scott Joplin.
PUBLISHER: Stark Music Co., St. Louis.
COPYRIGHT: Dec. 29, 1902, e; Jan. 8, 1903, r.
COMMENT: Performed in November 1899; in print by April 1902.

TITLE: *Rag-Time Dance. A Stop-Time Two Step* / *The Ragtime Dance*
PUBLISHER: Stark Music Co., New York and St. Louis.
COPYRIGHT: Dec. 21, 1906, e, r.

TITLE: *A Real Slow Drag*. Revised excerpt from *Treemonisha*
LIBRETTO: Scott Joplin.
PUBLISHER: Scott Joplin, New York.
COPYRIGHT: July 15, 1913, e, r.

TITLE: *Recitative Rag*
COMMENTS: Unpublished and lost.

TITLE: *Reflection Rag (Syncopated Musings)*
PUBLISHER: Stark Music Co., St. Louis.
COPYRIGHT: Dec. 4, 1917, e; Dec. 7, 1917, r.
COMMENT: Probably composed c. 1907–1908.

TITLE: *Rose Leaf Rag. A Ragtime Two Step* / *Rose Leaf Rag. A Ragtime Two-Step*
PUBLISHER: Daly Music Publisher, Boston.
COPYRIGHT: Nov. 15, 1907, e, r.

TITLE: *The Rose-bud March* / *Rosebud. Two-Step*
PUBLISHER: John Stark & Son, St. Louis, Mo.
COPYRIGHT: Not registered; 1905 notice on music.

TITLE: *Sarah Dear*
WORDS: Henry Jackson.
PUBLISHER: Bahnsen Music Co., St. Louis.
COPYRIGHT: Aug. 11, 1905, e, r.
COMMENTS: *Dal segno* symbols before vamp and after 2d ending of chorus, indicating repeated verse, not present in CWSJ.
  Announced in the *Indianapolis Freeman,* July 22, 1905. Written for Williams and Stevens, depicted on cover.

TITLE: *School of Ragtime* / *School of Ragtime. 6 Exercises for Piano*
PUBLISHER: Scott Joplin.
COPYRIGHT: Jan. 21, 1908, r; Jan. 29, 1908, e.

TITLE: *Scott Joplin's Dream*
COMMENT: Supposedly a collaboration between Lamb and Joplin, arranged by Eubie Blake; of doubtful authenticity. Copy in the Library of Congress.

TITLE: *Scott Joplin's New Rag*
PUBLISHER: Jos. W. Stern & Co., New York.
COPYRIGHT: May 1, 1912, p; May 2, 1912, e.

TITLE: *"Search-Light Rag"* / *Searchlight Rag*
PUBLISHER: Jos. W. Stern & Co., New York.
COPYRIGHT: Aug. 12, 1907, e, r (preliminary copy, without cover; 2d copy, with cover, Sept. 14, 1907).

TITLE: *Sensation*
COMPOSER: Joseph F. Lamb; arranged by Scott Joplin.
PUBLISHER: Stark Music Co., New York, St. Louis.

COPYRIGHT: Oct. 8, 1908, e, r.

COMMENTS: Scott Joplin had no hand in either the composition or arrangement. In advertising, Stark lists Lamb alone. Was published by May 18, 1908.

TITLE: *Silver Swan Rag*
PUBLISHER: New York Public Library, CWSJ.
COPYRIGHT: 1971, Estate of Lottie Joplin Thomas.
COMMENT: Existed only on piano roll until 1971 publication. Transcribed from piano roll by Dick Zimmerman and Donna McCluer.

TITLE: *"Sliding Jim" Rag*
COMMENT: Probably an erroneous citation. Listed in a concert program by the Indianapolis Military Band at the German House in Indianapolis, July 31, 1908. Included in the program printed in the *Indianapolis Star* of Friday, July 31, 1908, 9. The only copyrighted piece by this title is F. H. Losey's *Sliding Jim; Characteristic Reverie for Orchestra,* parts published by Carl Fischer, Nov. 2, 1904.

TITLE: *Snoring Sampson. A Quarrel in Ragtime*
WORDS & MUSIC: Harry La Mertha; arranged by Scott Joplin.
PUBLISHER: University Music Publishing Co., St. Louis.
COPYRIGHT: (Harry La Mertha) May 6, 1907, e, r.
COMMENTS: Not included in CWSJ.

TITLE: *Solace. A Mexican Serenade*
PUBLISHER: Seminary Music Co., New York.
COPYRIGHT: April 28, 1909, e, r.

TITLE: *"Something Doing." A Ragtime Two Step / Something Doing. Cake Walk March*
CO-COMPOSER: Scott Hayden.
PUBLISHER: Val A. Reis Music Co., St. Louis.
COPYRIGHT: Jan. 10, 1903, r; Feb. 24, 1903, e.

TITLE: *"Stoptime" Rag / Stoptime Rag*
PUBLISHER: Jos. W. Stern & Co., New York.
COPYRIGHT: Jan. 4, 1910, e, r (preliminary copy, with a generic cover).

TITLE: *The Strenuous Life. A Ragtime Two Step*
PUBLISHER: John Stark & Son, St. Louis.
COPYRIGHT: Not registered; 1902 notice on music.
COMMENTS: Probably issued by September 1902.

TITLE: *Sugar Cane. A Ragtime Two Step / Sugar Cane. A Ragtime Classic Twostep*
PUBLISHER: Seminary Music Co., New York.
COPYRIGHT: April 21, 1908, e, r.

TITLE: *Sun Flower Slow Drag. A Rag Time Two Step / Sunflower Slow Drag. Rag Time Two-Step*
CO-COMPOSER: Scott Hayden.
PUBLISHER: John Stark & Son, St. Louis.

COPYRIGHT: March 18, 1901, e, r.
COMMENT: Composed by 1899.

TITLE: *Swipesy Cake Walk*
CO-COMPOSER: Arthur Marshall.
PUBLISHER: John Stark & Son, St. Louis, Mo.
COPYRIGHT: July 21, 1900, e, r.

TITLE: *The Sycamore. A Concert Rag*
PUBLISHER: Will Rossiter, New York and Chicago.
COPYRIGHT: July 18, 1904, e, r.

TITLE: *Symphony No. 1.*
COMMENT: Unpublished and lost. Announced in the *New York Age,* Sept. 7, 1916, 6.

TITLE: *Syncopated Jamboree*
COMMENT: Unpublished and lost. Announced by Joplin in the *Indianapolis Freeman*
on Sept. 18, 1915.

TITLE: *Treemonisha—Opera in Three Acts*
LIBRETTO: Scott Joplin.
PUBLISHER: Scott Joplin, New York.
COPYRIGHT: May 19, 1911, p; May 22, 1911, e, r.
COMMENT: Pagination in CWSJ is 4 higher than in the original.

TITLE: *Wall Street "Rag"* / *Wall Street Rag*
PUBLISHER: Seminary Music Co., New York.
COPYRIGHT: Feb. 23, 1909, e, r.

TITLE: *Weeping Willow. A Rag Time Two Step* / *Weeping Willow. Ragtime Two Step*
PUBLISHER: Val. A. Reis Music Co., St. Louis.
COPYRIGHT: June 6, 1903, e, r.
COMMENT: Published by March 26, 1903.

TITLE: *When Your Hair Is Like the Snow*
WORDS: Owen Spendthrift.
PUBLISHER: Owen Spendthrift, St. Louis.
COPYRIGHT: May 18, 1907, e, r.
COMMENT: Joplin's name appears on the cover, but not on the music.

TITLE: *You Stand Good with Me, Babe*
COMMENT: Unpublished and lost. Completion announced in the *Indianapolis Freeman,* July 22, 1905, 5.

Chronological Listing

1895   Please Say You Will
         A Picture of Her Face

1896   The Crush Collision March
       Combination March
       Harmony Club Waltz
1899   Original Rags
       Maple Leaf Rag
1900   Swipesy Cake Walk
1901   Sunflower Slow Drag (completed 1899)
       The Augustan Club (completed Jan. 1900)
       Peacherine Rag
       The Easy Winners
       A Blizzard (lost)
1902   I Am Thinking of My Pickanniny Days
       Cleopha
       The Strenuous Life
       A Breeze from Alabama
       Elite Syncopations
       The Entertainer
       The Ragtime Dance. Song (performed Nov. 1899)
       March Majestic
1903   A Guest of Honor (lost)
       Dudes' Parade (lost; from A Guest of Honor)
       Patriotic Patrol (lost; from A Guest of Honor)
       Something Doing
       Weeping Willow
       Little Black Baby
       Palm Leaf Rag
       Maple Leaf Rag. Song (copyright 1904)
1904   The Favorite
       The Sycamore
       The Chrysanthemum
       The Cascades
1905   Bethena
       Binks' Waltz
       Sarah Dear
       Leola
       You Stand Good with Me, Babe (lost)
       The Rose-bud March
1906   Eugenia
       Antoinette
       Rag-Time Dance (instrumental version of song)
       Good-bye Old Gal Good-bye
1907   Snoring Sampson
       When Your Hair Is Like the Snow
       The Nonpareil
       Searchlight Rag
       Gladiolus Rag
       Lily Queen (Arthur Marshall; edited by Joplin)
       Rose Leaf Rag
       Heliotrope Bouquet

1908	School of Ragtime (advertised in Dec. 1907)
	Fig Leaf Rag
	Sugar Cane
	Sensation (composed by Joe Lamb)
	Pine Apple Rag
1909	Wall Street Rag
	Solace
	Pleasant Moments
	Country Club
	Euphonic Sounds
	Paragon Rag
1910	Stoptime Rag
	Pine Apple Rag. Song
1911	Treemonisha
	Felicity Rag (with Scott Hayden; probably composed by 1903)
	Lovin' Babe (arranged by Joplin; composed by Al. R. Turner)
1912	Scott Joplin's New Rag
1913	Kismet Rag (with Scott Hayden; probably composed by 1903)
	A Real Slow Drag (revised excerpt from Treemonisha)
	Prelude to Act 3 (revised excerpt from Treemonisha)
1914	Magnetic Rag
	Silver Swan Rag
1915	Frolic of the Bears (revised excerpt from Treemonisha)
	Morning Glories (lost)
	Syncopated Jamboree (lost)
	Pretty Pansy (lost; date approximate)
	Recitative Rag (lost; date approximate)
	For the Sake of All (lost; date approximate)
1916	If (lost)
	Symphony No. 1 (lost)
	Piano Concerto (lost; date approximate)
1917	Reflection Rag (probably composed c. 1907–08)

Spurious

Carnation
Mint Leaf Rag
Pepper Rag
A Princeton Tiger
Scott Joplin's Dream
"Sliding Jim" Rag

Appendix B: Three Songs

In 1971, the New York Public Library's Joplin edition was issued as *The Collected Works of Scott Joplin* rather than "The Complete Works" because reprint permission was denied for three rags—*Searchlight Rag, Fig Leaf Rag,* and *Rose Leaf Rag.* Permission to include these rags was obtained for the new edition in 1981, which was renamed *The Complete Works of Scott Joplin.* However, the editor missed the opportunity to include three songs arranged by Scott Joplin.

Snoring Sampson is listed in the index of *The Complete Works* with the notation "No copy found." The other two songs are not mentioned.

Snoring Sampson was reprinted in *Rag Times,* July 1974. *Good-bye Old Gal Good-bye* was reprinted there in May 1977. All three songs were reprinted in *Three Ragtime Songs,* compiled by Dick Zimmerman (Rockville Centre, N.Y.: Belwin Mills, 1985), but that edition is now out-of-print. To make the songs available to the reader, we include them here, with the following corrections:

Good-bye Old Gal Good-bye

Introduction, measure 6 (second measure of vamp), beat 2: changed tied eighth in r.h. to a tied quarter.

Initial word of lyric in verse corrected from "Its" to "It's".

Chorus, measure 14, r.h., second eighth: changed F to F-sharp.

Lovin' Babe

Verse, measure 6, voice: over the word "sweet," dotted quarter-note changed to a dotted eighth-note.

Good-bye Old Gal Good-bye

Words by H. CARROLL TAYLOR.

Music by MAC DARDEN.

Arr. by Scott Joplin.

hid from me and feed 'em to some oth-er coon.__ So take one last fond
au-to-mo-bile and rent-ed__ a big swell flat.__ In a week he

lov - ing look just gaze like on__ the dead, 'Cause I'm
bought__ a yacht and start-ed to sail a - way, When his

go - ing so far I can't find my-self__ so re - mem-ber these words I
old__ gal said don't I go too__ these words she heard him

CHORUS

said:— I say good-bye, bye, bye, old gal— A good, good-bye to
say:—

you———— Just say good-bye— to all my friends And

say that we are through.— Now don't think I'll come back to you— For

Snoring Sampson

A Quarrel in Ragtime

HARRY LA MERTHA

Arr. by Scott Joplin.

Composer of "Maple Leaf Rag,"
"Antoinette," etc.

To a lit-tle dusk-y maid-en that he__ made his choice in__
hon - y when I wakes you in the__ mid - dle of the night with a

nine - teen - two.__ Ev-'ry time they went to bed there was a big up-roar, For__
bass ta - too,__ Ev-'ry time I el - o-cutes that honk__ honk__ snore, Put your

when it came to snor - ing, Samp - son had__ the floor.__ One
feet a - gin ma back and shove me out on the floor. But, Ma -

Arranged by   **Lovin' Babe**   Words and Music

SCOTT JOPLIN   by AL. R. TURNER

Strol - ling thro' the park,
We had a long chat, there

out on a lark, I hap-pened to meet Miss De - li - lah, I
as we sat In___ the shade of the trees,___ I

Dress'd so ve-ry neat, man-ners ve-ry sweet, To cap-ture her was my de-
asked to know, to be her stea-dy beau, And these words seemed to

sire._____ She gim-me a glance, I took a chance, To
please her. She be-gan to blush, said, "Man, you hush, You

sit down by her side; I told her what I'd do, and that
sho is work-in' fas'; I likes you, 'deed I

Notes

Major References

BeRa   Berlin, Edward A. *Ragtime: A Musical and Cultural History.*
BeRR   Berlin, Edward A. *Reflections and Research on Ragtime.*
BlSJ   Blesh, Rudi, "Scott Joplin: Black-American Classicist."
BlJa   Blesh, Rudi and Harriet Janis, *They All Played Ragtime.*
CWSJ   *The Complete Works of Scott Joplin.*
HaBe   Haskins, James with Kathleen Benson, *Scott Joplin.*
JaNo   Janis, Harriet, Unpublished notes for *They All Played Ragtime.*
ThLJ   Thompson, Kay C. "Lottie Joplin."

Major Historic Newspaper and Periodical References

AMAJ   *American Musician and Art Journal*
ChRR   *Christensen's Ragtime Review*
IFr   *Indianapolis Freeman*
NYA   *New York Age*
SLP   *St. Louis Palladium*
SeCa   *Sedalia Capital*
SeCo   *Sedalia Conservator*
SeD   *Sedalia Democrat*
SeS   *Sedalia Sentinel*
SeT   *Sedalia Times*

Preface

1. Rudi Blesh, *Shining Trumpets. A History of Jazz* (New York: Knopf, 1946; 2d ed., 1958).

Chapter 1

1. Respectively: ThLJ, 18; Edward B. Marks, as told to Abbott J. Liebling, *They All Sang, from Tony Pastor to Rudy Vallee* (New York: Viking, 1934), 159–160.

2. Roy Carew and Don E. Fowler, "Scott Joplin. Overlooked Genius," *Record Changer* (Sept. 1944), 12–14, 59; (Oct. 1944), 10–12; (Dec. 1944), 10–11.

3. *The ASCAP Biographical Dictionary of Composers, Authors and Publishers,* ed. Daniel I. McNamara (New York: Thomas Y. Crowell, 1948).

4. The name is spelled "Givins" in HaBe, 199–200, note 6. It is spelled "Givens" in BlJa. Joplin's death certificate spells it "Givens," and the 1870 census lists a Milton and Susan Givens, ages 60 and 70, just before the Joplins. They may have been Florence's parents.

5. The spelling of his name is inconsistent: it is "Giles" in the 1880 census, in the Texarkana directories of 1899–1900 and 1922, and on Scott's death certificate; it is "Jiles" in the 1870 census, in the Texarkana directory for 1906, and on his own death certificate. Haskins and Benson show impressive research skills in efforts to locate Giles during his childhood and adolescence. They point to a young slave living in northeastern Texas, owned by a family that had come from South Carolina. He was last owned by a Josiah Joplin, from whom he would have adopted his family name (HaBe, 20–26). The evidence is not conclusive, but the argument can be persuasive if one accepts the authors' assumptions. Moreover, it is an enlightening demonstration of how enslaved African Americans can be traced through official records.

6. Ann Vanderlee and John Vanderlee, [part 1] "Scott Joplin's Childhood Days in Texas," *Rag Times* 7/4 (Nov. 1973), 5–7; and [part 2] "The Early Life of Scott Joplin," *Rag Times* 7/5 (Jan. 1974), 2–3. The testimony of Alexander Ford, as with that of most others interviewed, appears to be honest and mostly correct, but is demonstrably untrue in some specifics. The age sequence that Ford gave for the Joplin children is contradicted by the census listings of 1870 and 1880.

7. Theodore Albrecht, "Julius Weiss: Scott Joplin's First Piano Teacher," *College Music Symposium* 19/2 (Fall 1979), 92.

8. BlJa, 37.

9. Vanderlee and Vanderlee, Jan. 1974, 2.

10. This point was originally noted by Albrecht, 103. The reference in *Treemonisha* is page 7. Page numbers for the opera are given according to the edition published by the New York Public Library; subtract 4 to determine the pagination in the original edition.

11. Vanderlee and Vanderlee, Nov. 1973, 6; Jan. 1974, 2.

12. Jerry L. Atkins, "Early Days in Texas," *Rag Times* 6/3 (Sept. 1972), 1.

13. Vanderlee and Vanderlee, Jan. 1974, 3.

14. Albrecht, 89–105. The identification is not conclusive, but the study is strongly reasoned and convincing.

15. Albrecht, 93–97.

16. BlJa, 37.

17. Vanderlee and Vanderlee, Jan. 1974, 3.

18. Vanderlee and Vanderlee, Jan. 1974, 2.

19. HaBe, 61–62.

20. BlJa, 37; Brun Campbell, "From Rags to Ragtime and Riches," *Jazz Journal* 2/7 (July 1949): 13.

21. Campbell, July 1949, 13; BlJa, 38, 40–41.

22. "Scott Joplin is Dead," ChRR (July 1917), 13; repr. *Rag Times* (March 1987), 8. Stark was Joplin's first important publisher and a close associate.

23. BlJa, 40–41.

24. "Finds First 'Rag' Author," *Philadelphia Tribune,* July 6, 1912, 3. *Harlem Rag* remained unpublished until 1897, then becoming the first rag published by a black composer. Since this published work is a fully mature rag, as opposed to some of the undeveloped and still primitive examples that appeared in the same year, the report of his being active in the field since 1892 gains credence.

25. JaNo, 21.

26. *1883–84 Simmons & Kernodle's Pettis County and Sedalia City Directory* . . . (Sedalia: Simmons & Kernodle, c. 1883); *Hoye's Sedalia City Directory, for 1884–5,* . . . (Kansas City, Mo., c. 1885); *Stone, Davidson & Co.'s Sedalia City Directory 1886–'87* (Sedalia: Stone, Davidson, 1886). Another black Joplin in town was Jennie Joplin, but she was not listed until the 1888–89 directory.

27. "Scott Joplin a King," SeT, June 13, 1903, 1.

28. E.R.V., "One Deplorable State," IFr, July 25, 1891, 10; repr. Aug. 1, 1891, 2. For this and other newspapers that did not number their pages, page numbers have been supplied.

29. Mrs. Alice E. Albert, "A Monument for Jeff. Davis," *Southwestern Christian Advocate,* July 30, 1891, 1.

30. "The Texarkana Minstrel Company and the Jefferson Davis Monument Fund," *Southwestern Christian Advocate,* Aug. 13, 1891, 4. (Mr. Lynn Abbott discovered the episode in the New Orleans sources cited and generously supplied me with copies of the articles and additional background information.) In another reply from the minstrels, they claimed not to have received any money; see "Awfully Dissatisfied," IFr, Aug. 8, 1891, 10; repr. Aug. 15, 1891, 7. See also "The Southern States," *Times-Democrat* (New Orleans), July 21, 1891, 8; July 30, 1891, 8; and "Had Nothing To Do with It," IFr, Aug. 8, 1891, 5. I have been unable to find any record of the monument fund and assume the producer simply invoked the name of Jefferson Davis in an effort to play upon the sentiment of the Confederate veterans, thereby boosting sales.

31. Campbell, July 1949, 13; BlJa, 41.

32. Respectively: "Questions and Answers," *Etude* 16 (Dec. 1898), 349; "'Coon Songs' on the Wane," AMAJ 22 (June 12, 1906), 26a; Natalie Curtis, "The Negro's Contribution to the Music of America," *Craftsman* 23 (March 15, 1913), 662; Isidore Witmark and Isaac Goldberg, *The Story of the House of Witmark: From Ragtime to Swingtime* (New York: Lee Furman, 1939), 169. See also discussions in BeRa, 21, 25–26, 64–66, and BeRR, 1–3.

33. Performances by Blake and Johnson are transcribed by Riccardo Scivales in *Harlem Stride Piano Solos* (Katonah, N.Y.: Ekay Music, 1990).

34. BlJa, 25, 28, 41; Campbell, July 1949, 13.

Chapter 2

1. Most of the background information was taken from Sedalia town directories, especially, *E. L. Russel & Co.'s Sedalia City Directory, 1888–9* (Fort Scott, Kans.: E. L. Russell, [c. 1888]), 6–30. See also, HaBe, 87–88, and Rose M. Nolen, *Lost on the Prairie: George R. Smith College, Methodist School for Blacks, Sedalia, Missouri 1888–1925* (Sedalia, Mo.: RoseMark Communications; nd [c. 1986]), 7.

2. Bureau of the Census, as reported in *The First One Hundred Years: A History of the City of Sedalia, Missouri, 1860–1960.* (Sedalia, Mo.: Sedalia Chamber of Commerce, 1960), 123.

3. *1883–84 Simmons & Kernodle's Pettis County and Sedalia City Directory* . . . (Sedalia: Simmons & Kernodle, *c.* 1883, 192, 197–198, 212–220.

4. Russell, 30; *Stone, Davidson & Co.'s Sedalia City Directory, 1886–87* (Stone, Davidson & Co.: Sedalia, Mo., March 1886), 198.

5. Russell, 43, 306.

6. See, for example, in SeT: "Off and Gone," Aug. 31, 1901, 1; "Baseball Sunday," Sept. 22, 1901, 1.

7. "He Objected to Losing," SeS, Dec. 4, 1899, 8. The local newspapers routinely reported the daily court actions.

8. For an example of efforts to clean up the town, see "After the Bawds," SeD, Sept. 24, 1899, 1; "Sedalia, Mo. It is Taking Campaign Calmly," *Life,* Oct. 21, 1940, 92.

9. JaNo, 22.

10. "The Wages of Sin," SeCa, Nov. 29, 1898, 1; "Ed Gravitt Fined," SeD, Nov. 30, 1898, 5.

11. "Lottie Wright Arrested," SeS, Sept. 25, 1897, 4; "In the Justice Courts," SeCa, Aug. 10, 1898, 8.

12. "Back to Her Home," SeD, Jan. 15, 1899, 1.

13. "Shut Out Again," SeT, Aug. 31, 1901, 1.

14. "To Move the Tough Element," SeS, July 27, 1899, 1.

15. That there were restrictions in arrest powers is intimated in "Snap Shots," SeCa, June 15, 1898, 5, although the exact nature of restrictions is not specified. Probably, black officers could not arrest white offenders. For blacks in other municipal positions, see Editorial, SeT, April 12, 1902, 2.

16. Editorial, SeT, Jan. 25, 1902, 2.

17. Editorial, SeT, May 30, 1903, 2.

18. Editorial, SeT, Aug. 16, 1902, 2.

19. "Davis Day Detail," SeCa, Sept. 22, 1898, 4.

20. See the informative 26-page booklet by Rose M. Nolen, *Lost on the Prairie: George R. Smith College. Methodist School for Blacks, Sedalia, Missouri, 1888–1925,* Sedalia, Mo.: RoseMark Communications [nd; c.1986]; see also Rose M. Nolen, "Where Joplin Studied," *Pettis County Times,* June 8, 1983, Sect. B, 6; and "George R. Smith College" [advertisement], IFr, Jan. 4, 1896, 5.

21. *Annual Catalogue,* 1896–97, 43.

22. S. Brunson Campbell, "A Silver Half-Dollar and the 'Ragtime Kid,'" 3.

23. Untitled, SeT, Feb. 14, 1903, 1.

24. JaNo, 27; see also BlJa, 19–23.

25. "Off and Gone," SeT, Aug. 31, 1901, 1.

26. "Band Entertainment," SeT, Nov. 16, 1901, 1; "Society Notes," SeCo, Sept. 25, 1903, 3.

27. "Personal," SeT, Sept. 22, 1901, 2. The frequent assertion that Ireland was owner and editor of the newspaper *The Western World* (1891–1894) is untrue. *Western World* was a white industrial and labor newspaper. Ireland may have had another position with the paper, such as running the presses, which he did for other newspapers in later years.

28. "Wanted at Once," SeT, Nov. 28, 1903, 2.

29. Untitled, SeT, Sept. 7, 1901, 1.

30. Untitled, SeT, Sept. 22, 1901, 1.

31. In SeD: "Will Sell the Instruments," Oct. 22, 1899, 1; "Struck a Snag," Oct. 26, 1899, 1; "The Negro Band Instruments," Nov. 2, 1899, 1; "Sale of Instru-

ments," Nov. 3, 1899, 8; "Got Their Instruments Back," Nov. 29, 1899, 1. In SeS: untitled, Oct. 23, 1899, 7; "Will Be Music on the Bowery," Nov. 3, 1899, 8; "Serenaded the Justices," Nov. 29, 1899, 1. In SeC: "Snap Shots," Nov. 29, 1899, 4; Nov. 30, 1899, 4.

Chapter 3

1. JaNo, 24.

2. Tom Ireland, letter to S. Brunson Campbell, April 19, 1947.

3. "An Interview with Arthur Marshall," *Rag Times* 21/3 (Sept. 1987), 2; BlJa, 19, 26. Both Arthur Marshall and his daughter Mildred Steward cited their address as 117 West Henry (JaNo, 28). This was the same house that had been 135 West Henry, renumbered several years after Joplin left town. In the earlier days, there were two different east-west borders: south of the railroad, the border was Ohio; in the north, it was Lamine. Another unusual address feature in the north side was that the two-block span from Lamine to Osage had house numbers all in the 100 series, ascending from east to west. The confusion is evident in conflicting reports of house addresses that adjoined the border. Take, for example, Solomon Dixon's home, where Joplin lived for a while. The town directories were correct in listing it as 124 West Cooper. That the newspapers cited it as 124 East Cooper indicates that the house was in the block between Lamine and Ohio.

Neither Julius Walter Joplin nor Jennie Joplin, with whom Scott Joplin might have stayed if he had been in Sedalia during the 1880s, were any longer listed in the town directories during the 1890s.

4. JaNo, 22.

5. Blesh and Janis were told that Wesley lived on the corner of Lamine and Cooper, but this was a later address, listed in the 1898–1899 directory as 502 North Lamine. Wesley originally had two houses at Lamine and Morgan, one of which he rented. Both were destroyed by fire in 1897 ("A North Side Fire," SeD, Oct. 31, 1897, 10). Since Joplin moved to Wesley's home after six months with the Marshalls, it would have been to one of these two houses. He was not living there at the time of the 1897 fire.

6. Hazel N. Lang, *Life in Pettis County—1815–1973.* [Sedalia, Mo.]: Hazel N. Lang, 1975, 477, 737. This account is suspect, for the book is replete with errors. I also doubt that the Maple Leaf Club existed in 1894. Still, the author had access to Tom Ireland, so the account may contain elements of fact.

7. Ireland, letter to Campbell, April 19, 1947.

8. BlJa, 27. Marshall said, "Joplin played call jobs in daytime at some places at odd occasions" (JaNo, 29). Both Wright and Hall are identified as prostitutes in local newspapers. The 1903 town directory lists Hall at 208 West Main Street.

9. JaNo, 28. It is possible that this membership was from a few years later, as Will and Robert were in Sedalia in 1899. Emmett Cook was also a drummer with the Queen City Concert Band and is in the photograph referred to above. Richard Smith is mentioned in a 1904 newspaper as a baritone soloist performing with the Sedalia Quartet. Leonard Williams was probably the same person identified as Lynn Williams, also with the Sedalia Quartet.

10. All of the Joplin's music, unless otherwise indicated, can be found in CWSJ. Appendix A in the current book contains copyright and publication information and other comments about the music.

11. I thank Patricia Finley and Jean Palmer, of the Onondaga County Public Library, for assistance in researching Joplin's Syracuse publishers. They examined several months of the *Syracuse Herald* and the *Syracuse Standard* and also checked issues of the city directory for 1894, 1895, and 1896. They found no direct evidence of Joplin being in Syracuse. See also, Edward A. Berlin, "Scott Joplin's First Two Publishers," *Rag Times* 22/2 (July 1988), 6–7.

12. JaNo, 55.

13. *McCoy's Hannibal Business Directory, 1895–1896* (Keokuk, Iowa: William H. McCoy, [1895]): "S. M. Walker, Pianos and Organs, 405 Broadway." I thank Roberta Hagood, historian from Hannibal, for her assistance on this search.

14. Henry Jackson provided the music for a cakewalk exhibition in 1899; see "News in Colored Circles," SeS, Dec. 16, 1899, 5. Jackson was described as "the veteran M. K. & T. train porter" in SeT, Jan. 25, 1902, 1.

15. William H. Goetzmann, "Tying the Nation Together," a chapter in *We Americans* (Washington, D.C.: National Geographic Society, 1975), 182; reprinted as "The Crush Collision Story," *Rag Times* 12/4 (Nov. 1978), 3. Also see Roy Carew and Don E. Fowler, "Scott Joplin," *Record Changer* (Oct. 1944), 10.

16. The march, from which the formal structure of the rag was derived, is comprised of successive strains, each usually of 16 measures and independent thematic material. Typically, after two strains in the tonic key, the subdominant key is introduced, this being the "trio." The trio may have one or more 16-measure strains, frequently with an interlude—of 4, 8, 12, 16, 24, or even 32 measures—placed between the strains or their repeats. For a fuller general description of the interlude, see BeRa, 91–94.

In *Crush Collision March,* there are three strains (designated for analytical purposes as A, B, and C) before the trio. Since the A strain is in a minor key (d minor), the trio (section D), following the convention for the form, is in the subdominant of the relative major. (The relative major of d minor is F major; the subdominant of F is B-flat.) The interlude is between the D section and its repeat.

17. "An Interview with Arthur Marshall!!" *Rag Times* 21/3 (Sept. 1987), 2.

18. Trebor Tichenor, "Chestnut Valley Days," *Rag Times* 5/4 (Nov. 1971), 3.

19. Campbell's story is recounted briefly in BlJa, 28–31. See also the impressionistic sketch in the collection of Campbell's music, *Brun Campbell. The Music of "The Ragtime Kid,"* transcribed and with introduction by Richard A. Egan, Jr. (St. Louis, Mo.: Morgan, 1993).

20. Paul Affeldt, "The Saga of S. Brun Campbell," *Mississippi Rag* 15/3 (Jan. 1988), 1–4. As an extreme example of how Campbell allowed his imagination to take charge, I offer his description of Joplin's funeral:

> His funeral procession was one of the most unusual in American musical history, for every carriage in it carried the name of one of his ragtime hits. "Maple Leaf" was on the first. It presented a unique sight as it slowly made its way through the streets of Harlem on Good Friday of 1917. (S. Brun Campbell, "From Rags to Ragtime and Riches," *Jazz Journal,* 12/7 (July 1949), 13.)

This description is pure fantasy. There are no confirmed reports of a procession at the funeral, and since Joplin was buried in an unmarked grave, the funeral must have been exceedingly modest.

I also strongly doubt that Campbell's style of playing derived from Scott Joplin. More likely, his playing reflected the styles of the essentially untrained "folk pianists" of the period.

21. Campbell (July 1949), 13. The copyright date of *Original Rags* is March 15, 1899; this date, though, does not rule out an earlier publication.

22. BlJa, 19. JaNo, 23. Robert Allen Bradford, "Arthur Marshall—Last of the Sedalia Ragtimers", *Rag Times* 2/1 (May 1968), 5.

23. BlJa, 23–24, 19.

24. For example, see "A Dancing Party," SeCa, Oct. 17, 1899, 8. "The Maple Leaf club, one of the popular social organizations of this city, gave a dancing party last night in honor of Miss Lena Hume, of Fulton, Mo. Miss Hattie Gross presided at the punch bowl and served cooling drinks throughout the evening . . ."

25. JaNo, 22.

26. "The 'Black 400'," SeS, March 30, 1898, 1.

27. A picture of "Doc" Brown appears in the sheet music publication. See reprint in Trebor Jay Tichenor, *Ragtime Rarities* (New York: Dover, 1975). An oil painting of "Doc" Brown is displayed in the Kansas City Museum. See Kevin Sanders, "Doc Brown, Cakewalker," *Rag Times* 20/3 (Sept. 1986), 1–2.

28. Amusements, SeD, May 22, 1898, 2. "Ernest Is a Winner," SeCa, May 27, 1898, 1.

29. "A Coming Event," SeCa, Oct. 22, 1898, 1.

30. The name of the saloon below is given in "Black 400 Club Room," SeCa, Oct. 1, 1898, 5; another saloon, J. W. Lopp, is listed at that address in 1898–99 directory; in early May, J. H. Wilkerson and J. W. Brown opened a new saloon there, calling it "The Nonpareil" (see SeS, May 4 1899, 1, and SeCa May 17, 1899). Sedalia legend that the Black 400 Club was situated above Archia's Seed Store is not quite correct. Archia's opened next door at 108 East Main on Sept. 14, 1899. It was not enlarged to include the adjoining address of 106 until after the club's demise.

31. In SeCa: "Black 400 Club Room," Oct. 1, 1898, 5; "'Black 400' Ball," Oct. 6, 1898, 4.

32. S. Brunson Campbell, "A Silver Half-Dollar and the 'Ragtime Kid'," 4.

33. "Colored Society Dance," SeCa, Oct. 14, 1898, 1. "Colored 400," SeS, Nov. 25, 1898, 2. In SeD: "The Cake Walk," Nov. 25, 1898, 1; "The Cake Walk," Dec. 27, 1898, 7. All male instrumentalists were given the honorary title of "Professor."

34. "Maple Leaf Club," SeCa, Nov. 25, 1898, 1.

35. Articles of Agreement of the Maple Leaf Club of Sedalia, Missouri, Dec. 13, 1898.

36. "Will Give a Masked Ball," SeD, Dec. 7, 1898, 6.

37. The card was discovered by Larry Melton and donated to the State Fair Community College Ragtime Archives, Sedalia, Missouri. "W. J. Williams," on the face of the card, was Will Williams rather than Walker Williams, for the 1897 directory lists both W. J. Williams and Walker Williams. See *Sedalia City Directory, and Pettis County Directory, for 1897* (Sedalia: Capan and Bowman, [1897]).

38. "Some Clippings," SeT, Oct. 26, 1901, 1.

39. "Black Diamond in the City," SeS, June 13, 1899, 4.

40. In SeCa: "Closing the Clubs," Jan. 7, 1899, 1; "The Club Will Run," Jan. 8, 1899, 1.

41. "Fighting the Clubs," SeCa, Jan. 17, 1899, 5.

42. "Want 'Em Closed," SeD, Jan. 17, 1899, 8.

43. "That Cake Walk," SeCa, Jan. 19, 1899, 1. "Circuit Court," SeS, Feb. 7, 1899, 5. "The Circuit Court," SeCa, Oct. 11, 1899, p. 1.

44. "The Blind Tiger," SeCa, Feb. 14, 1899, 1.

45. All in SeCa: "For the Poor," Feb. 16, 1899, 1; "Concern for Charity," Feb.

19, 1899, 1; "Fancy Dress Ball," Feb. 22, 1899, 5; "Charity Benefit Off," Feb. 24, 1899, 8.

46. In SeD: "Fred Mack in Hock," Feb. 28, 1899, 1; "Took Him to Clinton," March 3, 1899, 1; "He, Too, Gave Bond," March 7, 1899, 2. In SeCa: "Fred Mack Arrested," March 3, 1899, 5, and "At the Black 400," 5; "Drinks Are Served," March 6, 1899, 9; "Tony Williams Arrested," March 7, 1899, 4; "Card from Tony Williams," March 8, 1899; "'Tony' and Fred Acquitted," March 22, 1899, 4. The performances are all described in SeD: "The Unlucky Coons" [Advertisement], March 29, 1899, 2; "Rehearsed Last Night," April 2, 1899, 9; "The Weather Interfered," April 4, 1899, 3; "Will Give a Parade," April 9, 1899, 11; "A Hot Time Tonight," April 11, 1899, 7.

47. In SeS: "Black '400' Will Reorganize," May 6, 1899, 3; "Closing Dance," May 30, 1899, 1.

48. In SeCa: "The Four Hundred Club," Aug. 4, 1899, 1; "Tony Williams at Joplin," Dec. 26, 1899, 1.

49. D. O. H. is Deutsche Orden der Harugari, i.e., the German Order of Harugari, "Harugari" referring to ancient warriors. This was an organization of German immigrants, established in New York in 1864. The Sedalia chapter was formed in 1869.

50. "News Notes," SeS, Oct. 26, 1899, 4.

51. "Battle Row Makes a Record," SeS, Oct. 2, 1899, 1; "The Courts," SeCa, Oct. 3, 1899, 8.

52. "Gathering 'em In," SeCa, Oct. 25, 1899, 1. In SeS: "The Police Were Busy," Oct. 25, 1899, 4; "Jots of Justice," Oct. 25, 1899, 1.

53. In SeS: untitled, Dec. 21, 1899, 6, and "Women Jailed for Fighting," Dec. 22, 1899, 4. "Got Three Days Each," SeD, Dec. 22, 1899, 4.

54. In the SeCa, "Raided the Club," Jan. 14, 1900, 5; "That Club Mess," Jan. 16, 1900, 1. In SeD: "He Used His Gun," Jan. 15, 1900; "Justice Jots," Jan. 17, 1900, 4. In SeS: "Tough Colored Clubs," Jan. 15, 1900, 5; "What Could He Expect?" Jan. 17, 1900, 5. The newspapers all refer to the invaded club as the Black 400; this may have been because it occupied the hall vacated by the Black 400. Notices of the previous three months clearly refer to the new organization as the Social Club.

55. In SeD: "Clubs Must Close," Jan. 25, 1900, 1; "Club Remained Closed," Jan. 30, 1900, 2. In SeS: untitled, Jan. 25, 1900, 5; "News Notes," March 16, 1900, 5. In SeCa: "The Clubs Closed," Jan. 26, 1900, 1; "Snap Shots," Jan. 30, 1900, 5. A single notice in March indicates that an illegal prize fight was held in the Maple Leaf club rooms, but it is unclear whether the club itself still existed.

56. "Court an Investigation," SeS, Jan. 25, 1900, 8. In "The Stage," IFr: Aug. 4, 1900, 5; Sept. 15, 1900, 5. JaNo, 27.

57. Ireland letter to Campbell, April 19, 1947.

Chapter 4

1. "Scott-Joplin," AMAJ 23/23, Dec. 13, 1907, 5.

2. For a more detailed examination of ragtime rhythms, see BeRa, 82–89, 128–134, 147–152. For excerpts of ragtime-like music from the minstrel stage of the 1880s, see *Patrol Comique* and *The Hottentots,* 107.

3. The early appearances of ragtime and the connection with coon songs is treated in more detail in BeRa, 20–31, 63–66.

4. "Decadence of the Waltz: Sousa's Marches Held Responsible by Dancing Masters for the Reign of the Two-Step," *New York Times,* Sept. 10, 1899, 16; Paul E. Bierly, *John Philip Sousa: American Phenomenon* (New York: Appleton-Century-Crofts, 1973) 7, 48. For a discussion of the musical sources of ragtime, see BeRa, 99–122.

5. For a more detailed discussion of ragtime form, see BeRa, 89–98, 134–146.

6. The report of Turpin's *Harlem Rag* being composed in 1892 is in "Finds First 'Rag' Author," *Philadelphia Tribune,* July 6, 1912, 3. Brun Campbell dates *Harlem Rag* from 1896. See S. Brun Campbell, "Ragtime Begins," *Record Changer* 7/3 (March 1948), 8; repr. *The Ragtime Society [Newsletter]* 2/9 (Nov. 1963), 4. For a discussion of the fair's place in ragtime history, see BeRa, 25–26, 134–135, and BeRR, 1–2.

7. Krell's *Mississippi Rag* is reprinted in *The Best of Ragtime Favorites* (New York: Charles Hansen, n.d.); *Golden Encyclopedia of Ragtime, 1900 to 1974* (New York: Charles Hansen, 1974); *34 Ragtime Jazz Classics for Piano* (New York: Melrose Music, 1964); *One Hundred Ragtime Classics,* ed. Max Morath (Denver: Donn Printing, 1963). The 1897 version of Turpin's *Harlem Rag* is in *Classic Piano Rags* (New York: Dover, 1973).

8. For discussion of types of early rags, see BeRa, 81–98.

9. S. Brunson Campbell, "A Silver Half-Dollar and the 'Ragtime Kid'," 2.

10. BlJa, 24. John Stark provided the information that Joplin had tried selling *Maple Leaf* and *Sunflower Slow Drag* as well as *Original Rags.* See in ChRR: "A Ragtime Pioneer" (Sept. 1915), 8; "Scott Joplin is Dead" (July 1917), 13.

11. *W. H. McCoy's Sedalia, Mo., City Directory, for 1898–1899* (Keokuk, Iowa: W. H. McCoy, [1898]), 256.

12. Unless we have more precise information, I use the copyright entry date as an indication of when a work was published. It should be understood that this is not an infallible guide: a work may be published either prior to or after copyright registration. In many cases there is no registration at all, despite a copyright notice on the title page of the music, and in other instances the registration remained incomplete for failure to submit two copies of the music. In the case of *Original Rags,* application was made on March 15, 1899, but apparently the requisite two copies of the score were never submitted to complete the registration. The Claimant Card at the Copyright Office is stamped "card filed Jan 28 1905 without credit," meaning that the music was never received.

13. See, for example, advertisement in the *Kansas City Star,* May 12, 1899.

14. For more on Daniels, see David A. Jasen and Trebor Jay Tichenor, *Rags and Ragtime* (New York: Seabury, 1978), 135–137. For a brief description of the Indian phase in American popular music, see William J. Schafer and Johannes Riedel, *The Art of Ragtime* (Baton Rouge: Louisiana State University Press, 1973), 118–127.

15. I am indebted to Trebor Tichenor for calling this to my attention and giving me a copy of the music.

16. See BeRa for discussions of the convention of octaves (95, 140–144) and blues elements in early ragtime (154–161).

17. L. Edgar Settle's *X. L. Rag* is reprinted in Trebor Jay Tichenor, ed., *Ragtime Raritites* (New York: Dover, 1975). Will H. Etter's *Whoa! Maude* is reprinted in *101 Rare Rags* (Los Angeles: Dick Zimmerman, 1988).

18. The contract for the *Maple Leaf Rag* was dated August 10, 1899. A published copy of the music was received by the Copyright Office on September 20.

19. BlJa, 28.

20. BlJa, 33.

21. BlJa, 32–33. Three decades after the publication of this book, Blesh retold the story, but without any hedging as to its reliability. He cited Tom Ireland as the story's source. See BlSJ, xxii.

22. "An Interview with Mildred Steward," Part II, *Rag Times* 24/5 (Jan. 1991), 4.

23. Bartlett D. Simms, and Ernest Borneman, "Ragtime History and Analysis," *Record Changer* (Oct. 1945), 5.

24. Dorothy Brockhoff, "Missouri was the Birthplace of Ragtime," *St. Louis Post-Dispatch,* Jan. 1960; repr. *Rag Times* 25/2 (July 1991), 10.

25. "Scott Joplin is Dead," ChRR (July 1917), 13; repr. *Rag Times* 20/6 (March 1987), 8. This is essentially the same story Stark told Christensen, as published two years earlier in Axel Christensen, "A Ragtime Pioneer," ChRR 1/9 (Sept. 1915), 8.

26. "The Maple Leaf Rag Contract," *Rag Times* 9/3 (Sept. 75), 8.

27. "R. A. Higdon," *Rag Times* 9/4 (Nov. 1975), 11; Edward A. Berlin, "More on R. A. Higdon," *Rag Times* 23/3 (Sept. 1989), 6–7. Sedalia newspapers make frequent mention of the numerous dance clubs. Will Stark's association with Higdon is documented in SeCa: "Old People's Day," July 8, 1899, 1; "Some Social Happenings," Dec. 8, 1899, 8. An example of Higdon's advertisements for loans in SeD, Sept. 17, 1899, 5.

28. Ernest Hogan, a star comedian and composer of ragtime songs, most notably *All Coons Look Alike to Me,* reported in 1897 that he was receiving almost $400 a month in royalties. See "The Stage," IFr, Sept. 11, 1897, 5.

29. Respectively: J. Russel Robinson, "Dixieland Piano," *Record Changer* (Aug. 1947), 7; Alan Lomax, *Mister Jelly Roll,* 2nd ed. (Berkeley: University of California Press, 1973), 149; Tom Davin, "Conversations with James P. Johnson," *Jazz Review* 2 (July 1959), 11, repr. in John Edward Hasse, ed., *Ragtime: Its History, Composers, and Music* (New York: Schirmer, 1985), 172; Edward R. Winn, undated advertisement, reprinted in *Rag Times* 7 (Jan. 1974), 8; Axel Christensen, "Chicago Syncopations: John Stark, Pioneer Publisher" *Melody* 2 (Oct. 1918), 8; John N. Burk, "Ragtime and Its Possibilities," *Harvard Musical Review* 2 (Jan. 1914), 11–13. "To Play Ragtime in Europe," *St. Louis Post-Dispatch,* Feb. 28, 1901, 8.

30. Respectively: Monroe H. Rosenfeld, "The King of Ragtime Composers is Scott Joplin, a Colored St. Louisan." *St. Louis Globe-Democrat,* June 7, 1903, Sporting Section, 5. Advertisement entitled "Famous and Lasting Rags," back covers of many of his publications in the mid-teens, such as Joplin's *Easy Winners* (1902) and Joseph F. Lamb's *Cleopatra* (1915) and *Top Liner* (1916).

31. Respectively: Trebor Jay Tichenor, "John Stillwell Stark, Piano Ragtime Publisher: Readings from *The Intermezzo* and His Personal Ledgers, 1905–1908," *Black Music Research Journal,* 9/2 (Fall 1989), 196. Roy Carew and Don E. Fowler, "Scott Joplin: Overlooked Genius," *Record Changer,* Oct. 1944, 10.

32. "Ragtime Music (Invented in St. Louis) is Dead," *St. Louis Post-Dispatch,* April 4, 1909, Sunday Magazine, 1.

33. Dick Zimmerman, "An Interview with John Stark," *Rag Times* 10/3 (Sept. 1976), 5.

34. Cf. listings in David A. Jasen, *Recorded Ragtime 1897–1958* (Hamden, Conn.: Archon Books, 1973).

35. Trebor Jay Tichenor, the owner of these ledger books, has kindly given me copies and granted permission for their use.

36. Annual salaries in the manufacturing industries from 1900–1909 ranged between $490 and $595. These figures are calculated from table Series D 780, in U.S. Bureau of Census, *Historical Statistics of the United States. Colonial Times to 1970,* Part 1 (Washington, D.C.: U.S. Government Printing Office, 1975), 168. Postal salaries are mentioned in "Salaries Have Been Raised," SeD, Dec. 17, 1899, 10. Real estate figures are in "For Rent," SeD, Sept. 15, 1899, 1.

37. JaNo, 56.

38. Respectively: "An Interview with Mildred Steward," *Rag Times* 24/5 (Jan. 1991), 4; JaNo, 28.

39. "An Interview with Arthur Marshall," *Rag Times* 12/3 (Sept. 1987), 2.

40. Cf. cover reprint in CWSJ I, 239.

41. Dick Zimmerman, "Joe Jordan and Scott Joplin," *Rag Times* 2/4 (Nov. 1968), 5.

42. See Alternate Covers in CWSJ I, xlii–xliii.

43. Respectively: "Scott Joplin a King," SeT, Oct. 13, 1903, 1; JaNo, 22; letter from Tom Ireland to S. Brunson Campbell, July 19, 1947.

44. A picture of the residence is in the *E. L. Russell & Co.'s Sedalia City Directory, 1888–1889* (Fort Scott, Kans.: Russell, [c. 1888], 141.

45. This is not the current Maplewood in the St. Louis area.

46. Scott DeVeaux, "The Music of James Scott," in Scott DeVeaux and William Howland Kenney, eds., *The Music of James Scott* (Washington and London: Smithsonian Institution Press, 1992), 45.

47. I should like to thank two eagle-eyed collectors for bringing the more obscure examples to my attention: Elliott Adams, for *Latonia Rag,* and Michael Montgomery, for *The Tantalizer* and *That's Goin' Some.*

Chapter 5

1. These advertisements were widely reprinted on Stark publications and in magazines. It is impractical to try to pinpoint their first printings. "A Fierce Tragedy" was reprinted in ChRR 1/1 (Dec. 1914), 2.

2. Alan Lomax, *Mister Jelly Roll,* 2d ed. (Berkeley: University of California Press, 1973), 149.

3. Respectively: JaNo, 26; "An Interview with Arthur Marshall," *Rag Times* 12/3 (Sept. 1987), 2.

4. This incident was described by Trebor Jay Tichenor at the Scott Joplin Festival Seminars in June 1992. Tichenor's evidence consisted of four letters in his possession, written by Stark to Marshall.

5. Stark must have been in Sedalia by 1882 because he is in the *1883–84 Simmons & Kernodle's Pettis County and Sedalia Directory* (Simmons & Kernodle, c. 1883), which has a preface dated February 1883. He has a full page advertisement on page 185.

6. Advertisement, SeCa, June 5, 1898, 2.

7. "Pianos, Pianos," SeS, Dec. 5, 1898, 1.

8. "Glimpses of Local Happenings," SeS, Dec. 24, 1898, 8. I have been unable to locate a copy of the *Music Trades* issue in question, which was that of Dec. 17, 1898.

9. The progress of the Stark family is traced in Sedalia newspapers, which were especially willing to report on the town's most celebrated musician. For Eleanor's

appearance with the Busch orchestra, see in SeS: "In Society," Jan. 3, 1899, 5; "A Complete Triumph," Jan. 9, 1899, 8. For her and her family's move to St. Louis, see in SeS: "In Society," Jan. 28, 1899, 8; Feb. 2, 1899, 5; March 17, 1899, 8.

10. Quoted from excerpts of letter from Ted Browne to Karl Kramer, about 1966. Copy supplied by Max Morath.

11. BlJa, 69.

12. Letter owned by Trebor Jay Tichenor.

13. For example, see the statement by Will Stark's widow in 1960: Dorothy Brockhoff, "Missouri was the Birthplace of Ragtime," *St. Louis Post-Dispatch,* Jan. 1960; repr. [as part 2] *Rag Times* 25/3 (Sept. 1991), 6.

14. "'Black American' Cake Walk," SeCa, Oct. 8, 1899, 1.

15. BlJa, 71.

16. "'Black American' Cake Walk," SeCa, Oct. 8, 1899, 1.

17. BlJa, 69, 71; JaNo, 29.

18. JaNo, 22; BlJa, 69.

19. The expectation that an event held by blacks would result in violence was common. Every year, after the Fourth of August Emancipation Day celebrations that drew black Missourians to Sedalia, the daily newspapers would comment that the crowd behaved better than expected. This expectation is apparent also in a newspaper comment regarding a Black 400 ball at which Joplin performed: "The attendance was large and there was not an unpleasant incident to mar the evening." "Colored 400," SeS, Nov. 25, 1898, 2.

20. "Made Sweet Music," SeCa, Oct. 11, 1899, 1.

21. Respectively: "Cake-Walk Tomorrow Night," SeD, Dec. 11, 1899, 7. Untitled, SeS, Dec. 13, 1899, 8. "News in Colored Circles," SeS, Dec. 16, 1899, 5. See also "Snap Shots," SeCa, Dec. 12, 1899, 5. Sedalia newspapers mention a Robert Joplin as early as January 1898, but this was not Scott Joplin's brother; it was a white youth.

22. "Lincoln Theatre, Knoxville, Tenn." IFr, Feb. 15, 1908, 5.

23. A Cake Walk," SeD, Nov. 19, 1899, 4. See also Untitled, SeS, Nov. 20, 1899, 7; and Advertisement, SeCa, Nov. 24, 1899, 5. The "Pork Chops Greasy Quartette" was probably a takeoff on a well-known minstrel song of the time, which has as the opening lines of its chorus: "I'm livin' easy, eatin' pork chops greazy" (Irving Jones, *I'm Livin' Easy,* New York: F. A. Mills, 1899). Composer Charles Ives recalled hearing this sung by minstrels in Connecticut as far back as 1893–94. See Charles E. Ives, *Memos,* ed. John Kirkpatrick (New York: W. W. Norton, 1972), 56.

24. In SeS: untitled, Nov. 25, 1899, 2; untitled, SeS, Nov. 28, 1899, 2. In SeCa: "Last Night's Entertainment," Nov. 25, 1899, 5; "Snap Shots," Nov. 28, 1899, 8.

25. Words by Gene Jefferson, music by Leo Friedman (Chicago: Sol Bloom, 1900). The reference to the Queen City Cornet Band using this music is "Band Concert," SeT, May 10, 1902, 1.

26. Words and music by Will Marion [Will Marion Cook], *Darktown Is Out Tonight.* New York: M. Witmark, 1898. In another version, Cook's collaborator Paul Laurence Dunbar is credited with the lyric.

27. For further discussion, see my article "On Ragtime: Understanding the Language," *CBMR Digest* 3/3 (Fall 1990), 6–7.

28. "Will Honor a Local Club," SeD, Jan. 2, 1900, 1. See also, in the "Snap Shots" column of SeCa, Jan. 3, 1900, 8, and Jan. 31, 1900, 5.

29. In SeD: "Will Give a Masque Ball," Feb. 4, 1900, 5; and "The Masqued Ball," Feb. 20, 1900, 2.

30. Respectively: "Tony's Cake Walk," SeCa, Dec. 27, 1898, 1; unititled, SeS, Dec. 13, 1899, 8 (my emphasis); "Snap Shots," SeCa, Jan. 31, 1900, 5; "Snap Shots," SeCa, Jan. 3, 1900, 8.

31. BlJa, 26–27.

32. BlJa, 52–53.

33. "An Interview with Arthur Marshall!!" *Rag Times* 21/3 (Sept. 1987), 1; "Arthur Marshall: His Daughter's Memories," *Rag Times* 24/4 (Nov. 1990), 2. Sedalia resident Ollie Martin, whose mother Beatrice Martin knew both Joplin and Marshall, insists the title was always pronounced with a short "i"—"Swippsy." This pronunciation would not support the story of the title's origin.

34. Cassidy, Russell E., "Joseph F. Lamb: A Biography," *Ragtime Society* 5/4 (Summer 1966), 31–32. *Dynamite Rag* was renamed *Joe Lamb's Old Rag* and copyrighted by Bob Darch. Publication is planned by the Smithsonian Institution Press in a collection of Lamb's works, edited by Joseph R. Scotti.

35. Letter from John Stark to Arthur Marshall, Aug. 24, 1906. Letter is owned by Trebor Tichenor.

36. JaNo, 13.

37. BlJa, 26–27. Marshall confirmed this in a letter to Robert Allen Bradford, printed as "Arthur Marshall—Last of the Sedalia Ragtimers," *Rag Times* 2/1 (May 1968), 5.

38. S. Brunson Campbell, "From Rags to Ragtime and Riches," *Jazz Journal,* 2/7 (July 1949), 13.

39. "Scott Joplin is Dead," ChRR, July 1917, 13; "Scott Joplin a King," SeT, Oct. 13, 1903, 1.

40. "Scott Joplin a King," SeT, Oct. 13, 1903, 1; "Band Entertainment," SeT, Nov. 16, 1901, 1; Trebor Tichenor, "Missouri Ragtime Revival," *Rag Times* 4/5 (Jan. 1971), 3.

41. "Stage," IFr, July 20, 1901, 5; Nov. 2, 1901, 5.

42. BlJa, 53.

43. For a more detailed analysis, see BeRR, 8–10.

44. I thank Dr. Elliott Adams for bringing this Carter piece to my attention.

45. Discussions with Sedalian Bob Ault have led me to doubt the accuracy of this census listing. Comparing the listing with town directories shows a serious pattern of discrepancies. Though the address numbers for specific individuals are frequently the same in both sources, street names are different. For example, John Barlow, whose name follows Joplin's in the census, is listed at 212 Pacific, but in the directory he is at 212 St. Louis. Mary Lenox, line 15, is at 224 Pacific, but in the directory is 224 St. Louis. George Rich, in the census at an unnumbered house on Pacific, is in the directory at 414 W. Pettis. Michael Seethaler, listed before Joplin in the census at 801 Washington, is in the directory at 301 E. Main. Since the directory is consistent through several years, we assume the census is in error. We hypothesize that the census taker, rather than write directly on the census page, took notes and transcribed them incorrectly later.

Susannah Hawkins, listed in the census as a cook and as the head of the household in which Joplin lived, is never listed in the town directories. Conceivably she was the same person as Susan Hawkins, a cook listed in the 1903 directory at 119 W. Main. This information, though, does not help us confirm Joplin's address in 1900.

The significance of Joplin's census listing is that it shows he lived in Sedalia with Belle.

Chapter 6

1. Respectively: "Ragtime," *Times* (London), Feb. 8, 1913, 11; Hiram Kelly Moderwell, "Ragtime," *New Republic* 4 (Oct. 16, 1915), 286.

2. Antonin Dvorak, "Music in America," *Harper's New Monthly Magazine,* (Feb. 1895), 432.

3. Respectively: "Ragtime," *Times* (London), Feb. 8, 1913, 11; Hiram Kelly Moderwell, "A Modest Proposal," *Seven Arts* 2 (July 1917), 375–376.

4. Herbert Sachs-Hirsch, "Dangers That Lie in Ragtime," *Musical America* 16 (Sept. 21, 1912), 8.

5. Respectively: "Musical Impurity," *Etude* 18 (Jan. 1900), 16; Francis Toye, "Ragtime: The New Tarantism," *English Review* (March 1913), 655–658.

6. James Weldon Johnson, "Views and Reviews," NYA, Sept. 23, 1915, 4.

7. Walter Winston Kenilworth, "Demoralizing Rag Time Music," *Musical Courier* 46 (May 28, 1913), 22–23.

8. "What is American Music?" *Musical America* 3 (Feb. 24, 1906), 8.

9. "What the Concert-Goer Says of 'The Negro Music Journal,'" *Negro Music Journal* 1 (Oct. 1902), 28.

10. For a more thorough treatment of the reactions to ragtime, see Chapter Three of BeRa; Neil Leonard, "The Reactions to Ragtime," in John Edward Hasse, ed., *Ragtime: Its History, Composers and Music* (New York: Schirmer, 1985), 102–113; and Neil Leonard, *Jazz and the White Americans* (Chicago: University of Chicago Press, 1960).

11. There are many notices in Sedalia newspapers of train fares. As an example of the roundtrip excursions, see "Personal," SeT, Aug. 31, 1901, 2.

12. W. C. Handy, ed. Arna Bontemps, *Father of the Blues. An Autobiography* (New York: Macmillan, 1941), 30. Targee Street, no longer extant, was between Market and Clark, one block west of 14th Street.

13. *Gould's St. Louis Directory for 1891–92.*

14. "Finds First 'Rag' Author," *Philadelphia Tribune,* July 6, 1912, 3.

15. I discuss the variation aspect in BeRR, 19, 21–22.

16. All of these pieces, including the 1897 version of *Harlem Rag,* are reprinted in *Classic Piano Rags,* compiled by Rudi Blesh (New York: Dover Publications, 1973.)

17. BlJa, 54–55; JaNo, 19, 26.

18. Sylvester was a brother. The relationships of the others are not known. All were listed as musicians in various years of the city directory. Louis is listed at 1616 Gay Street (one block north of Morgan) and Abraham at 1111 Morgan in the 1900–01 directory. Peter and Sylvester were at 1927 Market in 1901–02, and Sylvester was there alone in 1903. Peter was at 2246 Lucas in 1904, and had a working address of 1613 Linden (one block south of Morgan) in 1905, 1906, and 1907; Louis and Sylvester were also at the last address in 1907.

19. BlJa, 56–57; JaNo, 12–15, 33.

20. JaNo, 16–19, 33. Beginning around 1906, the theatrical sections of the *New York Age* and the *Indianapolis Freeman*, both black newspapers, regularly carried items about Patterson.

21. JaNo, 13.

22. JaNo, 41. This description came from Artie Matthews, who was younger than those discussed above. The description most accurately reflects the district Matthews knew during the teens, which can reasonably be assumed to have been similar to the conditions ten to fifteen years earlier.

23. Karl Muck, "The Music of Democracy," *Craftsman* (Dec. 1915), 277.

24. "To Play Ragtime in Europe," *St. Louis Post-Dispatch,* Feb. 28, 1901, 8.

25. Ted Browne, letter to Karl Kramer, as quoted in letter to Max Morath, *c.* 1966.

26. "Scott Joplin," AMAJ, June 17, 1907, 35.

27. Monroe H. Rosenfeld, "The King of Ragtime Composers is Scott Joplin, a Colored St. Louisan," *St. Louis Globe-Democrat,* June 7, 1903, Sporting Section, 5.

28. Respectively: Edward A. Berlin, "On Ragtime: Echoes from the Past," *CBMR Digest* 2/2 (Fall 1989), 7; Edward A. Berlin, "On Ragtime: Scott Joplin the Educator," *CBMR Digest* 3/1 (Spring 1990), 4; as reported by Trebor Tichenor to the author on June 5, 1988.

29. Respectively: Trebor Tichenor, "Chestnut Valley Days," *Rag Times* 5/4 (Nov. 1971), 3. Rosenfeld, June 7, 1903. "Scott Joplin," AMAJ, June 17, 1907, 35. "Composer of Ragtime Now Writes Grand Opera," NYA, March 5, 1908, 6. [Harry] Bradford, "William H. Farrell, Composer and Producer," IFr, Sept. 18, 1909, 5. Beatrice Martin, taped interview with Dick Zimmerman, 1975.

30. Respectively: JaNo, 19; John Arpin, Interview with William Sullivan, as reported to me in June 1989; JaNo, 23; Beatrice Martin, taped interview; "An Interview with Arthur Marshall!!" *Rag Times* 21/3 (Sept. 1971), 2; JaNo, 23.

31. Respectively: Ted Browne; JaNo, 13.

32. Willie "The Lion" Smith, as told to George Hoefer, *Music on My Mind: The Memoirs of an American Pianist* (New York: Doubleday, 1964; repr. Da Capo, 1978), 53.

33. "Stage," IFr, July 20, 1901, 5; emphasis mine. John Stark's grandson William P. Stark said that Joplin moved to St. Louis in the middle of 1901; see Bartlett D. Simms and Ernest Borneman, "Ragtime: History and Analysis," *Record Changer* (Oct. 1945), 5.

34. *Gould's St. Louis Directory for 1902.* The building has been restored as part of the St. Louis Historic District. The street name is now Delmar.

35. JaNo, 28. No documentation of the marriage has been found.

36. Trebor Tichenor, "Chestnut Valley Days," *Rag Times* (Nov. 71), 3.

37. There was a Robert Joplin at 2617 Lawton in the 1901–02 directory and at 624A North Beaumont in 1903. While the addresses are in the neighborhood, we suspect this was not Scott Joplin's brother Robert *B.* Joplin. Both entries list Robert Joplin as a cook, and neither includes the middle initial that brother Robert habitually used.

38. "A Negro Exhibit at St. Louis," IFr, March 16, 1901, 4.

39. Information on Shattinger comes from Ernst C. Krohn, ed. and completed by J. Bunker Clark, *Music Publishing in St. Louis.* Bibliographies in American Music (The College Music Society and Harmonie Park Press, 1988), 82.

40. The editor of CWSJ introduced changes without editorial indications.

Since this edition is primarily intended for study and performance, it was judged more important that correct (as far as possible) music texts be presented rather than to perpetuate original errors for antiquarian interest. For this reason corrections have been incorporated into the facsimile pages (p. viii).

But changes to the music text, as this issue demonstrates, are of more than "antiquarian interest." Since the editor did not identify the changes, we have no way of knowing whether the tie was in the copy photographed for the edition, or resulted from an "editorial correction." For more discussion on this issue, see my review in *Notes* 40 (Sept. 1983), 147–149.

41. JaNo, 29.

42. "Our Trip to the World's Fair City," SeT, April 26, 1902, 1.

43. JaNo, 40.

44. JaNo, 26, citing Marshall.

45. "Tony's Cake Walk," SeCa, Dec. 27, 1898, 1.

46. Edward A. Berlin, "On Ragtime: Echoes from the Past," *CBMR Digest* 2/2 (Fall 1989), 7.

47. Edward A. Berlin, "On Ragtime: Scott Joplin the Educator," *CBMR Digest* 3/1 (Spring 1990), 4.

48. W. A. Corey, "Timely Tattle," AMAJ, 27/13, (July 8, 1911), 16.

49. Respectively: JaNo, 23, 26; Trebor Tichenor, unpublished and undated interview with Charlie Thompson, related to the author June 5, 1988; Dick Zimmerman, "Joe Jordan and Scott Joplin," *Rag Times* 2/4 (Nov. 1968), 5; John A. Fisher, Untitled interview with Joe Jordan, *Ragtime Society* 1/9 (Nov. 1962), 3; JaNo, 19, 40; and BlJa, 66.

50. Bartlett D. Simms and Ernest Borneman, "Ragtime: History and Analysis," *Record Changer* (Oct. 1945), 5.

51. Ted Browne, letter to Karl Kramer.

52. Walter L. Bruetsch, "Neurosyphilitic Conditions: General Paralysis, General Paresis, Dementia Paralytica," in *Handbook of Psychiatry,* ed. Sylvano Arieti (New York: Basic Books, 1975), 137–138.

53. In considering this issue, I have had the advice of Dr. Edith Schnall, a biologist with extensive professional experience with syphilis; dermatologist Dr. Elliott Adams; and psychiatrist Dr. Terry Parrish, the last two also being outstanding ragtime pianists.

54. JaNo, 19.

55. In SeT: "Local News," July 12, 1902, 3; "Deaths," Aug. 9, 1902, 1.

56. JaNo, 25. In JaNo, 29, Marshall says he was on the road until fall of 1903, but events at that time make the latter date unlikely. Evidence of Marshall's touring is contained in "Some Clippings," SeT, Oct. 26, 1901, 1. Other Sedalians who went with him were Jake Powell and Richard Smith.

57. "Stage," IFr, Dec. 6, 1902, 5.

58. BlJa, 68–69. Nothing is known of Bobby Kemp, to whom the work is dedicated. The Vasser Boys were W. C. (cornet) and M. B. (violin). In 1903 they were named as members of the World's Fair Band of St. Louis, reputed to be the foremost black band of that city. See "Stage," IFr, Nov. 14, 1903, 7.

59. "Stage," IFr, Dec. 6, 1902, 5.

60. A concert program has been found indicating that the work was performed at at least one band concert in St. Louis in 1902 or 1903. See Michael Montgomery and Trebor Tichenor, liner notes for *Scott Joplin, "Elite Syncopations": Classic Ragtime from Rare Piano Rolls,* Biograph BCD 102.

61. "Gov. Roosevelt in Chicago. Addresses the Hamilton Club on 'The Strenuous Life'," *New York Times,* April 11, 1899, 3; "Jealous of Gov. Roosevelt?" *New York Times,* April 12, 1899, 6.

62. New York: The Century Co.

63. In a 1904 biography of Roosevelt, the sociologist Jacob Riis referred to "the famous phrase 'the strenuous life.'" Jacob A. Riis, *Theodore Roosevelt the Citizen* (New York: Outlook, 1904; repr. New York: AMS Press, 1969), 420.

64. The Short work was copyrighted May 7, 1902, and published by C. H. Person Music House in Maynard, Mass. The cover, depicting Roosevelt, is reproduced in *Tin Pan Alley Goes to Sea: The U.S. Navy on Sheet Music Covers, 1895–1919,* Engagement Calendar for 1989, selections and commentary by Alfred C. Holden, Naval Institute Press, page facing Sept. 11–17. The other *Strenuous Life* marches are by W. Townsend (Lyon & Healy, March 11, 1903); J. Z. Fullblood (W. Z. Corbett, June 14, 1906); anonymous (Illinois Music Book House, June 29, 1906); T. W. Thurton (Francis Day & Hunter, March 4, 1907); and J. G. Boehme (Carl Fischer, Aug. 3, 1907).

65. SeS, Oct. 25, 1901, 1.

66. Rosenfeld, June 7, 1903.

67. "The Stage," IFr, Nov. 14, 1903, 7.

68. Reprinted in *Rag Times* 22/5 (Jan. 1989), 8.

69. For discussions on mandolin performance, I am indebted to Bob Ault.

70. I thank Dick Zimmerman for bringing this rare publication to my attention.

71. Related to me by Rose in June 1987.

72. "Stage," IFr, Sept. 6, 1902, 5.

73. Bartlett D. Simms and Ernest Borneman, "Ragtime: History and Analysis," *Record Changer* (Oct. 1945), 5.

74. David L. Joyner points out this connection in his *Southern Ragtime and Its Transition to Published Blues* (Ph.D. dissertation, Memphis State University, 1986), 150.

75. "Stage," IFr, Nov. 29, 1902, 5. Joplin's Sedalia friend Bob Henderson was with this company.

76. "Stage," IFr, Dec. 20, 1902, 5.

77. *Ibid.*

Chapter 7

1. These samples of operatic names are found in the following pages of IFr: July 12, 1902, 5; July 6, 1901, 5; July 25, 1903, 5; April 25, 1903, 5; Oct. 12, 1901, 5.

2. "Black Patti's Troubadours," IFr, April 26, 1902, 5.

3. Sylvester Russell, IFr, "A Review of the Stage," Feb. 15, 1902, 5; "Will Marion Cooke's [sic] Errors," Nov. 15, 1902, 5.

4. For sample advertisements, see SLP, Jan. 10, 1903, 8; Dec. 19, 1903, 8; and IFr, July 23, 1904, 3. The earliest advertisements for the Rosebud are from 1903, and the earliest reference to it in the city directory is in the 1903 edition. However, an article in SLP from February 22, 1904 states the Rosebud was having its third annual ball, suggesting that the saloon existed in 1902.

5. Advertisement, SLP, March 21, 1903, 1; repr. April 4, 1903, 4.

6. BlJa, 79–80.

7. *Ibid.* I could not locate a record of the child's death. Death records were not required in St. Louis prior to 1910.

8. John Edward Hasse, *Interview of Rev. Alonzo Hayden (Sedalia, Missouri, August 25, 1976),* 5. Unpublished tape and typescript on deposit at the Smithsonian Institution.

9. This letter was addressed to Bob Ault and is used with his permission. Webster Groves is a community just west of the city of St. Louis.

10. Untitled personals, SeT, April 11, 1903, 2.

11. Monroe Rosenfeld, "The King of Rag-Time Composers Is Scott Joplin, a Colored St. Louisan," *St. Louis Globe-Democrat*, June 7, 1903. The following week this article was reprinted in the SeT, and the editor, W. H. Carter, tells how they had acknowledged Joplin many years earlier.

12. BlJa, 71.

13. "Scott Joplin's Opera," SeCo, Aug. 22, 1903, 2.

14. "Stage," IFr, April 8, 1902, 5.

15. Tom Fletcher, *100 Years of the Negro in Show Business* (New York: Burdge, 1954; repr. Da Capo, 1984), 57–58.

16. "Stage," IFr, March 15, 1902, 5. The *Freeman* reported on March 8 (p. 5) that Wright had an impressive funeral and that white friends had retained a counsel to bring charges against those guilty of the lynching.

Two New Madrid newspapers reported on the incident. The *Weekly Record* of February 22, 1902, p. 4, presented a one-sided view that is easily gauged by its opening sentence: "After the play was over last Saturday night at the Opera House a number of prominent, good people narrowly escaped death at the hands of a 'bad nigger' with a pistol." The outline of the story is the same, but the details differ greatly and are inconsistent and illogical. The *Southeastern Missourian* of February 20, 1902, p. 1, gave a slightly less inflammatory report. It stated that the lynching "is to be regretted from a humanitarian standpoint, but is condoned by a majority of the citizens, as the crime for which the negro's life paid the penalty was a grave one." It also suggested that the minstrels "did not demean themselves as becomes their race" because they had become accustomed to treatment as equals in the north and in Canada.

Apparently some other newspapers in the area were unconvinced by the official version, for an editorial in the *Weekly Record* complains of "inaccurate and false rumors" appearing in unnamed publications.

17. Except for the East St. Louis and Springfield dates, the tour is outlined in notices printed in the *New York Dramatic Mirror* and other papers during September and October of 1903. The *Dramatic Mirror* notices were supplied by the theater managers to the newspaper and are arranged in the paper under the general heading "Correspondence," in subsections by state, city, and theater. Since entries include both recent performances and future schedules, there are frequently repeated citations for a given date and theater. The specific listings in the *New York Dramatic Mirror* are as follows:

Sept. 12, 1903: p. 4: Illinois, Galesburg: Auditorium . . . Scott Ragtime Opera co., [Sept.] 3; p. 5: Missouri, Webb City: New Blake Theatre . . . Joplin Ragtime Opera co., [Sept.] 12.

Sept. 19, 1903: p. 8: Iowa, Ottumwa: Grand Opera House . . . S. Joplin Opera co., [Sept.] 29; Kansas, Pittsburg: Opera House . . . Scott-Joplin Minstrels, [Sept.] 17.

Sept. 26, 1903: p. 5: Kansas, Pittsburg: Opera House . . . Scott-Johnson [*sic*] Minstrels, [Sept.] 17; Kansas, Parsons: Edwards' Opera House . . . Scott Joplin's Rag Time Opera, [Sept.] 18; Iowa, Ottumwa: Grand Opera House . . . Joplin Opera co., [Sept.] 29; Iowa, Cedar Rapids: Greene's Opera House . . . Scott-Joplin Opera co., [Sept.] 30; Iowa, Mason City: Wilson Theatre . . . Scott Joggins [*sic*] Rag Time Opera co., [Oct.] 12.

Oct. 3, 1903: p. 5: Iowa, Ottumwa: Grand Opera House . . . Joplin Opera co., [Sept.] 29; Iowa, Mason City: Wilson Theatre . . . Scott-Joggins [sic] Ragtime Opera co., [Oct.] 12; p. 6: Nebraska, Beatrice: Paddock Opera House . . . Scott's [sic] Joplin Ragtime Minstrels, [Oct.] 6; Nebraska, Fremont: New Larson Theatre . . . Joplin Ragtime Opera, [Oct.] 7.

Oct. 10, 1903: p. 5: Iowa, Mason City: Wilson Theatre . . . Scott-Joplin Opera co., [Oct.] 12; Iowa, Ottumwa: Grand Opera House . . . Joplin Opera co., [Sept.] 29.

Oct. 17, 1903: p. 4: Iowa, Ottumwa: Grand Opera House . . . Joplin Opera co. Sept. 29; failed to appear; reported disbanded; Iowa, Mason City: Wilson Theatre . . . Scott Joplin Opera co. [Oct.] 12; p. 5: Nebraska, Beatrice: Paddock Opera House . . . Scott's [sic] Joplin Ragtime Minstrels, [Oct.] 6; Nebraska, Fremont: NEWLARSON [sic] . . . Joplin Ragtime Minstrels, [Oct.] 7.

Oct. 24, 1903: p. 5: Iowa, Mason City: Wilson Theatre . . . Scott Joplin Opera co., [Oct.] 12, cancelled; p. 7: Nebraska, Fremont: Love's Theatre . . . Joplin Minstrels [Oct.] 7, cancelled.

In addition, I checked all issues in August and November without finding additional reference to Joplin. Nor could I locate any scheduled performances in Kentucky, as alluded to by Addison Reed in his dissertation *The Life and Works of Scott Joplin* (Ph.D. dissertation, University of North Carolina at Chapel Hill, 1973), 38. For the scheduled appearance on October 12, Haskins and Benson's listing of Parker's Opera House in Mason City is in error (HaBe, 138); Wilson Theatre is correct.

Information about the Springfield, Illinois, date on September 2 comes from IFr: "Stage," Sept. 26, 1903, 5. I also examined, or had examined for me, the following newspapers from towns on the scheduled tour route: *Illinois State Journal* and *Illinois State Register* (both from Springfield, Illinois); *Galesburg (Illinois) Daily Republican Register; Webb City (Missouri) Sentinel; Pittsburg (Kansas) Headlight, Kansan* and *Workers' Chronicle; Parsons (Kansas) Sun, Weekly Sun, Eclipse,* and *Evening Herald; Ottumwa (Iowa) Weekly Democrat* and *Morning Courier; Cedar Rapids (Iowa) Evening Gazette; Beatrice (Nebraska) Daily Sun* and *Daily Express; Fremont (Nebraska) Tribune; Mason City (Iowa) Times Herald.*

18. The advertisement for the Galesburg date is reprinted in my article "On the Trail of *A Guest of Honor:* In Search of Scott Joplin's Lost Opera," in *A Celebration of American Music* (Ann Arbor: University of Michigan Press, 1990), 58.

19. "Stage," IFr, Sept. 26, 1903, 5. The term "Bufay" is problematic. "Ofay," pig latin for "foe" and a term referring to whites, was in common use by blacks of this time. I suggest that "Bufay" means "black foe," as a black thief would have been in this case. See discussion in my article "On Ragtime: Understanding the Language," *CBMR Digest* 3/3 (Fall 1990), 6–7.

20. Cancellation notices for the performances scheduled on September 29 and October 7 and 12 appeared in the *New York Dramatic Mirror* in the following issues: Oct. 17, 4; Oct. 24, 5 and 7.

21. In announcing its double-sized Christmas issue in 1913, IFr set a five-day deadline for submissions. See "Gossip of the Stage," Dec. 13, 1913, 5.

22. BlSJ, xxvii.

23. Two individuals have claimed to own portions of the score, but have refused to show it, thereby encouraging skepticism toward their claims. I discuss this issue in "On the Trail of *A Guest of Honor:* In Search of Scott Joplin's Lost Opera," 55–56.

Nathan B. Young's *A Guest of Honor. A Recreation—1999 a.d.* (St. Louis: Warren H. Green, 1986) is a fictional account about the opera and other events in the black community of the time. Young, an African-American lawyer and judge who graduated from Yale Law School in 1918, moved to St. Louis in 1924 and dedicated himself to collecting materials on black life in that city. He asserts that his story is based on the testimony of a friend who saw the opera in 1904 at True Reformers' Hall. I have been unable to verify any part of the story.

24. Roy Carew and Don E. Fowler, "Scott Joplin: Overlooked Genius," *Record Changer* (Oct. 1944), 12.

25. "The Stage," IFr, Sept. 12, 1903, 6.

26. I choose the date 1906 because it is the year of the latest composition on the flier, James Scott's *Frog Legs Rag.* I am indebted to Trebor Tichenor for calling this flier to my attention and for providing me with a copy from his collection.

27. "Roosevelt-Washington Affair," IFr, Oct. 26, 1901, 4.

28. The information about Sydney Brown comes from Trebor Tichenor, Oct. 15, 1989.

29. Brian Rust, *Jazz Records, 1872–1942,* 5th ed. (Chigwell, Essex, England: Storyville Publications, 1983), c.v. Sweatman. Rust informs me that Sweatman gave this information in an interview in 1959.

30. "The Stage," IFr, Dec. 5, 1903, 5.

Chapter 8

1. Advertisement, SLP, Jan. 9, 1904, 5.

2. A 400 Bar listed at 1300 Morgan Street, probably the same as the 400 Social Club, was run by a Charles W. Williams, not to be confused with Tony's brother Charles E. Williams. See in SLP, "Notice," March 5, 1904, 1; "Business Directory," June 11, 1904, 7.

3. In SLP: "Announcement," Feb. 20, 1904, 1; "The Rose Bud Ball," Feb. 27, 1904, 1. The latter article is reprinted in full in David A. Jasen and Trebor Jay Tichenor, *Rags and Ragtime* (New York: Seabury, 1978), 102–103.

4. Advertisements in SLP: March 12, 1904, 1; March 19, 1904, 8; April 2, 1904, 3.

5. For example, see in IFr, "Shifting Scene," July 16, 1904, 1; "Thompson's Weekly Review," Oct. 15, 1904, 1.

6. JaNo, 18.

7. Robert Allen Bradford, "Arthur Marshall—Last of the Sedalia Ragtimers," *Rag Times* 2/1 (May 1968), 5.

8. Alan Lomax, *Mister Jelly Roll,* 2d ed. (Berkeley: University of California Press, 1973), 120. To Morton's description Blesh and Janis added that Alfred Wilson was from New Orleans and that Charlie Warfield, of Tennessee, came in second. They do not indicate the source of their information. See BlJa, 75.

9. "Scott Joplin Tonight," SeCa, July 28, 1903, 5.

10. The dedication and full subtitle appear on the cover of the copyright copies, received and registered by the Library of Congress on August 22, 1904. However, the yellow lettering on the cover shows up poorly against the pale blue background. Consequently, a new cover was made, with a new plate for the printed matter. On the new cover, the dedication and the "American" portion of the subtitle were omitted. This latter cover is the more common and is reproduced in SJCW.

The title "Chrysanthemum" is conceivably a reference to a Chrysanthemum Club, one having been formed in St. Louis in the spring of 1903. However, there is no known connection between this club and Joplin. See SLP, April 4, 1903, 1.

11. "Sedalian's New Music," SeD, April 4, 1901, 1. See also "Our Town Editor," SeCo, April 15, 1904, 1.

12. Rose M. Nolen, *Sedalia's Ragtime Man* ([Sedalia, Mo.]: RoseMark Communications, n.d. [*c.* 1986]), 9–10.

13. "Scott Joplin's Dance," SeD, April 12, 1904, 7; "Our Town Editor," SeCo, April 15, 1904, 1.

14. The success of *Cascades* at the fair was reported to me by Dick Zimmerman, who had learned of it from Steward.

15. "A Musical Prophecy," *New York Sun,* Sept. 14, 1905, 6.

16. Stark wrote in response to a letter printed on Sept. 9 in which a writer complained of piano music coming from a neighboring apartment: "Her music for the most part is the purest quality of ragtime. . . ."

17. "Scott Joplin's Concert," SeS, July 24, 1904, 7.

18. Brun Campbell, "Rags to Ragtime to Riches," *Jazz Journal* 2/7 (July 1949), 13.

19. "Scott Joplin Tonight," SeCa, July 28, 1903, 5.

20. "Sedalians Should Appreciate Qualified Talent," SeCo, July 15, 1904, 4.

21. "A Wonderful Musician," SeD, July 27, 1904, 8.

22. "All Out to Liberty Park, Thursday Night, July 28," SeS, July 26, 1904, 4. "A Correction," SeCo, July 29, 1904, 1; "Scott Joplin Musicale," SeCa, July 29, 1904, 4.

23. Advertisements and articles in the SeS, Aug. 1, 3, and 5, 1904, all on p. 5; SeD, Aug. 2, 3, and 4, 1904, pp. 8, 4, and 5; SeCo, Aug. 5, 1904, 1; SeCa, Aug. 4, 1904, 5.

24. "The Fourth of August Celebration Here Was Quiet and Creditable Day and Night," SeCa, Aug. 5, 1904, 4. The last clause, "a marked departure from former celebrations," may suggest that previous Fourth of August celebrations were unruly. However, appraisals for prior years invariably commented on the good behavior of the attendees. What the statement probably reflects was the expectation that crowds of black folk would be unruly.

25. "Colored Folks Have the Town," SeS, Aug. 4, 1899, 1. The Cook song was published by W. Witmark in 1898.

26. "To Celebrate the Fourth," SeCa, July 14, 1899, 4.

27. Untitled, SeCo, Aug. 12, 1904, 1; "Local and Personal," SeCo, Sept. 16, 1904, 4.

28. SeD, Sept. 11, 1904, p. 4. See also "Mrs. Scott Joplin Dead," SeS, Sept. 11, 1904, 8; "Mrs. Joplin Dead," SeCa, Sept. 11, 1904, 8. The address given in the newspapers was incorrect. The town directories for 1904 and preceding years all place Dixon at 124 *West* Cooper.

29. "Demise of Mrs. Joplin," SeCo, Sept. 16, 1904, 1.

30. Unpublished taped interview, Dick Zimmerman and Beatrice Martin. I spoke to Beatrice Martin in 1989, shortly before her death, but she no longer recalled the incident of Joplin's marriage.

31. Rose M. Nolen, "Local Woman Remembers Scott Joplin," *Pettis County Local Times-News,* June 6, 1984, 5. See also Rose M. Nolen, *Sedalia's Ragtime Man* (RoseMark Communications [*c.* 1986]), 9–10.

32. Reverend Marvin Albright of Sedalia advised me in efforts to locate church records. He also inquired from various church organizations about the possibility that the records had been moved to another institution.

33. The 1900 Arkansas Soundex (Census index) contains some entries that

appear intriguingly close. There is a listing for a young woman named *Eddie* Alexander born in September 1884, but she was white (head of household, Alexander, Joseph P.; city of Pine Bluff, vol. 19, E.D. 103, sheet 8, line 92). It is unlikely that Freddie was white because Joplin could not have traveled through Arkansas and Missouri with a white woman without serious risk of being lynched. Also, had Freddie been white, the newspapers would certainly have commented upon it.

A listing that could refer to Lovie was that of a black woman named Lovenia Alexander, born in 1877 and the head of a household in Little Rock (vol. 33, E.D. 93, sheet 13, line 2). While this could have been Freddie's sister, there is no way to link the two, and the listing tells us nothing more about Freddie.

34. My original speculations about the marriage, prior to finding the evidence in the newspapers, are in my article "Scott Joplin in Sedalia: New Perspectives," *Black Music Research Journal* 9/2 (Fall 1989), 209–210.

Chapter 9

1. "A Rosebud Novelty," SLP, Dec. 24, 1904, 1. This article is reprinted in David A. Jasen and Trebor Jay Tichenor, *Rags and Ragtime* (New York: Seabury, 1978), following p. 76.

2. Advertisement, SLP, Oct. 28, 1905, 8.

3. Advertisement, SLP, June 24, 1905, 1.

4. All in the "Stage" column in the IFr, March 11, 1905, 5; July 29, 1905, 5; Nov. 18, 1905, 5.

5. All in SeCo: "Personal," Jan. 13, 1905, 1; Advertisements, May 13, 1905, 3; and May 20, 1905, 3.

6. BlJa, 79.

7. This appraisal is made on an analysis of Stark's surviving business ledgers, which date from April 1907 to October 1908. They are owned by Trebor Tichenor, who has kindly permitted me to use them.

8. For studies of James Scott, see Scott DeVeaux and William Howland Kenney, eds., *The Music of James Scott* (Washington and London: Smithsonian Institution Press, 1992); William H. Kenney, "James Scott and the Culture of Classic Ragtime," *American Music* 9/2 (Summer 1991), 149–182.

9. James Allen "Bing" Morgens died around 1969. Information for this account comes from his children William, James, and June Morgens in telephone conversations on July 27, 1989. See my "Echoes from the Past," *CBMR Digest* 2/2 (Fall 1989), 6–7.

10. Incorporation papers, T. Bahnsen Piano Manufacturing Company, signed June 8, 1898, City of St. Louis.

11. "Stage," IFr, Nov. 18, 1905, 5.

12. In IFr: "Williams and Stevens," Dec. 27, 1902, 16; "Stage," June 4, 1904, 5.

13. *Cakewalk in the Sky* is reprinted in Richard Zimmerman, *A Tribute to Scott Joplin and the Giants of Ragtime* (New York: Shattinger-International, 1975); *St. Louis Tickle* is reprinted in Trebor Jay Tichenor, *Ragtime Rarities* (New York: Dover, 1975).

14. "Stage," IFr, July 22, 1905, 5.

15. "Stage," IFr, Nov. 18, 1905, 5.

16. The allocation of shares is described in Articles of Association for Stark

Music Printing and Publishing Company, signed January 22, 1904, City of St. Louis. The concerts are reported in SeD: "Miss Stark's Concert," Sept. 1, 1897, 3; "Sixteen Years," SeD, Jan. 17, 1899, 3; in SeS: "Miss Stark's Concert," Sept. 4, 1897, 4; "At the Opera House," Sept. 28, 1898, 1..

17. I am unable to locate business records for Stark in New York, so I cannot determine precisely when he opened his office there. A letter he wrote to the *New York Sun* on September 11, 1905, giving New York as his address, establishes his presence in the city by that date.

18. BlJa, 57.

19. Michael Montgomery [and Trebor Tichenor], liner notes for *Scott Joplin "Elite Syncopations" Classic Ragtime from Rare Piano Rolls*. Biograph BCD 102, 6. Further indication of the Stark connection with this company is in publications by Carrie Bruggeman Stark, Will's wife. Under the pseudonym Cal. Stark, she published *Sunset Waltz* (1914) and *Baby Blues* (1917) with American Music Syndicate. (See Elliott L. Adams, "They Gotta Quit Kickin' My Dawg Around," *Sacramento Ragtime Society*, July 1990, 7–10). I could not locate business papers for the American Music Syndicate in St. Louis archives.

20. The music has a notice also of British copyright, but no copy has been found at Stationer's Hall in London. See Edward S. Walker, "Scott Joplin in England," *Storyville* 68 (Dec. 1976–Jan. 1977), 66–68.

21. Trebor Tichenor, "'The Real Thing' As Recalled by Charles Thompson," *Ragtime Review* 2/2 (April 1963), 5–6.

22. Mayhew Lake, *Great Guys* (Grosse Pointe Woods, Mich.: Bovaco Press, 1983), 122.

23. Sources for these accounts are Elise K. Kirk, *Music at the White House* (Urbana: University of Illinois Press, 1986), 161, 183, and 390; "President Leads Xmas Cakewalk," *New York World,* Dec. 26, 1901, 1; and "Alice Roosevelt's Prank," *New York Sun,* Sept. 13, 1905, 1.

24. JaNo, 25; Robert Allen Bradford, "Arthur Marshall—Last of the Sedalia Ragtimers," *Rag Times* 2/1 (May 1968), 8.

25. JaNo, 29; *Chicago City Directory.*

26. BlJa, 231.

27. Trebor Tichenor and Michael Montgomery suggest that the title refers to Eugenia Street, located behind Tom Turpin's Rosebud Cafe. (Notes to *Scott Joplin "Elite Syncopations" Classic Ragtime from Rare Piano Rolls*. Biograph BCD 102, p. 5.) Today Eugenia Street is behind where Turpin's saloon was situated, but in 1904 Walnut Street was between Market and Eugenia. The intervening street removes any connection between Eugenia and the Rosebud.

28. Isabele Taliaferro Spiller Collection, Moorland-Spingarn Research Center, Howard University.

29. This discussion owes much to conversations I've had with ragtime performers and dancers. I've also adapted significant portions from "It's About Time: The Evidence on Ragtime Tempos," a seminar presentation given by Dick Zimmerman at the Scott Joplin Ragtime Festival in Sedalia, Missouri, in June 1992.

30. BlSJ, xxx.

31. For more detailed observations on this remarkable piece, see Peter Dickinson, "The Achievement of Ragtime: An Introductory Study with Some Implications for British Research in Popular Music," *Proceedings of the Royal Musical Association* 105 (1978–79), 69–75; and James Bennighof, "*Heliotrope Bouquet* and the Critical Analysis of American Music," *American Music* 10/4 (Winter 1992), 391–410.

32. This song and two others not included in CWSJ are reprinted in Appendix B.

33. "Personals," SLP, July 14, 1906, 5. SLP, Jan. 12, 1907, 5, indicates that Turpin was no longer in business.

34. Advertisement, SLP, July 28, 1906, 1.

35. Information on Patterson's joining with Spiller is found in IFr: "From New York City," Aug. 11, 1906, 5; "Initial Opening of the Dandy Dixie Minstrels," Aug. 25, 1906, 6. Spiller's family background was sketched by Phyllis Anderson in "William Newmeyer Spiller, Vaudeville Musician: The Hampton Years: 1876–1899," a presentation at the Annual Conference of the Sonneck Society for American Music, April 4, 1991.

36. "Stark Music Printing Company's New Home," AMAJ, June 26, 1906, 6.

37. These letters were preserved by Mildred Steward, Arthur Marshall's daughter. They are now owned by Trebor Tichenor. Our commentary makes use of Tichenor's presentation "The Stark-Marshall Letters," given in a symposium at the 1992 Scott Joplin Ragtime Festival in Sedalia, Missouri.

38. "An Interview with Arthur Marshall," *Rag Times* 21/3 (Sept. 1987), 2.

39. From the Stark's comments about the song, Trebor Tichenor theorizes that it is "Down in that Foreign Land" by Charles Warfield, arranged by Arthur Marshall, published in Chicago in 1907. Tichenor suggests also that the rag is probably *The Pippin,* which Stark published in 1908.

40. Advertisements: "Pythian Temple," SLP, Nov. 10, 1906, 1; Nov. 17, 1906, 1. The review was "Pythian Temple Crowded," Nov. 24, 1906, 1.

41. David A. Jasen, *Recorded Ragtime, 1897–1958* (Hamden, Conn.: Archon), 64. The 1902 *Maple Leaf* recording by Parke Hunter, listed on p. 63, is not Joplin's piece; it is a march.

42. The music may have been self-published by LaMertha. Information about LaMertha and his relationship to Joplin is in letters from his daughter Harriette Ballman to Trebor Tichenor (June 27, 1973) and Max Morath (Oct. 24, 1974). Additional information comes from the obituary "Harry LaMertha," *Overset* (April 1954), 6. The music is reproduced in Appendix B.

43. Respectively: "Purchase a Saloon," SeS, May 4, 1899, 1; "Stage . . . Notes from the Nonpariel of Minstrelsy, Billy Kersand's own . . ." IFr, Oct. 10, 1903, 5; "Stage . . . Notes from Nonpariel Jubilee Singers," IFr, Feb. 25, 1905, 5.

44. See also my speculation on the thematic connection between strains B and C: BeRa, 137–139.

45. "Scott Joplin," AMAJ, June 17, 1907, 35.

46. As reported in Addison W. Reed, *The Life and Works of Scott Joplin* (Ph.D. dissertation, University of North Carolina at Chapel Hill, 1973), 39, 41; and HaBe, 150–152.

47. Walker, 66–68.

Chapter 10

1. His date of arrival is not known precisely, but an article in NYA ("Composer of Ragtime Now Writes Grand Opera," March 5, 1908, 6) places it in the summer of 1907. It was probably by early July, for in August he had music issued by Jos. W. Stern, a New York publisher. His address on 29th Street is first cited in AMAJ, Nov.

8, 1907, in an item to be discussed below. The size of the boarding house is judged by the 1910 census, which shows eight lodgers, as well as the owner and his wife. Joplin's intention of a short stay is indicated in several articles cited below.

2. Entertainment, illegal establishments, and corruption were present also in other parts of the city, but the designation "Tenderloin" referred to the area patrolled by the 19th (later, the 23rd) Precinct, which had its station house on West 30th Street between Sixth and Seventh avenues. In 1907, the boundaries of the precinct were 23rd to 42nd streets between Fourth and Eighth avenues. It had previously extended down to 14th Street and was bounded on the west by Seventh Avenue. See "The Tenderloin Carved by Murphy," *New York World,* March 23, 1901, 2; "Tenderloin Divided Now," *New York Times,* July 6, 1907, 2.

3. " 'The Shoo-Fly Regiment' on Broadway," NYA, Aug. 1, 1907, 3; Juli Jones, "Chicago Life," IFr, Sept. 21, 1907, 5.

4. James Weldon Johnson, *The Autobiography of an Ex-Colored Man,* (Boston: Sherman, French, 1912; repr. in *Three Negro Classics* [New York: Avon Books, Discus Books, 1965], 446–452, from chapters VI and VII. Page numbers are from the reprint. I have changed sentence order here for continuity in the description.

Johnson identifies the club as Ike Hines's in two later books: the historical study *Black Manhattan* (New York: Knopf, 1930), 74, and his autobiography *Along This Way* (New York: Viking, 1933), 175.

Other addresses that Hines occupied in the Tenderloin were 122 West 27th Street and 133 West 26th Street. Hines's move to Harlem is noted in "Stage," IFr, Aug. 1, 1903, 5. For a sketch of the man, see Carle Browne Cooke, "Letter From New York," IFr, Jan. 27, 1906, 5.

5. Wilkins's establishment was also known as the Little Savoy Club. Other addresses that it occupied were 269 West 35th Street, 50 West 29th Street (1908–09) and 136 West 37th Street (1910–12). After the last date, he moved to Harlem. Noble Sissle's lengthy description of Wilkins is in "Show Business," NYA, Oct. 23, 1948, 13.

6. Tom Fletcher, *100 Years of the Negro in Show Business* (New York: Burdge & Co., 1954; repr. Da Capo Press, 1984), 173–175.

7. Sylvester Russell wrote "The Marshall House . . . caters for the very first class table board for those who can pay the highest prices." ("New York Notes," IFr, Aug. 12, 1905, 5.) A reproduction of one of Demuth's painting "Marshall's" (1915) is in Alvord L. Eisman, *Charles Demuth* (New York: Watson-Guptill, 1982), plate 6. A similar painting, showing the same group, is entitled "Negro Jazz Band" (1916), reproduced in Barbara Haskell, *Charles Demuth* (New York: Abrams, 1987), plate 19. "At Marshall's (Negro Dancing)" (1917) is reproduced in Emily Farnham, *Charles Demuth. Behind a Laughing Mask* (Norman, Okla.: University of Oklahoma Press, 1971), plate 12. The dates of these paintings present an unresolved problem: the Marshall Hotel closed in 1913 and was replaced by the Douglass Hotel. However, we could find no alternate Marshall Club during the later years. Possibly the Douglass Hotel retained the former name for its famous cabaret.

8. For more on the ragtime scene in New York, see my BeRR, 29ff.

9. "I have often sat in theatres and listened to beautiful ragtime melodies . . . ." See "Theatrical Comment: Use of Vulgar Words a Detriment to Ragtime," NYA, April 3, 1913, 6.

10. Both pieces were copyrighted with pre-publication copies, without covers. A second copy of *Searchlight,* with cover, was submitted on September 24, 1907, and

a cover copy of *Gladiolus* on March 30, 1908. The title *Searchlight* may have been a second tribute toward Joplin's friends the Turpins who, a quarter-century earlier, had worked a mine in Searchlight, Nevada. See BlSJ, xxxi.

11. The Cole and Johnson Brothers team began receiving royalties from Stern in 1901. See James Weldon Johnson, *Along This Way* (New York: Viking, 1933), 181. In 1903, black song writer Shep Edmonds brought suit against Stern for failing to pay agreed-upon royalties. ("The Stage," IFr, Jan. 10, 1903, 4.) In 1909, columnist Sylvester Russell confirmed that it was the practice of the major houses to give royalties to black writers as well as white. See Sylvester Russell, "Eighth Annual Review," IFr, Jan. 9, 1909, 5.

12. Edward B. Marks, as told to Abbott J. Liebling, *They All Sang: From Tony Pastor to Rudy Vallee* (New York: Viking, 1934), 159–160.

13. Letter to Blesh and Janis, Jan. 27, 1950, as quoted in JaNo, 28, and David A. Jasen and Trebor Jay Tichenor, *Rags and Ragtime* (New York, Seabury, 1978), 108.

14. Russ Cassidy, "Joseph Lamb—Last of the Ragtime Composers," [part 3] *Jazz Monthly* 7 (Nov. 1961), 9.

15. "Scott-Joplin," AMAJ 23/22, Nov. 8, 1907, 12.

16. "Scott Joplin," AMAJ 23/23, Dec. 13, 1907, 5.

17. Russ Cassidy, "Joseph Lamb: Last of the Ragtime Composers," [part 1] *Jazz Monthly* (Aug. 1961), 6; Joseph Scotti, *Joe Lamb: A Study of Ragtime's Paradox* (Ph.D. dissertation, University of Cincinnati, 1977), 43, 52. *Dynamite Rag* was renamed *Joe Lamb's Old Rag* and copyrighted by Bob Darch. To date, it has not been published. *Old Home Rag* was copyrighted in 1959 and was published for the first time in *Ragtime Treasures. Piano Solos by Joseph F. Lamb* (Rockville Centre, NY: Belwin Mills, 1964).

18. In BlJa, 236 as well as several other accounts, after Lamb performed *Sensation* for Joplin it was Joplin who exclaimed "That's a good rag—a regular Negro rag." On the recording *Joseph Lamb: A Study in Classic Ragtime* (Folkways FG-3562), however, Lamb narrates the story and has another black man making that statement to Joplin. As recounted by Scotti, 44–45, this assumption was confirmed by Lamb in a letter to Michael Montgomery, dated January 25, 1958.

19. The copyright date is almost five months later, October 8, 1908.

20. BlJa, 236–237.

21. The story of Lamb's meeting with Joplin is told in greater detail in many sources, among them BlJa, 235–239; "The Ragtime Game," *New Yorker,* July 2, 1960, 3; repr. *The Ragtime Society* 2/7 (Sept. 1963), 3; Russell E. Cassidy, "Joseph Lamb: Last of the Ragtime Composers," *Jazz Monthly* (Aug. 1961), 4–7; (Oct. 1961), 13–15; (Nov. 1961), 9–10; (Dec. 1961), 15–16; repr. as "Joseph F. Lamb: A Biography," *Ragtime Society* 5/4 (Summer 1966), 29–41; and Scotti, 43–45. On Lamb's arrangements for Helf, see also Elliott L. Adams, "The Joseph F. Lamb Story, Part V," *Sacramento Ragtime Society* (Feb. 1990), 3–11.

22. Scotti, 47–48.

23. Prices are in advertisements in the *St. Louis Post-Dispatch,* March 26, 1903, 5; and the *New York World,* May 22, 1904, M7.

24. The sheet music war is discussed in greater detail in Russell Sanjek, *American Popular Music and Its Business: The First Four Hundred Years,* vol. 2, *From 1790 to 1909* (New York: Oxford University Press, 1988), 417–420.

25. "Odds and Ends," AMAJ 24/2 (Jan. 24, 1908), 16.

26. David A. Jasen, *Recorded Ragtime, 1897–1958* (Hamden, Conn.: Archon,

1973), c.v. The recording may be heard on *They All Played the Maple Leaf Rag*, Archive Productions CD1600.

27. Trebor Jay Tichenor, "Missouri Ragtime Revival," *Rag Times* 4/5 (Jan. 1971), 5.

28. "Composer of Ragtime Now Writes Grand Opera," NYA, March 5, 1908, 6.

29. Notices for Robert Joplin are all in IFr: "The Stage," Dec. 1, 1906, 5; advertisement, Dec. 7, 1907, 5; "Lincoln Theatre, Knoxville, Tenn.," Feb. 15, 1908, 5; "The Lincoln Theatre," Feb. 22, 1908, 5; "Lincoln Theatre, Knoxville, Tennessee," Feb. 29, 1908, 5; "Lincoln Theatre, Knoxville, Tennessee," March 7, 1908, 5; "Clark's Theatre at Columbus, Ohio," March 21, 1908, 5.

30. JaNo, 30; BlJa, 62. An advertisement in SeCo, April 5, 1908, 4, indicates that Marshall was performing in Sedalia.

31. The music bears a 1908 copyright and was registered on January 29, but it may have been issued at the end of the previous month.

32. Jasen and Tichenor, *Rags and Ragtime*, 86.

33. Advertisement. AMAJ 23/24, Dec. 22, 1907, 13; repeated Jan. 10, 1908, 41; Jan. 24, 1908, 14; Feb. 14, 1908, 23; and Feb. 28, 1908, 17.

34. For example, on James Scott's *Quality Rag*.

35. Page 6 for both dates.

36. Company and Incorporation papers, dated Aug. 21, 1901, and May 8, 1903. The papers include Jeremiah Luschein as a partner, but his name is never mentioned in the trade press. See also, in AMAJ: Monroe Rosenfeld, "Along the Great White Way," July 24, 1908, 16, and "The Interviewer Talks with . . . Herman Snyder," Oct. 9, 1908, 17.

37. Company papers for Seminary Music. The firm was not incorporated. We assume that the Seminary owners were married to, or otherwise related to, the Crown owners.

38. On the company papers of July 10, 1908, a Max Josephson is also listed as a partner. Josephson is not mentioned in the incorporation papers of Nov. 24, 1908. Of Waterson's role in setting up Ted Snyder Music, Monroe Rosenfeld wrote: ". . . Mr. Watterson [*sic*] has furnished him [Snyder] with very cozy quarters." See "Along the Great White Way," July 24, 1908, 16. See also in the same periodical: "Along the Great White Way," Sept. 11, 1908, 18; "The Interviewer Talks with . . . .," Oct. 9, 1908, 15; "The Interviewer Talks with . . . Ted Snyder," Jan. 8, 1909, 23. It was Ian Whitcomb's observation that Seminary Music and Ted Snyder Music were at the same address which prompted me to investigate the business and incorporation papers. See Whitcomb's *Irving Berlin and Ragtime America* (New York: Limelight, 1988), 34.

39. I thank Dr. Phyllis Anderson for sending me a copy of these notes, which are part of the Isabele Taliaferro Spiller Collection, Moorland-Spingarn Research Center, Howard University. For more on Spiller, see Eileen Southern, *Biographical Dictionary of Afro-American and African Musicians* (Westport, Conn.: Greenwood Press, 1982; Phyllis Wynn Anderson, *Isabele Taliaferro Spiller: Harlem Music Educator, 1925–1958* (doctoral dissertation, University of Georgia, 1988); Edward A. Berlin, "On Ragtime: The Musical Spillers." *CBMR Digest* (Spring 1989), 8–9.

40. The passage is on p. 107 of one of the ledger books, designated by owner Trebor Tichenor as Ledger #1. Similar versions are on pages 105, 106, and 108. The draft for *School of Ragtime* is on p. 103.

41. *Ibid.,* 106.

42. "The Interviewer Talks with ————," AMAJ 24/23, Dec. 11, 1908, 13.

43. Rudi Blesh, *Classic Piano Rags* (New York: Dover, 1973).

44. "A Musical Novelty," AMAJ 27/12 (June 24, 1911), 7.

45. "The Interviewer Talks with ————," AMAJ 24/23 (Dec. 11, 1908), 13.

46. For discussion, see BeRR, 13–15.

47. Advertisement, ChRR 1/1 (Jan. 1915), 23.

48. "'Classic Rags' Composed by May Aufderheide," AMAJ 25 (June 24, 1909), 7. A selection of Aufderheide's compositions have been reprinted in the following folios: Trebor Jay Tichenor, *Ragtime Rarities* (New York: Dover, 1975); Richard Zimmerman, *A Tribute to Scott Joplin and the Giants of Ragtime* (New York: Shattinger-International Music, Charles Hansen, 1975); Richard Zimmerman, *101 Rare Rags* (Los Angeles: Richard Zimmerman, 1988); David A. Jasen, *Ragtime. 100 Authentic Rags* (New York: Big 3 Music Corporation, 1979).

49. For further discussion of Classic Ragtime, see BeRa, 185–193.

50. BlJa, 239, 242.

51. BlJa, 239; Cassidy, Russell E., "Joseph F. Lamb: A Biography," *Ragtime Society* 5/4 (Summer, 1966), 32; Scotti, 46–48, 52. A so-called Joplin-Lamb collaboration, *Scott Joplin's Dream,* has been copyrighted by Robert Darch and is on deposit at the Library of Congress. It is labeled "arranged by Eubie Blake" and is in Blake's handwriting, but the authenticity of this work as a Joplin-Lamb collaboration is seriously doubted. The music does not reflect the styles of either Joplin or Lamb.

52. The information comes from Leonard Kunstadt, editor of *Record Research Magazine* and co-author (with Samuel Charters) of *Jazz: The New York Scene.* Kunstadt told me he learned of this in the early 1940s from a pianist named Muller, who had studied with Joplin. Kunstadt could not recall the name of the restaurant, but said it was the same one that had been bombed by the FALN (a militant organization for Puerto Rican independence). Such a bombing had occurred on January 24, 1975, at Fraunces' Tavern, the inn at which George Washington had made his farewell address to his officers. I consulted with Miriam Friedman, curator of the museum attached to the restaurant. She reported, after some research, that there is no record of live music being performed at the Tavern during this period, but that record of such activities would not necessarily have been kept.

53. Introduction to *Ben Harney's Ragtime Instructor.* See discussion in BeRa, 47, 115–118.

54. *Whoa! Maud* is reprinted in *101 Rare Rags.* For further discussion of the music, see David Lee Joyner, *Southern Ragtime and Its Transition to Published Blues* (Ph.D. dissertation, Memphis State University, 1986), 155–158.

55. [Harry] Bradford, "C.V.B.A. Special Notes," IFr, Nov. 20, 1909, 6.

56. Lester A. Walton, "Colored Vaudevillans Organize," NYA, June 10, 1909, 6. Businessman John Nail, described above, guided the formation of the organization, using both his business skills and his connections in city government. For this, he was made an honorary member. See Fletcher, 175.

57. "Ragtime Music (Invented in St. Louis) Is Dead," *St. Louis Post-Dispatch,* April 4, 1909, Sunday Magazine, 1.

58. "Theatrical Jottings," NYA, Aug. 12, 1909, 6.

59. All by Harry Bradford in IFr: "William H. Farrell, Composer and Producer," Sept. 18, 1909, 5. "What the Colored Vaudevillians Are Doing in New York City and the East," Oct. 16, 1909, 6; Oct. 30, 1909, 6; Nov. 6, 1909, 6; Nov. 13, 1909, 5. "C.V.B.A. Special Notes," Nov. 13, 1909, 6.

60. "Death of Harry Bradford," IFr, Nov. 2, 1909, 5. It was first announced that

Bradford had died of brain fever after an illness of one week. An announcement a week later revealed "Harry Bradford's death was not caused solely by his theatrical and literary work. When he discovered that a well known actor was paying undue attention to his wife he was driven insane by grief." See "The Stage," IFr, Dec. 11, 1909, 5.

Chapter 11

1. ThLJ, 18.

2. Harry A. Brown, "New York City and C.V.B.A. Notes," IFr, Jan. 1, 1910, 5. Lester A. Walton, "C.V.B.A. Entertainment," NYA, Feb. 3, 1910, 6.

3. Harry A. Brown, "New York City and C.V.B.A. Notes," IFr, Feb. 12, 1910, 5.

4. The song was published in New York by the black company Gotham-Attucks Music Co., 1909. It was not the first song based on this idea. Five years earlier Orville Jones had a song entitled *And This Shall Be His Name*. The conclusion of the song has the father saying to the mother: "Take this pencil, do the 'riting, While I do the 'citing, And this shall be his name: George Washington, Pierpont Morgan Cornelius Vanderbilt Gould, Napoleon Bonaparte Mark Hanna Julius Caeser [sic] Williams Walker Ernest Hogan Alexander, Martin Luther Mark Twain Booker T. Washington Jack Johnson Abraham Lincoln Jones." (See untitled, SeCo, June 10, 1904, 4.)

A later generation of Yale students created still another variant of the song under the title *George Jones*.

For a contemporaneous comment on a supposed practice of African Americans to use inappropriately ponderous names, see "People Want Simpler Name for New York Theater," IFr, April 6, 1912, 5.

I thank Wayne Shirley for bringing the Smith-Mack song to my attention.

5. JaNo, 26.

6. In NYA: "Theatrical Jottings," Jan. 27, 1910, 6; "Charles H. Turpin Elected Constable," Nov. 17, 1910, 1; "Theatrical Jottings," April 6, 1911, 6. In IFr: "Chas. H. Turpin," June 4, 1910, 10; "Items of Race Interest," May 13, 1911, 6; "Turpins [sic], of St. Louis, Wins in His Contest for Constable," Aug. 14, 1915, 6.

7. "Luna Park Theater, Atlanta, Georgia," IFr, Jan. 29, 1910, 6.

8. "The Stage," IFr, June 18, 1910, 5; "Theatrical Jottings," NYA, June 23, 1910, 6; April 13, 1911, 6.

9. "Stage Notes," IFr, Aug. 13, 1910, 5.

10. John Stark Ledgers, volume 1, page 96.

11. Information about Eleanor was given to me on December 19, 1992, by Hilda L. Stark, widow of John Stark III, William's son.

12. In NYA: "The Lure of the Cabaret," Sept. 12, 1912, 6; James W. Johnson, "Views and Reviews," Sept. 23, 1915, 4.

13. In the NYA: advertisement, Oct. 20, 1910, 6; "The Clef Club Concert," Oct. 27, 1910, 6.

14. "Negro's Place in Music," *New York Evening Post*, March 13, 1914, as reprinted in Robert Kimball and William Bolcom, *Reminiscing with Sissle and Blake* (New York: Viking, 1973), 61. See also R. Reid Badger, "James Reese Europe and the Prehistory of Jazz," *American Music* 7/1 (Spring 1989), 48–67.

The instrumentation for the October 27 concert includes "first mandolins and bandoris." The mandolins, having the same tuning as a violin, could have easily played second violin parts, as indicated in the 1914 interview. But the meaning of "bandoris" is not certain. It is probably the bandurria, a small guitar that has paired strings, making it sound similar to the mandolin.

15. Lester A. Walton, "Is Ragtime Dead?" NYA, April 8, 1909, 6.

16. "Negro's Place in Music."

17. "Enumeration of Mulattoes by Census Bureau Causes Genealogy Mix-Up," NYA, July 25, 1912, 1.

18. David A. Jasen, *Recorded Ragtime 1897–1958* (Hamden, Conn.: Archon, 1973), 103.

19. "New Music. Seminary Music Co." AMAJ 26/10, May 28, 1910, 34; "3 Great Rags," NYA, May 19, 1910, 6; also, May 26, 1910, 6.

20. Lester A. Walton, "Music Publishers Drawing Line," NYA, Feb. 9, 1911, 6.

21. "Theatrical Jottings," NYA, Feb. 2, 1911, 6.

22. "Theatrical Jottings," NYA, April 6, 1911, 6.

23. The score was stamped by the British Museum with the date May 22, 1911. See Edward S. Walker, "Scott Joplin in England," *Storyville* 68 (Dec. 1976–Jan. 1977), 67.

24. "Latest Negro Opera," NYA, May 25, 1911, 6.

25. "Scott Joplin Honored," NYA, June 27, 1911, 6.

26. "A Musical Novelty," AMAJ, June 24, 1911, 7.

27. JaNo, 19.

28. For an excellent discussion of the black origins of this quartet style, see Lynn Abbott, "'Play That Barber Shop Chord': A Case for the African-American Origin of Barbershop Harmony," *American Music* 10/3 (Fall 1992), 289–325. For a discussion of folklore elements in the opera, see Ann Charters, "Treemonisha," *Jazz Monthly* Aug. 1962, 7–11.

29. Page numbers refer to the reprint in CWSJ; these are four higher than the pagination in Joplin's original publication.

30. Monroe Rosenfeld, "The King of Rag-Time Composers is Scott Joplin, A Colored St. Louisan," *St. Louis Globe Democrat,* June 7, 1903, Sporting Section, 5.

31. Willie the Lion Smith, with George Hoefer, *Music on My Mind, The Memoirs of an American Pianist* (New York: Doubleday, 1964; repr. Da Capo, 1978), 25–26.

32. "Negro Minstrel Should Go," IFr, March 13, 1897, 2.

33. "The Church and the Stage," IFr, Jan. 2, 1909, 4.

34. Theodore Albrecht, "Julius Weiss: Scott Joplin's First Piano Teacher," *College Music Society Symposium* 19 (Fall 1979), 104–105.

35. ThLJ, 18.

36. JaNo, 3, 47. "Mayflower Rag" is not a known piece.

37. Interview with Trebor Tichenor at his home on June 5, 1988. Blesh and Janis evidently did not believe the story and attributed the claim to Joplin's declining mental health. They wrote in their book (p. 242):

> The old Joplin, trusting and confident was changing, becoming the prey of gnawing suspicions . . . that his friends were betraying him and that the world was against him. He began to believe that his compositions were being stolen and that the kingpins of Tin Pan Alley were waxing rich on piracies from his work.

38. W. A. Corey, "Timely Tattle," AMAJ 27/21 (Nov. 11, 1911), 20.

39. Irving Berlin, with Justus Dickinson, "'Love-Interest' As a Commodity," *Green Book Magazine* (April 1916), 695–698.

40. Lawrence T. Carter, *Eubie Blake: Keys of Memory* (Detroit: Balamp, 1979), 53. "Eubie . . . questioned Lukie about it. 'Does Irving Berlin buy his tunes from you?' he asked. 'Did he write "Alexander's Ragtime Band" or did you write it?' 'I wish to God I *had* written that song,' answered Lukie. 'Irving Berlin don't buy no tunes from me. He writes them himself.'"

41. For example, in 1928 Ira B. Arnstein brought suit against Irving Berlin with a charge strikingly similar to Joplin's. He asserted that, in hopes of having his song *Alone* published, he brought it to Berlin in 1927. A few months later, it was published as Berlin's hit song *A Russian Lullaby.* ("Says Berlin 'Pirated' Song," *New York Times,* March 24, 1928, 10.)

Chapter 12

1. This account was provided by John Arpin, an outstanding ragtime and jazz pianist and president of the Canadian-based Ragtime Society. He interviewed Sullivan in 1975 and I confirmed much of the information with Sullivan's son Paul on June 23, 1990. See also, my article "On Ragtime: Scott Joplin the Educator," *CBMR Digest* 3/1 (Spring 1990), 4.

2. "C.V.B.A. Entertainment," NYA, Aug. 17, 1911, 6.

3. All by W. A. Corey, "Timely Tattle," AMAJ 27 (July 8, 1911), 16; (Nov. 25, 1911), 17; (Dec. 11, 1911), 18.

4. For speculation on Stark's attitude toward Hayden, see Dick Zimmerman, "Did Stark Snub Hayden?" *Rag Times* 12/5 (Jan. 1979), 3.

5. The piece is very rare and was not discovered until 1974. It was omitted from CWSJ, and we reprint it in Appendix B.

6. "To Produce 'Treemonisha,'" NYA, Oct. 5, 1911, 6.

7. BlJa, 248–249. Eubie Blake's report is contained in numerous sources, one being in Al Rose, *Eubie Blake* (New York: Schirmer, 1979), 149–150. Lottie's information comes from the *ASCAP Biographical Dictionary of Composers, Authors and Publishers,* ed. Daniel I. McNamara (New York: Thomas Y. Crowell, 1948); repeated in the 1952 edition. Later editions omit mention of the performance.

8. Newspaper items reporting on the opposition to the new dances were almost a daily occurrence in 1912. See, for example, in the *New York Times:* "Welfare Inspector at Society Dance," Jan. 4, 1912, 1; "Philadelphia Bans the Trot," Jan. 5, 1912, 9; "Bars 'Grizzly Bear' at Dance at Astor," Jan. 16, 1912, 13; "To Bar Turkey Trot in Their Ballrooms," Feb. 3, 1912, 20; "Barred the 'Rag' Dances," Feb. 4, 1912, part 2, 5. In IFr: "Gossip of the Stage," June 15, 1912, 4; June 29, 1912, 4; "Trying to Put Out Grizzly Bear et al." Oct. 5, 1912, 6.

9. "Theatrical Jottings," NYA, Sept. 6, 1912, 6; "Chicago Stage Notes and Stray News," IFr, Sept. 14, 1912, 5.

10. "The Lincoln Theater," IFr, Oct. 12, 1912, 5.

11. Lawrence T. Carter, *Eubie Blake: Keys of Memory,* (Detroit: Balamp, 1979), 74–75.

12. Advertisement, NYA, Feb. 1, 1912, 4.

13. Tom Davin, "Conversations with James P. Johnson," in *Ragtime: Its History, Composers, and Music,* ed. John Edward Hasse (New York: Schirmer, 1985), 175; Willie "The Lion" Smith, with George Hoefer, *Music on My Mind: The Memoirs of an American Pianist* (New York: Doubleday, 1964; repr. New York: Da Capo, 1978), 53.

14. "Barron D. Wilkins Slain," *New York Times,* May 25, 1924, 1.

15. "Notes," IFr, March 2, 1912, 5.

16. All in Sylvester Russell, "Chicago Weekly Review," IFr: March 23, 1912, 5; July 6, 1912, 5; July 13, 1912, 5.

17. In IFr: advertisement, Aug. 10, 1912, 5; "Booker Washington Airdrome," Sept. 14, 1912, 3.

18. All in IFr: "Gossip of the Stage," April 6, 1912, 5; July 27, 1912, 5; Nov. 2, 1912, 6. Jas. H. Price, "Ruby Theater, Louisville, Ky." May 4, 1912, 5; May 25, 1912, 4; June 22, 1912, 4.

19. David A. Jasen, *Recorded Ragtime,* (Hamden, Conn.: Archon, 1973), c.v.

20. Lester A. Walton, "Things Theatrical," NYA, April 5, 1917, 6.

21. Addison W. Reed, "Scott Joplin, Pioneer," in *Ragtime: Its History, Composers, and Music,* ed. John Edward Hasse (New York: Schirmer, 1985), 132.

22. "Theatrical Jottings," NYA, June 16, 1910, 6.

23. In the NYA: "C.V.B.A. Masquerade," Feb. 22, 1912, 6; Lester A. Walton, "C.V.B.A. Entertainment," Aug. 22, 1912, 6; "New C. V. B. A Committees," June 6, 1912, 6.

24. Perry Bradford, *Born with the Blues: Perry Bradford's Own Story* (New York: Oak, 1965), 95. In IFr: (Patterson) "The Monogram," April 19, 1913, 5; (Jordan) "Elmwood Cabaret," April 26, 1913, 5; "Joe Jordan and Evylin Joiner Stop the Bill," May 31, 1913, 5; "The States," Aug. 9, 1913, 5; (Saunders) Sylvester Russell, "Chicago Weekly Review," Oct. 11, 1913, 4.

25. In IFr: Joe Golphin, "Vaudeville in St. Louis," May 10, 1913, 5; Walter Fearance, "At the Booker Washington Theater," July 19, 1913, 4. "Theatrical Jottings," NYA, May 15, 1913, 6. The Charles Hunter mentioned here was not the blind white ragtime composer of the same name.

26. Lester A. Walton, "A Bit of Biography," NYA, Oct. 9, 1913, 6.

27. Newspapers have many notices about tango teas and other dances. See, for example, in the *New York Times:* "Bustanoby Starts Sunday Tango Tea," April 7, 1913, 18; "Mayor Won't Bar Proper Tea Dances," April 8, 1913, Part 1, 22; "Stop 'Tea Room' Dancing," April 9, 1913, 20. "Race is Dancing Itself to Death," NYA, Jan. 8, 1914, 1.

28. In NYA: "Afternoon Teas Now the Rage in Harlem," April 16, 1914, 1; "The Maxixe," May 7, 1914, 6; "Theatrical Jottings," May 21, 1914, 6; "Haynes Dance Contest" and "A. O. W. Tango Dance," June 25, 1914, 6.

29. Lester A. Walton, "Use of Vulgar Words a Detriment to Ragtime," NYA, April 3, 1913, 6.

30. In the IFr: "Gossip of the Stage," June 14, 1913, 5; "New York News," June 21, 1913, 5, and July 5, 1913, 5; "Notes," July 12, 1913, 5, and July 19, 1913, 5.

31. "Joplius [*sic*] Comedians at Washington Park," *The Evening Review (Bayonne, N. J.),* July 14, 1913, 2.

32. The same address is listed in the 1913 edition ("year ending August 1," 1913) of the city directory. This is the first New York directory that contains Joplin's name.

33. Real estate advertisements indicate the number of rooms in the building. See NYA, August 19, 1909, 2. Judging by rates on similar apartments in the area, the monthly rental probably ranged between $15 and $20. See NYA, June 12, 1913, 7. A photograph of the building is reproduced in HaBe, following p. 152.

34. ThLJ, 8.

35. BlJa, 233.

36. Since there is no 1910 Soundex for New York City or for Washington, D.C., where Lottie lived prior to moving to New York, we could not locate her address in 1910. Her earliest listing in the New York City directories (and there is no way to know if this was actually she) is in 1915 as "Lettie Stokes."

37. Brun Campbell, "From Rags to Ragtime and Riches," *Jazz Journal* 2/2 (July 1949), 14.

38. JaNo, 19.

39. BlJa, 233.

40. In IFr: "Notes," July 26, 1913, 5; "New York Notes," and Billy Jones, "Eastern Theatrical Notes," Aug. 16, 1913, 5. "Theatrical Jottings," NYA, Aug. 7, 1913, 6.

41. "Gossip of the Stage," Aug. 9, 1913, 5.

42. "Theatrical Jottings," NYA, Aug. 14, 1913, 6.

43. "New York Notes," IFr, Oct. 4, 1913, 5.

44. Advertisement, NYA, Jan. 1, 1914, 6.

45. "Negro Show Made Stir on Broadway," *New York World,* Feb. 19, 1903, 3.

46. "Color Line for a Play," *New York Times,* Feb. 22, 1903, 3.

47. In the NYA: "New Theatre for Harlem," Nov. 30, 1911, 6; Lester A. Walton, "Walker-Hogan-Cole Theatre," March 7, 1912, 6; Lester A. Walton, "Theatrical Comment," Dec. 21, 1911, 6. In IFr: "Gossip of the Stage," March 16, 1912, 5; "Comments on the Name Given to the Colored Theater . . . ," March 30, 1912, 6.

48. All in NYA: "Crescent Theatre Sold," Aug. 4, 1911, 6; "Warranted Progress," Aug. 31, 1911, 3; Lester A. Walton, "The Crescent as a Stepping-Stone," Nov. 7, 1912, 6. Walton identifies Nibur's first name as "Martin." This is probably an error, for in other articles and in advertisements for the Lafayette, the first name is "Benjamin."

49. In NYA: "Drawing Color Line is Resented," Nov. 14, 1912, 1; "Theatrical Jottings," Dec. 5, 1912, 6; Lester A. Walton, "The Theatres in Harlem," Jan. 9, 1913, 6; "Acquire Control of the Lafayette Theatre," Feb. 6, 1913, 1; Advertisement, Feb. 6, 1913, 2; Lester A. Walton, "An Evening at the Lafayette," Feb. 20, 1913, 6.

50. "The Lafayette Theater, New York," IFr, Aug. 8, 1914, 6. "Colored Vaudeville Circuit," NYA, Aug. 20, 1914, 6.

51. In NYA: "'Jim' Europe Out," Jan. 1, 1914, 6; "Tempo Club Plans to Tour the Country," March 26, 1913, 1; "Tempo Club," April 2, 1914, 6. Billy E. Jones, "Eastern Theatrical Notes," IFr, Jan. 24, 1914, 6.

52. "Clef Club Triumph," NYA, June 11, 1914, 6.

53. "New York News," IFr, March 7, 1914, 5. Len Kunstadt and Bob Colton, "In Retrospect: Wilbur Sweatman. Daddy of the Clarinet," *Black Perspective in Music,* Fall 1988, 227–241.

54. Leonard Kunstadt reported seeing a few pages of Sweatman's autobiography, which is now lost. "He is also busily engaged in writing his autobiography. We've seen some of it that was typed by his secretary, and all we say at this moment is *wow!* The golden information pours forth! You read about the great ragtimer Scott Joplin. . . ." Kunstadt and Coulton, 229.

55. ThLJ, 8.

56. James Dapogny, ed., *Ferdinand "Jelly Roll" Morton. The Collected Piano Music* (Washington, D.C., and New York: Smithsonian Institution Press and G. Schirmer, 1982), 495–496.

57. "Theatrical Jottings," NYA, Oct. 29, 1914, 6. Billy E. Jones, "New York News," IFr, Aug. 14, 1915, 6.

58. Telephone interview with Bertha Niederhofer, Martin Niederhofer's widow, June 26, 1989.

59. "Theatrical Jottings," NYA, July 16, 1914, 6. Advertisement, NYA, Oct. 22, 1914, 6. The address of Crown is now 1437 Broadway, the same address as the enlarged publishing firm of Waterson, Berlin, and Snyder. In IFr: untitled, Dec. 26, 1914, 4; Billy E. Jones, "New York News," Jan. 2, 1915, 5; and Jan. 16, 1915, 5.

60. QRS 31533. National Music Roll Company, Master Record 1239. See Michael Montgomery and Trebor Jay Tichenor, liner notes for both Biograph BCD 102, *Scott Joplin. Elite Syncopations* and Biograph BLP-1010Q, *Scott Joplin*. See also in *Rag Times*: "Rare Rag Roll Found," 4/1 (May 1970), 1; "Saga of the Silver Swan," 4/2 (July 1970), 1.

61. The transcription was done by Richard Zimmerman and Donna McCluer, and edited by William Bolcom and Vera Brodsky Lawrence. See the "Editor's Note," 10–11, in either edition of CWSJ. The copyright is held by the Trust of Lottie Joplin Thomas.

62. This account is based on an unpublished essay distributed by Michael Montgomery. Others who had contributed significantly to the search were Thornton Hagert and Edward J. Sprankle.

63. "Theatrical Jottings," NYA Dec. 17, 1914, 6. The notice is the first concrete evidence that Joplin played the violin. It thereby supports assertions that Joplin had learned the instrument as a child and that he tried to teach it to Belle.

64. IFr, Feb. 6, 1915, 5; also Feb. 13, 1915, 5.

65. Songwriter Harrison Smith referred to it as "Scott's massive Steinway." He had used it while staying at Lottie's boarding house in the late 1920s. See Harrison Smith, "Wilbur C. Sweatman, 'Original Jazz King,'" in Kunstadt and Colton, 232.

Sylvia Ogden Lee, a concert pianist and opera coach, described the piano to me in a telephone interview on July 1, 1987. She played on it daily for several months while she and her husband, conductor Everett Lee, boarded in Sam Patterson's home in 1942.

The piano still exists and is in private hands; I have not been permitted to see it. The serial number indicates it had been purchased new—not by Joplin—in 1908.

66. Advertisement, "David I. Martin Announces the Eighth Annual Recital of His Violin Pupils," and "Recital by Martin's Pupils," both in the NYA, April 29, 1915, 6.

67. "Martin-Smith Recital," NYA, May 13, 1915, 6.

68. Advertisement, "Frolic of the Bears," NYA, May 20, 1915, 8. In IFr: Billy E. Jones, "New York News," May 29, 1915, 5; "The Frolic of the Bears," June 19, 1915, 6, and July 3, 1915, 5.

69. Billy E. Jones, "New York News," IFr, Sept. 4, 1915, 6.

70. BlSJ, xxxix.

71. Billy E. Jones, "New York News," IFr, Sept. 18, 1915, 5.

72. Peter Gammond, *Scott Joplin and the Ragtime Era* (New York: St. Martin's Press, 1975), 8; Reed, 132; Carter, 24; Al Rose, 1979), 149–150; HaBe, 158.

73. Rose, 149–150.

74. Gammond, 8.

75. The date of the interview was June 26, 1989.

76. BlSJ, xxxix.

77. Actually, Joplin did use his home, or another apartment, as a studio. Mary Wormley, Lottie's niece and the ultimate beneficiary of the Joplin estate, was in this studio once as a child, the only occasion she had ever seen him. She recalled that

Joplin, while teaching a young black boy, became impatient with his student and chased him out of the house. (Gerald Kearney, Wormley's lawyer, told me this on August 14, 1986.) The 1916/1917 city directory lists Joplin as a music teacher at 133 West 138th Street and has him also at 212 West 138th Street. The 1917 directory has him as a music teacher at 160 West 133rd Street. Lottie had two additional listings in the latter directory: Lottie Joplin, furnished rooms, 163 West 131st Street, and Lottie Stokes, boarding house, 259 West 137th Street. Since they lived at the 131st Street address, Joplin may have had a separate studio in the 133rd Street building.

78. BlSJ, xxxix.

79. JaNo, quoting ragtimer Charlie Thompson.

80. Biograph has issued CD recordings of the Connorized rolls on *Scott Joplin, "The Entertainer,"* BCD 101; *Scott Joplin, "Elite Syncopations,"* BCD 102 (including, also, the Master Record roll of *Silver Swan Rag*); and *The Greatest Ragtime of the Century,* BCD 103. The Uni-Record performance is available on a Biograph LP recording, BLP 1006Q.

81. Untitled, NYA, Sept. 7, 1916, 6.

82. Jack Trotter, "New York. Notes of Stage and Sport," IFr, Oct. 21, 1916, 4.

83. ThLJ, 8.

84. In IFr: Jan. 20, 1917, 4; Feb. 10, 1917, 4; March 24, 1917. "Theatrical Jottings," NYA, March 29, 1917, 6. The date of his confinement is indicated in the State Supreme Court document headed, "In the matter of the appointment of a committee of the estate of Scott Joplin, an alleged incompetent person . . . ," April 1, 1917. Joplin's estate consisted of the plates of his opera, valued at $300.

85. Walter L. Bruetsch, "Neurosyphilitic Conditions: General Paralysis, General Paresis, Dementia Paralytica," 134–143. In *Handbook of Psychiatry,* ed. Sylvano Arieti (New York: Basic Books), 1975.

86. "Scott Joplin Dies of Mental Trouble," NYA, April 5, 1917, 1.

87. Lester A. Walton, "Things Theatrical," NYA, April 5, 1917, 6. This item was the basis for a similar notice that appeared in IFr, April 14, 1917, 5: "Real Cause of Scott Joplin's Death at New York." Joplin was interred on April 5, 1917, at St. Michael's Cemetery in East Elmhurst, New York. This is a nonsectarian cemetery in Queens County. The location of his grave is plot 5, row 2, grave 5.

88. "Scott Joplin is Dead," ChRR, July 1917, 13. Stark does not have a byline for the item, but identifies himself with the verb "came" in the opening sentence of the next paragraph, which discusses his initial association with Joplin:

> When he first came into the office of the Stark Music Company some years ago, with the manuscripts of Maple-Leaf Rag and the Sunflower Slow Drag, he had tried other publishers, but had failed to sell them. Stark quickly discerned their quality, bought them and made a five-year contract with Joplin to write only for his firm, which firm has all of his great compositions.

Chapter 13

1. Edward B. Marks, as told to Abbott J. Liebling, *They All Sang: From Tony Pastor to Rudy Vallee* (New York: Viking, 1934), 159–160.

2. Maude Cuny-Hare, *Negro Musicians and Their Music* (Washington, D.C.: Associated Publishers, 1936). The two chapters are "Negro Idiom and Rhythm" (131–156) and "Musical Comedy" (157–177). Showing her undisguised animosity

toward ragtime, she wrote: "So far did the Rag craze and Jazz spread, that in traveling and visiting many institutions of learning, the author found that the musical taste of the youth was being poisoned" (133).

3. Alain Locke, *The Negro and His Music* (Washington, D.C.: The Associates in Negro Folk Education, 1936; repr. in *The American Negro: His History and Literature* (New York: Arno, 1969), 62.

4. Locke, 82.

5. ThLJ, 8.

6. Jasen, *Recorded Ragtime, 1897–1958,* (Hamden, Conn: Archon, 1973), c.v. Maple Leaf Rag. The recording has been re-released on *They All Played the Maple Leaf Rag,* Archive Productions CD1600.

7. Dick Zimmerman, "An Interview with John Stark," *Rag Times* 10/3 (Sept. 1976), 5. In a letter in November 1926, Stark wrote that the *Maple Leaf* was his chief source of income. See BlJa, 267.

8. We discuss this version in Chapter 10; see Ex. 10-5. Also, Jasen, c.v. Fuzzy Wuzzy Rag. Handy's recording can be heard on *They All Played the Maple Leaf Rag,* Archive Productions CD1600.

9. IFr, May 12, 1917, 4.

10. Len Kunstadt and Bob Colton, "In Retrospect: Wilbur Sweatman. Daddy of the Clarinet," *Black Perspective in Music* 16/2 (Fall 1988), 229.

11. Willie the Lion Smith, with George Hoefer, *Music on My Mind, The Memoirs of an American Pianist* (New York: Doubleday, 1964; repr. Da Capo, 1978), 90.

12. Copyrights were good for twenty-eight years and could be renewed for a second term of the same length.

13. Roy Carew and Don E. Fowler, "Scott Joplin: Overlooked Genius," *Record Changer* (Sept. 1944), 12–14; (Oct. 1944), 10–12; (Dec. 1944), 10–11. This was the third ragtime article published by Carew, and he was to write about a dozen more through the next twenty years.

14. (New York: Duell, Sloan and Pearce, 1950; 2nd ed., Berkeley: University of California Press, 1973.)

15. *The Complete Works of Scott Joplin*. Audiophile AP 71–72.

16. "Sedalia Rediscovers Joplin," *Rag Times* 2/4 (Nov. 1968), 13.

17. Folkway Records FG 3562.

18. Charters recorded the Utah performance on Portents 3. See in *The Ragtime Society:* "Treemonisha on Record!!!" July–Aug. 1965, 32–33, and "Record Reviews" (Sept.–Oct. 1965), 43. The selections, all on side one, are: Overture, Prelude to Act 3, The Sacred Tree, We're Goin' Around, We Will Rest Awhile, Aunt Dinah Has Blowed de Horn, A Real Slow Drag. Ann Charters performed Joplin rags on side two.

19. For a listing of Campbell's articles, see BeRa, 227.

20. ThLJ, 18.

21. BlSJ, xxxix.

22. John Edward Hasse, "Rudi Blesh and the Ragtime Revivalists," 185. In John Edward Hasse, ed., *Ragtime* (New York: Schirmer, 1985).

23. "Thomas" comes from a subsequent marriage, of which we have no information. This information comes from the Surrogate Court record, File No. A-873/1953; "Petition for Letters of Administration, . . . Lottie Joplin Thomas, aka Lottie Joplin and Lottie T. Joplin," March 20, 1953.

24. Telephone conversation with Pete Clute on May 5, 1991.

25. Kunstadt and Colton, 229; telephone conversation with Leonard Kunstadt on May 9, 1991.

26. This information on Harry G. Bragg was supplied by his son, also named Harry G. Bragg, in a telephone conversation in April 1991.

27. "Examination of Barbara Doris Sweatman, Respondent, In the Matter of the Application of Harry G. Bragg, as Administrator of the Estate of Wilbur C. Sweatman, Deceased, to Discover Property Withheld." Surrogate Court, County of New York, A.851/1961. Sept. 22, 1961, 15–16, 19–20, 23, 24, 28–29, 32.

28. Eva Sweatman made a Last Will and Testament on July 21, 1962 (file #96251, Kansas City, MO, Probate Court); she died on July 21, 1964. An Inventory and Appraisal of her estate done on May 11, 1965, indicated royalties of $10,250 and no real property. She left the bulk of her estate to Sweeney, the only items specified being the royalties. She left to Barbara Sweatman, "who is said to be a daughter of my brother Wilbur Sweatman, deceased, and was born out of wedlock, the sum of One Dollar ($1.00)." In the Finding and Order Discharging Administration, dated August 8, 1965, is an indication that Barbara Sweatman could not be located. One other person, Mattie James, received a small amount of money. We could not locate Robert L. Sweeney and there is no record of him in the Kansas City Probate Court.

29. The amount of the ASCAP payments is indicated in the Surrogate Court record, file No. A-873/1953, "In the Matter of the Petition of Mary L. Warmley [sic]," Nov. 3, 1962.

30. Fifteen of Bolcom's rags are collected in William Bolcom, *Piano Rags* (New York: Edward B. Marks Music, 1991). Several recorded samples are included in *Heliotrope Bouquet: Piano Rags 1900–1970,* Nonesuch H-71257, 1971; *Bolcom Plays His Own Rags,* Jazzology JCE-72; and *Paul Jacobs Plays Blues, Ballads, and Rags,* Nonesuch D-79006.

31. Samples of Albright's ragtime can be heard on *Albright Plays Albright,* Musical Heritage Society MHS 4253, 1973.

32. Speight Jenkins, "Perspective on Joplin: A Talk with Vera Brodsky Lawrence," *Record World,* July 6, 1974, 20.

33. Nonesuch H-71248, 1970. "Nonesuch and the Scott Joplin Craze," *Record World,* July 6, 1974, 22.

34. H. Wiley Hitchcock, "Music of the Higher Class," *Stereo Review,* April 1971, 84.

35. *Piano Rags by Scott Joplin* volumes II and III, H-71264, 1972; H-71305, 1974. Selections from Volumes I–III have been reissued on CD as *Scott Joplin Piano Rags* on Nonesuch 9 79159-2, 1987. Bolcom's recordings at this time include *Pastimes and Piano Rags,* Nonesuch H-71299.

36. *Birth of the Cool,* Capitol DT 1974.

37. *Scott Joplin: The Red Back Book,* Angel S-36060, 1973; CDC-47193.

The actual title of the collection of orchestrations is *Fifteen Standard High Class Rags,* published by John Stark, no date. It has been called "The Red Back Book" because of its red cover. See John Kronenberger, "The Ragtime Revival—A Belated Ode to Composer Scott Joplin," *New York Times,* Aug. 11, 1974, part 2, 4. Robert Adels, "Joplin Dominates Classical Sales," *Record World,* July 6, 1974, 20.

38. "MCA and 'The Sting,'" *Record World,* July 6, 1974, 20.

39. "4 Grammys Given to Stevie Wonder," *New York Times,* March 3, 1975, 37.

40. Robert Adels, "Joplin Dominates Classical Sales," *Record World,* July 6, 1974, 20.

41. Harold C. Schonberg, "Music: 'Treemonisha,'" *New York Times,* Jan. 30, 1972, 51. A summary of reviews nationwide is contained in Dick Zimmerman, "Treemonisha Charms the Critics," *Rag Times* 5/6 (March 1972), 7.

42. Raymond Ericson, "This Orchestra Has 24 Conductors," *New York Times,* Dec. 15, 1974, II, 22.

43. Donal Henahan, "'Treemonisha,' the Legend Arrives," *New York Times,* Oct. 22, 1975, 40; "Kathleen Battle is Standout as 'Treemonisha' Stand-in," *New York Times* Oct. 23, 1975, 47; Walter Kerr, "The Rag King Comes to Broadway," *New York Times,* Nov. 2, 1975, II, 1.

44. Peter Kihss, "Pulitzer Prizes to Bellow, 'Chorus Line,' 2 on Times," *New York Times,* May 4, 1976, 1.

45. I interviewed Kearney on August 14, 1986. This description of Mrs. Wormley agrees with a report given by Kearney's predecessor, Robert Rosborne; see Jack Viertel, "Scott Joplin: His Rags, Whose Riches?" *New Times,* Nov. 29, 1974, 58.

46. T. J. Anderson, "Scott Joplin and the Opera *Treemonisha,*" address to the Association for the Study of Afro-American Life and History, Oct. 30, 1976.

47. *Ibid.*

48. Murray Hill 931079.

49. The record of Lottie Joplin Thomas Trust v. Crown Publishers is contained in 456 F.Supp. 539 (*Federal Supplement,* Vol. 456, 1979, pp. 531–539); the appeal is in 592 F.2d 651 (1978) (*Federal Reporter,* Second Series, Vol. 592, 1979, pp. 651–658). For other views of the ragtime and Scott Joplin revivals, see Terry Waldo, *This Is Ragtime* (New York: Hawthorne, 1976), chapters 7–10; David A. Jasen and Trebor Jay Tichenor, *Rags and Ragtime* (New York: Seabury, 1978), 259–266. Waldo has particularly strong comments on the litigation aspects of the revival.

50. *One Hundred Ragtime Classics,* Denver: Donn Printing, 1963.

51. See my article "On Ragtime: The Ragtime Clubs," *Black Music Research Newsletter* 9/2 (Fall 1987), 13–15.

52. Dick Zimmerman, "Joplin Returns to Sedalia," *Rag Times* 8/3 (Sept. 1974), 1–3.

53. "Sedalia Festival A Hit," *Rag Times* 17/2 (July 1983), 1.

54. *New York Times,* Jan. 24, 1971, II, 15.

55. I exclude from this count three nonscholarly books for children and adolescents: Barbara Mitchell, *Raggin': A Story about Scott Joplin* (Minneapolis: Carolrhoda, 1987), for very young children; Mark Evans, *Scott Joplin and the Ragtime Years* (New York: Dodd, Mead, 1976); Katherine Preston, *Scott Joplin* (New York: Chelsea House). The last of these is a particularly attractive, nicely illustrated, and sensibly written book for older children. The scholarly notes are listed in the bibliography of the present volume.

56. Peter Williams, "Four for Festival," *Dance and Dancers* 25/7 (July 1974), 23.

57. "A Musical Novelty," AMAJ, June 24, 1911, 7.

Select Bibliography

Albrecht, Theodore, "Julius Weiss: Scott Joplin's First Piano Teacher," *College Music Society Symposium* 19 (Fall 1979), 89–105.

Bennighof, James, "*Heliotrope Bouquet* and the Critical Analysis of American Music," *American Music* 10/4 (Winter 1992), 391–410.

Berlin, Edward A., "On Ragtime: A Different Perspective on Tin Pan Alley," *CBMR Digest* 4/1 (Spring 1991), 4–5.

———, "On Ragtime: Echoes from the Past," *CBMR Digest* 2/2 (Fall 1989), 6–7.

———, "On Ragtime: Ragtime and the Church," *CBMR Digest* 4/2 (Fall 1991), 6–7.

———, "On Ragtime: Scott Joplin the Educator," *CBMR Digest* 3/1 (Spring 1990), 3–4.

———, "On Ragtime: The Musical Spillers," *CBMR Digest,* 2/1 (Spring 1989), 8–9.

———, "On Ragtime: Understanding the Language," *CBMR Digest,* 3/3 (Fall 1990), 6–7.

———, "On the Trail of *A Guest of Honor:* In Search of Scott Joplin's Lost Opera," in *A Celebration of American Music: Words and Music in Honor of H. Wiley Hitchcock,* ed. Richard Crawford, R. Allen Lott, and Carol J. Oja (Ann Arbor: University of Michigan Press, 1990), 51–65.

———, "Scott Joplin and the Jefferson Davis Monument Fund," *I.S.A.M. Newsletter* 19/1 (November 1989), 8.

———, "Scott Joplin in Sedalia: New Perspectives," *Black Music Research Journal* 9/2 (Fall 1989), 205–223.

———, *Ragtime: A Musical and Cultural History* (Berkeley: University of California Press, 1980; 1984).

———, *Reflections and Research on Ragtime* (Brooklyn: Institute for Studies in American Music, 1987).

———, "Scott Joplin's *Treemonisha* Years," *American Music* 9/3 (Fall 1991), 260–276.

Blesh, Rudi, "Scott Joplin: Black-American Classicist," in *The Complete Works of Scott Joplin,* (New York Public Library, 1971; 1981), both volumes, xiii–xl.

Blesh, Rudi and Harriet Janis, *They All Played Ragtime,* (New York: Kmopf, 1950; 4th ed. New York: Oak, 1971).

Bradford, Perry, *Born with the Blues: Perry Bradford's Own Story* (New York: Oak, 1965).

Campbell, S. Brunson, "From Rags to Ragtime and Riches," *Jazz Journal* 2/7 (July 1949), 13.

———, "A Silver Half-Dollar and the 'Ragtime Kid'." Typescript on deposit at Western Historical Manuscript Collection, Ellis Library, University of Missouri-Columbia.

Carew, Roy, and Don E. Fowler, "Scott Joplin. Overlooked Genius," *Record Changer* (Sept. 1944), 12–14, 59; (Oct. 1944), 10–12; (Dec. 1944), 10–11.

Carter, Lawrence T., *Eubie Blake: Keys of Memory* (Detroit: Balamp Publishing, 1979).

Charters, Ann, "Treemonisha," *Jazz Monthly* (Aug. 1962), 7–11.

Charters, Samuel B. and Leonard Kunstad, *Jazz, A History of the New York Scene* (Garden City: Doubleday, 1962).

Davin, Tom, "Conversations with James P. Johnson," *Jazz Review* 2 (June 1959, July 1959, Aug. 1959, Sept. 1959, March–April 1960). Repr. in Hasse, *Ragtime: Its History, Composers, and Music,* 1985.

Dickinson, Peter, "The Achievement of Ragtime: An Introductory Study with Some Implications for British Research in Popular Music," *Proceedings of the Royal Musical Association* 105 (1978–79), 69–75.

Fletcher, Tom, *100 Years of the Negro in Show Business* (New York: Burdge, 1954; repr. Da Capo, 1984).

Gammond, Peter, *Scott Joplin and the Ragtime Era* (New York: St. Martin's, 1975).

Haskins, James with Kathleen Benson, *Scott Joplin* (Garden City: Doubleday, 1978).

Hasse, John Edward, ed., *Ragtime: Its History, Composers, and Music* (New York: Schirmer, 1985).

Hasse, John Edward, *The Creation and Dissemination of Indianapolis Ragtime, 1897–1930* (Ph.D. Dissertation, Indiana University, 1981).

Hebert, Rubye Nell, *A Study of the Composition and Performance of Scott Joplin's Opera "Treemonisha"* (D.M.A. Dissertation, Ohio State University, 1976).

Ireland, Tom, Letter to S. Brunson Campbell, April 19, 1947. Folder D, Letter 2, Fisk University Library's Special Collections.

Janis, Harriet, Unpublished notes for *They All Played Ragtime,* c. 1949. On deposit in the Fisk University Library's Special Collections.

Jasen, David A., *Recorded Ragtime 1897–1958* (Hamden, Conn: Archon, 1973).

Jasen, David A., and Trebor Jay Tichenor, *Rags and Ragtime* (New York: Seabury, 1978).

Johnson, James Weldon, *The Autobiography of an Ex-Colored Man* (Boston: Sherman, French, 1912).

Joyner, David Lee, *Southern Ragtime and Its Transition to Published Blues* (Ph.D. Dissertation, Memphis State University, 1986).

Kenney, William H., "James Scott and the Culture of Classic Ragtime," *American Music* 9/2 (Summer 1991), 149–182.

Kimball, Robert and William Bolcom, *Reminiscing with Sissle and Blake* (New York: Viking, 1973).

Kunstadt, Len and Bob Colton, "In Retrospect: Wilbur Sweatman. Daddy of the Clarinet," *Black Perspective in Music* 16/2 (Fall 1988), 227–241.

Lomax, Alan, *Mister Jelly Roll. The Fortunes of Jelly Roll Morton, New Orleans Creole and "Inventor of Jazz",* 2d ed. (New York: Duell, Sloan and Pearce, 1950; Berkeley: University of California Press, 1973).

Marks, Edward B. as told to Abbott J. Liebling, *They All Sang, from Tony Pastor to Rudy Valee* (New York: Viking, 1934).

Reed, Addison, *The Life and Works of Scott Joplin* (Ph.D. Dissertation, University of North Carolina at Chapel Hill, 1973).

Rose, Al. *Eubie Blake* (New York: Schirmer, 1979).

Schafer, William J., and Johannes Riedel, *The Art of Ragtime* (Baton Rouge: Louisiana State University Press, 1973).

Scotti, Joseph, *Joe Lamb: A Study of Ragtime's Paradox* (Ph.D. Dissertation, University of Cincinnati, 1977).

Smith, Willie "The Lion," as told to George Hoefer, *Music on My Mind: The Memoirs of an American Pianist* (New York: Doubleday, 1964; repr. Da Capo, 1978).

Thompson, Kay C. "Lottie Joplin," *Record Changer* 9 (Oct. 1949), 8.

Tichenor, Trebor Jay, "John Stillwell Stark, Piano Ragtime Publisher: Readings from *The Intermezzo* and His Personal Ledgers, 1905–1908," *Black Music Research Journal* 9/2 (Fall 1989), 193–204.

Waldo, Terry, *This Is Ragtime* (New York: Hawthorne, 1976).

Witmark, Isidore and Isaac Goldberg, *The Story of the House of Witmark: From Ragtime to Swingtime* (New York: Furman, 1939).

Music Collections

Brun Campbell. The Music of "The Ragtime Kid," transcr. with intro. by Richard A. Egan, Jr. (St. Louis: Morgan, 1993).

Classic Piano Rags, compiled by Rudi Blesh (New York: Dover, 1973).

The Complete Works of Scott Joplin, Vera Brodsky Lawrence, ed., 2 vols., 2d ed. (New York Public Library, 1981).

Ferdinand "Jelly Roll" Morton. The Collected Piano Music, James Dapogny, ed. (Washington, D.C. and New York: Smithsonian Institution Press and Schirmer, 1982).

The Music of James Scott, Scott DeVeaux and William Howland Kenney, eds. (Washington: Smithsonian Institution Press, 1992).

101 Rare Rags, Dick Zimmerman (Los Angeles: Dick Zimmerman, 1988).

Ragtime Rarities, Trebor Jay Tichenor, ed. (New York: Dover, 1975).

Ragtime Rediscoveries, Trebor Jay Tichenor, ed. (New York: Dover, 1979).

Index